Water and Environmental Challenges in a Changing World

In Focus – Special Book Series

Water and Environmental Challenges in a Changing World

Editors

Mauro De Marchis and Luigi Berardi

IWA PUBLISHING

Published by **IWA Publishing**
Republic – Export Building, 1st Floor
2 Clove Crescent
London E14 2BE, UK
Telephone: +44 (0)20 7654 5500
Fax: +44 (0)20 7654 5555
Email: publications@iwap.co.uk
Web: www.iwapublishing.com

First published 2020

British Library Cataloguing in Publication Data
A CIP catalogue record for this book is available from the British Library

ISBN: 9781789062069 (paperback)
ISBN: 9781789062076 (eBook)
ISBN: 9781789062083 (ePUB)

Contents

Editorial: Water and environmental challenges in a changing world: the perspective of the 13th International Conference on Hydroinformatics HIC 2018

The IAHR/IWA Joint Committee on Hydroinformatics is a community of scientists working on the application of information and communications technologies (ICTs) in addressing the increasingly serious problems of the equitable and efficient use of water in its various and really different forms. The community is thus composed of particularly wide and lively research areas widening from numerical modelling to the development of new sensors and tools, from the interpretation of increasing amounts of data, to the analysis and management of complex systems requiring the integration of physically based models, data and controls for practical applications. Today, multi-disciplinarity is the key to tackle challenges for sustainable management of water resources, especially considering the increasing complexity in the *water world*. It is thus fundamental to link the mathematical laws describing the water bodies and the new technologies given by the ICT and information science scientific communities.

In this framework, the 13th International Conference on Hydroinformatics 2018 (HIC 2018) held in Palermo, Italy, was the concrete location where researchers, working in very different areas related to water problems, over an entire week. The attention of the scientific community on HIC 2018 was demonstrated by presence of several International Water Association Specialist Groups. More than 400 attendees came to Palermo from about 40 countries all over the world (with the largest delegations from China, UK, Italy and South Korea). Figure 1 shows the distribution of delegates' nationalities.

The conference covered seven main topics: technologies for water management and monitoring; remote sensing; big-data; knowledge and water data management; hydraulic and hydrologic modelling; climate change impacts; environmental and costal hydroinformatics. Moreover, the International Scientific Committee proposed 12 special sessions on topics of special interest for the technical-scientific community:

1. Data assimilation of spatial information for hydrologic and hydraulic models
2. Complex Network Theory and applications to water systems
3. Climate change impacts on urban water systems: flooding forecasting and warning
4. Integrated use of the water reservoirs
5. IA techniques for Smart Water Systems: Smart Sensors, Smart Networks and Serious Gaming
6. Model predictive control for water management
7. Development and application of the next generation of shallow flow models
8. Long-term resilience of water systems: input data analysis
9. Monitoring network optimization and model choice: information for predictions and value for decisions
10. Simulation of fluvial eco-hydraulic and morphodynamic processes
11. Advance in uncertainty estimation of hydro-science in a changing environment
12. Time series analysis for climate change detection

This special issue includes 16 papers selected among 412 presented at HIC 2018 in various themes related to hydroinformatics. The articles followed the standard peer-review process of the *Journal of Hydroinformatics* with three/four independent reviews for each paper. The 16 papers can be subdivided into five main topics.

The first three articles deal with numerical models to predict the water hydrodynamic in terms of flow fields and contaminant dispersion. Di Cristo *et al.* (2020) present a two-phase shallow-water morphodynamical model, particularly suited to the analysis of fast geomorphic transients, applied to the numerical simulation of the propagation of a dam-break wave over an erodible floodplain in the presence of a rigid obstacle. Wang *et al.* (2020) focus attention on the investigation of the wake flow around a circular patch of cylinders with different solid volume fraction (SVF), analyzing the spatial evolution of the wake flow behind the patch and the correlation between the flow pattern and the wake flow stability. Yang *et al.* (2020) propose a generic sampling method to interpret the output of the

doi: 10.2166/hydro.2019.200

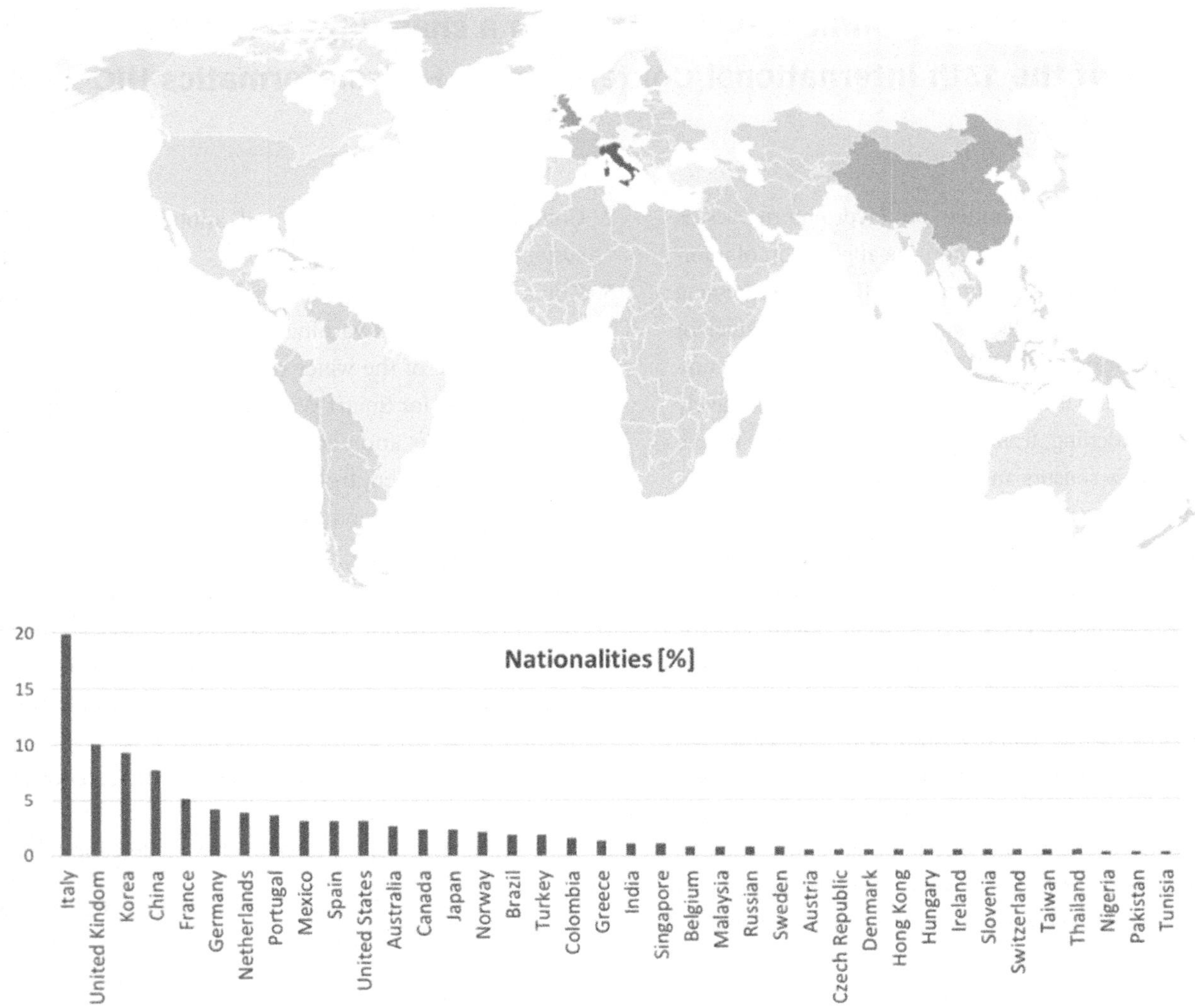

Figure 1 | Nationalities of delegates at HIC 2018.

random walk method, a meshfree scheme, and highlight the importance of accurately taking diffusion into account in studying the mixing phenomena.

Further, four articles investigate problems related to floods in urban and coastal areas and to wastewater treatments. Sambito *et al.* (2020) analyze an approach for positioning water quality sensors based on the Bayesian decision network (BDN). The analysis is focused on soluble conservative pollutants, such as metals. The methodology incorporates several sources of information, including network topology, flows and non-formal information about the possible locations of contamination sources. The methodology was tested using two sewer systems with increasing complexity. Pinho *et al.* (2020) apply hydroinformatics tools for the identification and characterization of the key factors responsible for flood events; moreover, they propose a flood early warning system resulting from the studied flood events. Gibson *et al.* (2020) propose an integrated and participatory methodological approach to assess the risk and enhance the resilience of interconnected critical infrastructures to urban flooding under climate change. The proposed methodology has been applied to protect coastal communities.

Visanji *et al.* (2020) present a new methodology for a decision support tool (WiSDOM), which focuses on producing treatment solutions suited to treating water for reuse to Indian Water Quality Standards. Basically, the study aims at finding optimal wastewater treatment solutions to calculate the removal of different contaminants of emerging concern (CECs) found in developing countries.

Three studies are focused on problems concerning water distribution networks. Braun *et al.* (2020) develop a mathematical method to investigate water distribution networks with an alternative spectral approach that uses polynomial chaos expansion and has the potential to give results of comparable accuracy to Monte Carlo sampling through the definition of a stochastic model. This approach is applied in order to evaluate the influence of uncertain demands on the water age. Simone *et al.* (2020) propose the use of edge betweenness metrics, tailored for water distribution networks from Complex Network Theory, to analyze the network domain of several real systems. Results show that the metric can represent a useful tool for supporting analysis, design and management tasks since it captures the role of topological network features on its emergent hydraulic behavior. Fellini *et al.* (2020) present an algorithm for real-time fault detection in the SCADA system of a modern water supply system (WSS) in an Italian alpine valley. A coupled model of the hydraulic and remote-control system is developed to test the performance of the WSS. Simulations show that the occurrence of errors in the sensors causes significant worsening in the economic, energy and mechanical performance of the infrastructure.

Four studies are related to hydrological and climate change problems. Grasso *et al.* (2020) develop a tool, named MultiRain, to compute depth-duration-frequency (DDF) curves, both related to a point and integrated over an area, from multiple regional statistical analyses. The MultiRain procedure is based on Python scripting, geographic information system (GIS) functions and web technologies, and can be performed via web browser or in a desktop GIS environment. A demonstration version has been built with four different regional analyses proposed in 20 years for the north west of Italy. Dimitriadis & Koutsoyiannis (2020) observe that geophysical processes are often characterized by long-term persistence inducing large statistical bias, thus propose to examine the most probable value (i.e. mode) of the estimator of variance to adjust the model for statistical bias. Particularly, they performed an extensive Monte Carlo analysis based on the climatogram (i.e. variance of the average process vs scale) of the simple scaling (Gaussian Hurst-Kolmogorov) process, and show that its classical estimator is highly skewed especially in large scales. Real-Rangel *et al.* (2020) introduce a transparent framework for monitoring the spatiotemporal distribution of drought hazard, based on uni- and multi-variate standardized drought indices that use reanalysis datasets of hydrological variables available freely and globally. The framework provides a significant extension to the capabilities for national drought monitoring and it is being used by the Mexican water authority in the decision-making process related to drought severity assessment. Chen *et al.* (2020) analyze the problem of vegetation change in China regions, attributing growing season vegetation activity to climate change and human activities and investigating the interactions among different driving variables. The authors performed Mann-Kendall trend analysis, Pearson correlation analysis, and partial correlation analysis. They observed that dynamic relationship between vegetation activity and the driving variables reflected the acclimatization and adaption processes of vegetation, which needs further investigation.

Lastly, two studies are in the fields of groundwater and erosion. Tapoglou *et al.* (2020) propose a study to examine the uncertainty of various aspects of a combined artificial neural network (ANN), kriging and fuzzy logic methodology, which can be used for the spatial and temporal simulation of hydraulic head in an aquifer. This simulation algorithm was applied in a study area in Miami, USA. Francipane *et al.* (2020) develop a geographic object-based image analysis to detect and map gullies based on very high resolution (VHR) imageries, considering that mapping of gullies through conventional field surveying can be an intensive and expensive activity. The authors used a one-meter resolution LIDAR DEM (digital elevation model) to identify gullies. The tool has been calibrated for two relatively large gullies surveyed in the Calhoun Critical Zone Observatory (CCZO) area in the southeastern United States.

The complexity of the problems and the different approaches used within the articles published in this special issue clearly shows the multi-disciplinary feature of the researches focused on water bodies. Thanks to the vivacity of the hydroinformatics community, day by day fundamental advances are made in understanding complex phenomena related to water physics as well as in novel methodologies and tools to drive sustainable and efficient management of natural and artificial water systems.

Guest Editors
Mauro De Marchis
Faculty of Engineering and Architecture, University of Enna Kore, Enna, Italy

Luigi Berardi IWA
Engineering and Geology Department, University 'G. d'Annunzio' of Chieti Pescara, Pescara, Italy

REFERENCES

Braun, M., Piller, O., Deuerlein, J., Mortazavi, I. & Iollo, A. 2020 Uncertainty quantification of water age in water supply systems by use of spectral propagation. *Journal of Hydroinformatics* **22** (1), 111–120.

Chen, C., Li, T., Sivakumar, B., Li, J. & Wang, G. 2020 Attribution of growing season vegetation activity to climate change and human activities in the three-river headwaters region, China. *Journal of Hydroinformatics* **22** (1), 186–204.

Di Cristo, C., Greco, M., Iervolino, M. & Vacca, A. 2020 Interaction of a dam-break wave with an obstacle over an erodible floodplain. *Journal of Hydroinformatics* **22** (1), 5–19.

Dimitriadis, P. & Koutsoyiannis, D. 2020 The mode of the climacogram estimator for a Gaussian Hurst-Kolmogorov process. *Journal of Hydroinformatics* **22** (1), 160–169.

Fellini, S., Vesipa, R., Boano, F. & Ridolfi, L. 2020 Fault detection in level and flow rate sensors for safe and performant remote-control in a water supply system. *Journal of Hydroinformatics* **22** (1), 132–147.

Francipane, A., Cipolla, G., Maltese, A., La Loggia, G. & Noto, L. V. 2020 Using very high resolution (VHR) imageries within a GEOBIA framework for gully mapping: an application to the calhoun critical zone observator. *Journal of Hydroinformatics* **22** (1), 219–234.

Gibson, M. J., Chen, A. S., Khoury, M., Vamvakeridou-Lyroudia, L. S., Stewart, D., Wood, M., Savic, D. & Djordjevic, S. 2020 A case study of the cascading effects on critical infrastructure in Torbay coastal/pluvial flooding with climate change and 3D visualisation. *Journal of Hydroinformatics* **22** (1), 77–92.

Grasso, S., Libertino, A. & Claps, P. 2020 MULTIRAIN: A GIS-based tool for multi-model estimation of regional design rainfall for scientists and practitioners. *Journal of Hydroinformatics* **22** (1), 148–159.

Pinho, J., Vieira, L., Vieira, J., Venâncio, S., Simões, N., Marques, J. & Seabra Santos, F. 2020 Assessing causes and associated water levels for an urban flood recurring to hydroinformatic tools. *Journal of Hydroinformatics* **22** (1), 61–76.

Real-Rangel, R. A., Pedrozo-Acuña, A., Breña-Naranjo, J. A. & Alcocer-Yamanaka, V. H. 2020 A drought monitoring framework for data-scarce regions. *Journal of Hydroinformatics* **22** (1), 170–185.

Sambito, M., Di Cristo, C., Freni, G. & Leopardi, A. 2020 Optimal water quality sensor positioning in urban drainage systems for illicit intrusion identification. *Journal of Hydroinformatics* **22** (1), 46–60.

Simone, A., Ciliberti, F. G., Laucelli, D. B., Berardi, L. & Giustolisi, O. 2020 Edge betweenness for water distribution networks domain analysis. *Journal of Hydroinformatics* **22** (1), 121–131.

Tapoglou, E., Varouchakis, E. A., Trichakis, I. C. & Karatzas, G. P. 2020 Hydraulic head uncertainty estimations of a complex artificial intelligence model using multiple methodologies. *Journal of Hydroinformatics* **22** (1), 205–218.

Visanji, Z., Seyed, M. K. S., Johns, M. B., Savic, D. & Memon, F. A. 2020 Optimising wastewater treatment solutions for the removal of contaminants of emerging concern (CECs): a case study for application in India. *Journal of Hydroinformatics* **22** (1), 93–110.

Wang, J., Zhang, J., Liang, D. & Gan, L. 2020 Non-hydrostatic modelling of shallow water flow around a circular array of emergent cylinders. *Journal of Hydroinformatics* **22** (1), 20–32.

Yang, F., Liang, D., Wu, X. & Xiao, Y. 2020 On the application of the depth-averaged random walk method to solute transport simulations. *Journal of Hydroinformatics* **22** (1), 33–45.

Interaction of a dam-break wave with an obstacle over an erodible floodplain

Cristiana Di Cristo, Massimo Greco, Michele Iervolino and Andrea Vacca

ABSTRACT

Because of climate change, flood-prone areas are more and more frequently exposed to potential casualties and damage. The capability of the flow to carry relevant quantities of sediments during these critical events adds to the further complexity of the resulting scenarios. The interaction between the flow and the obstacles in flood-inundated areas contributes to an increase in the hazard level and constitutes a relevant concern in the framework of risk analysis. Despite this relevance, the existing literature on the topic is relatively scarce, especially for the estimation of the forces acting on rigid obstacles in the presence of a mobile bed. In the present paper, a recent two-phase shallow-water morphodynamical model, particularly suited for the analysis of fast geomorphic transients, is applied for the numerical simulation of the propagation of a dam-break wave over an erodible floodplain in the presence of a rigid obstacle. The geometry of the test-case is inspired by a recent fixed-bed study reported in the literature, for which extensive experimental and numerical data concerning the flow field and the dynamic loading against the obstacle are available. Results of the numerical simulations contribute to highlight the effects of the obstacle on the changes in the bottom topography.

Key words | dam break, impact force, mobile bed, two-phase model

Cristiana Di Cristo
Massimo Greco
Andrea Vacca
Università di Napoli 'Federico II',
via Claudio 21, Napoli 80125,
Italy

Michele Iervolino (corresponding author)
Università della Campania 'Luigi Vanvitelli',
via Roma 9, Aversa (CE) 81031,
Italy
E-mail: *michele.iervolino@unicampania.it*

INTRODUCTION

The presence of obstacles, such as bridge piers, is common in rivers and channels, but they may also be encountered in potentially flooded areas. The magnitude of floods caused by intense rainfall is becoming more and more severe owing to climate change (Hoerling *et al.* 2001). In turn, the exposition of initially dry areas to flood waves arising from the failure of embankments or dikes has increased. In these cases, the wave impacts on a variety of obstacles such as buildings, piles, industrial and commercial structures, with damaging consequences. The presence of structures in flooding areas has many aspects that require adequate investigation, especially in case of a severe flood like that following a dam break. First, an appropriate risk control strategy, essential to prevent damage and to reduce the losses, requires the individuation of risk areas and the selection of adequate measures for its mitigation, such as structural countermeasures including longitudinal walls, deflection and redirecting structures (Hung *et al.* 1984). Second, obstacles may considerably affect the flood wave propagation, especially when they are close to the dam and in the first moments after the break (Soares-Frazão & Zech 2007). Third, the impacting force on the obstacles may produce structural damage even compromising their stability.

Finally, the capability of the flood to carry relevant quantities of sediment in the flow is another important aspect. In this case, the sediment concentration influences the characteristic of the mixture flow, moving from the bed and suspended load to debris transport. The interaction between the liquid–solid mixture and the presence of obstacles in the inundated areas both contribute to affect the flood wave propagation. On the one hand, the obstacles

doi: 10.2166/hydro.2019.014

deviate the flow pattern, and, on the other hand, the morphological change induced by the flow may expose the obstacle to unexpected loading condition, therefore contributing to its structural damage. For these reasons, the propagation of flood waves over loose sediment bed and its interaction with obstacles constitute a relevant concern in the framework of risk analysis.

In conclusion, the study of the complex interaction between the flood wave and the obstacles is extremely important to evaluate the hazard levels and to support the design of structural countermeasures. This requires the knowledge of both the hydro- and sediment dynamics and the force acting on the structure by the flow (Thieken *et al.* 2005).

The dynamic impact of a flood wave on a structure and the consequent damage is, in some cases, evaluated by empirical formulas or adopting simplified analysis (e.g. Kelman & Spence 2004). However, these approaches may produce incorrect indications since they are usually derived through site-specific studies. Alternatively, the principal kinematics characteristics of the dynamic impact and the acting force may be numerically reproduced.

The study of a wave impacting on an obstacle has been widely investigated both experimentally and numerically considering clear water flowing on non-erodible beds (e.g. Bukreev 2009; Chen *et al.* 2014); debris flows (e.g. Canelli *et al.* 2012; Scheidl *et al.* 2013), and snow and dry avalanches (e.g. Tai *et al.* 2001; Chiou *et al.* 2005; Teufelsbauer *et al.* 2009; Cui & Gray 2013; Faug 2015).

Among the numerical studies concerning debris flows, Kattel *et al.* (2018) modelled debris impacting tetrahedral obstacles of different dimensions, number and orientation, reproducing the behaviour of the liquid–solid mixture with the quasi-three-dimensional two-phase model of Pudasaini (2012). The presence of obstacles may increase the solid- and fluid-phase separation, and it strongly influences the wave propagation. Debris flow propagation in urban areas has been recently simulated by Gao *et al.* (2017) with a depth-integrated continuum model, demonstrating that the obstacles increase the depth and the velocity of the flow and the impact pressure. More recently, Di Cristo *et al.* (2019) reproduced the impact force of a mud flow on a rigid obstacle with both a single-phase, shear-thinning power-law model and a two-phase model, which separately considers the liquid and solid phases. The comparison suggests that in the situations where the liquid and solid phases remain mixed, the essential features of the interaction with the obstacles and the maximum impact force are similarly predicted by the two models, while some differences are registered in the cases characterized by phase separation.

Concerning clear water flows over non-erodible beds, among the different experimental studies, only few evaluated the impact force. Lobovsky *et al.* (2014) performed laboratory tests measuring the load produced from a dam break on a vertical wall using miniaturized pressure sensors, while Aureli *et al.* (2015) evaluated the impact force on a single rigid obstacle.

With regard to numerical studies, Wang *et al.* (2000) used Saint Venant Equations with a second-order total variation diminishing finite-difference method for solving 1D and 2D dam-break in the presence of obstacles, also discussed the effect of bed slope, bottom friction and the tailwater/reservoir depth ratio. Successively, many other works considered the shallow-water formulation as a reasonable approach for the simulation of these complex processes. For example, Shige-eda & Akiyama (2003) investigated two-dimensional (2D) flood flows and the hydrodynamic force acting on structures, showing that the prediction accuracy of the shallow-water model is reasonable for both hydrodynamics and force on structures. Bukreev (2009) successfully applied a shallow-water model to predict the dynamic action exerted by a dam-break wave on a vertical wall. Soares-Frazao & Zech (2008) and El Kadi Abderrezzak *et al.* (2009) investigated the potentiality of a depth-averaged shallow-water model in reproducing the propagation in urban areas of both flash floods and dam-break waves. Numerical results, in terms of both flow depths and velocities, fairly agree with experimental ones, although some discrepancies were observed around buildings, where the flow is strongly 3D. More recently, Aureli *et al.* (2015) compared the predictions of a 2D depth-averaged model, a 3D Eulerian two-phase model and a 3D smoothed particle hydrodynamics model, in the simulation of the forces exerted by a dam-break wave on a rigid squat structure. Comparing the numerical and the experimental results, the authors found that the error in the peak load for the

2D model based on the shallow-water approach was in the order of 10%.

As far as the simultaneous presence of obstacles and of an erodible floodplain are concerned, from the experimental point of view, the tests by Palumbo *et al.* (2008) and the test-cases studied under the NSF-Pire project (Soares-Frazão *et al.* 2012), regarding a dam-break in an erodible channel with a sudden enlargement, contributed to shed some light on the basic elements of the interaction of the flood wave with rigid walls. However, they neither considered any obstacle in the floodplain nor evaluated the impact forces.

The numerical analyses which were subsequently developed, support the idea that for these kinds of processes, the shallow water approximation represents a suitable framework, provided that adequate modelling of the non-equilibrium sediment transport is accounted for. To match this latter requirement, several non-equilibrium models have been proposed, which can be grouped into mixture single layer (e.g. Wu & Wang 2007), multi-layer (e.g. Capart & Young 2002; Savary & Zech 2007; Li *et al.* 2013; Swartenbroekx *et al.* 2013) or multi-phase (e.g. Dewals *et al.* 2011; Greco *et al.* 2012; Rosatti & Begnudelli 2013; Di Cristo *et al.* 2016).

The objective of this paper is to numerically analyse the propagation of a dam-break wave over an erodible floodplain in the presence of a rigid obstacle. The results of both fixed and erodible bed conditions are compared for understanding the role of the sediment mobility on the hydrodynamics and the impact force. The shallow-water two-phase model of Di Cristo *et al.* (2016) is adopted for reproducing the impact of a dam break considering the same geometry of the fixed-bed test case by Aureli *et al.* (2015). In this study, both fixed bed and mobile bed conditions are reproduced, and in the latter case, two different tests characterized by two different sediments constituting the loose bed are considered. The preliminary study of Di Cristo *et al.* (2018a) shows a comparison between the fixed and mobile bed tests. In the present paper, the comparison is completed with the results presented in terms of the temporal evolution of the flow field, the bottom deformation and the impact force on the obstacle.

The article is organized as follows. In the 'Material and Methods' section, both the test-case and the morphodynamical model adopted for the numerical simulations are described. Some details concerning the numerical solution method are also given. In the section 'The Considered Test-Case', results of the numerical simulations are presented and discussed. Finally, the main conclusions are drawn.

MATERIAL AND METHODS

The geometry of the test-case considered for the numerical simulations has been selected from the existing literature on clear water dam-break waves impacting against rigid obstacles over a fixed bed. Namely, the study carried out by Aureli *et al.* (2015) has been considered. In what follows, both the test-case and the morphodynamical model adopted for the numerical simulations are described.

THE CONSIDERED TEST-CASE

Figure 1 reproduces the plan view of the test-case by Aureli *et al.* (2015). The flow is caused by the sudden removal of the gate, which triggers the propagation of a dam-break wave over the downstream floodable area.

In the numerical simulations, the floodplain can be assumed as non-erodible or constituted by loose sediment, namely uniformly graded sand. In the former case, the resulting layout exactly coincides with the scheme investigated by Aureli *et al.* (2015), while in the latter case two scenarios have been considered by changing the sediment diameter.

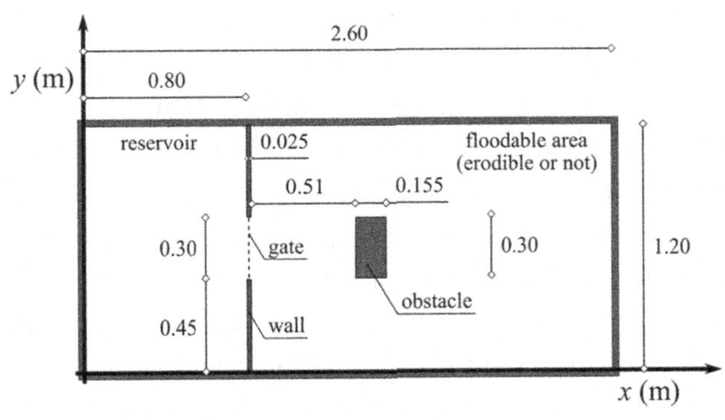

Figure 1 | Sketch of the considered test-case (dimensions are in metres).

THE MATHEMATICAL MODEL

A brief description of the model presented in Di Cristo *et al.* (2016) is provided below, whereas the reader is referred to the original reference for further details about the underlying assumptions and the subsequent limitations, along with a more exhaustive description of the numerical method employed for the simulation. Since in the present application the suspended load has been neglected, the mathematical model is constituted of the equations expressing the mass and momentum conservation for the liquid (Equations (1) and (3)) and the solid phase transported as bedload (Equations (2) and (4)). Finally, Equation (5) describes the evolution of the loose bed. They read as follows:

$$\frac{\partial \delta_l}{\partial t} + \nabla \cdot (\delta_l \mathbf{U}_l) - p e_B = 0 \tag{1}$$

$$\frac{\partial \delta_s}{\partial t} + \nabla \cdot (\delta_s \mathbf{U}_s) - (1-p) e_B = 0 \tag{2}$$

$$\frac{\partial \delta_l \mathbf{U}_l}{\partial t} + \nabla \cdot (\delta_l \mathbf{U}_l \mathbf{U}_l) + \nabla\left(\frac{gh^2}{2}\right) + gh\nabla(z_B) + \mathbf{S}_l = 0 \tag{3}$$

$$\frac{\partial \delta_s \mathbf{U}_s}{\partial t} + \nabla \cdot (\delta_s \mathbf{U}_s \mathbf{U}_s) + \frac{r}{r+1}\nabla\left(\frac{g\delta_s^2}{2C_s}\right) + g\delta_s\frac{r}{r+1}\nabla(z_B) + \mathbf{S}_s = 0 \tag{4}$$

$$\frac{\partial z_B}{\partial t} + e_B = 0 \tag{5}$$

in which t is the time, g denotes the gravity acceleration, $r = (\rho_s - \rho_l)/\rho_l$ with ρ_l and ρ_s are constant liquid and solid densities, respectively. As far as the flow variables are concerned, δ_l denotes the liquid-phase volume for the unit bottom surface and δ_s is the solid-phase volume transported as bedload for the unit bottom surface. Hence, the free surface and bottom elevation are denoted by z_w and z_B, respectively, $h = z_w - z_B = \delta_l + \delta_s$. $\mathbf{U}$ is the phase-averaged velocity vector, with the subscript l and s used to denote the water and the solid phase, respectively. For both phases, the second-order tensor $\mathbf{UU}$ denotes the diadic product of the phase-averaged velocity with itself. C_s is the bedload volume concentration and e_B is the bottom erosion/deposition rate.

The source terms of momentum equations $\mathbf{S}_l$ and $\mathbf{S}_s$ are given by the following formulas:

$$\mathbf{S}_l = \frac{\mathbf{U}_l}{C_{Ch}^2}|\mathbf{U}_l| - (r+1)\mu_d g\delta_s \frac{r}{r+1}\frac{\mathbf{U}_s}{|\mathbf{U}_s|} - \alpha(r+1)\mathbf{U}_s|\mathbf{U}_s| + g\delta_s s_B + \frac{\mathbf{D}}{\rho_l} \tag{6}$$

$$\mathbf{S}_s = \mu_d g\delta_s \frac{r}{r+1}\frac{\mathbf{U}_s}{|\mathbf{U}_s|} + \alpha\mathbf{U}_s|\mathbf{U}_s| - \frac{\mathbf{D}}{\rho_l} \tag{7}$$

where μ_d is the dynamic friction coefficient, C_{Ch} denotes the dimensionless Chezy coefficient, s_B is the bottom slope, $\mathbf{D}$ is the drag force of the liquid on the solid particle and α is the coefficient needed to evaluate the collisional stresses. In Equation (7), the first two terms represent the ratio $\tau_{B,s}/\rho_s$, where the bottom shear stress, $\tau_{B,s}$ accounts for both frictional and interparticle collisional stresses. In Equation (6), the first four terms represent the ratio $\tau_{B,l}/\rho_l$, where the bottom shear stress on the liquid phase, $\tau_{B,l}$, is expressed as the difference between the shear stress which would act upon the bottom in the absence of sediment transport and the momentum transferred to the solid phase $\tau_{B,s}$. The drag force $\mathbf{D}$ is evaluated as follows:

$$\mathbf{D} = \rho_l C_D \frac{\delta_s}{d}(\mathbf{U}_l - \mathbf{U}_s)|\mathbf{U}_l - \mathbf{U}_s| \tag{8}$$

where C_D is a bulk drag coefficient and d is the sediment particle diameter. The bottom entrainment/deposition is evaluated by means of the following formula (Pontillo *et al.* 2010):

$$e_B = w_s \frac{T^{3/2} - C_s}{1-p} \tag{9}$$

in which w_s denotes the sediment settling velocity, corrected to account for its dependence on C_s through the well-known semi-empirical formula by Richardson & Zaki (1954). The dimensionless mobility parameter T, which accounts for the excess of the mobilizing stresses onto the bottom surface with respect to the resisting ones (Van Rijn 1984), is evaluated through the following formula:

$$T = \frac{|\tau_{B,l} + \tau_{B,s} - \tau_c - \tau_B|}{|\tau_c + \tau_B|} \quad (10)$$

where τ_c is the threshold shear stress for particle motion and τ_B is the magnitude of Mohr–Coulomb stress at the bottom, depending on the static friction coefficient μ_s. Even if the coefficients α and C_D could be estimated from the existing empirical formulas, they are evaluated in the model by means of the following relations, deduced to comply with the equilibrium empirical formulas for transport at capacity (Di Cristo *et al.* 2016):

$$\alpha = \frac{(1 - c_1) - k_1(\mu_d - s_B)}{(r+1)k_2^2}, \qquad C_D = \frac{1 - c_1}{k_1} \frac{\rho_l g d r}{\left[C_{Ch}\tau_0^{1/2} - k_2(\tau_0 - \tau_c)^{1/2}\right]^2} \quad (11)$$

where k_1 and k_2 are the following two dimensionless coefficients:

$$k_1 = \frac{1}{2\mu_s} \frac{2 + (1-p)^{2/3}}{1 + (1-p)^{2/3}} \quad (12)$$

$$k_2 = \frac{K_{MPM}}{k_1} \quad (13)$$

In Equation (13), K_{MPM} is the Meyer–Peter and Müller formula coefficient and c_1 is a dimensionless model parameter. With the above closures, the model application requires the definition of only three independent dimensionless parameters. Two of them, C_{Ch} and K_{MPM}, may be evaluated from the standard literature. The remaining free model parameter c_1 is bounded between the lower and upper limits theoretically deduced, with very little sensitivity of the results to its variation (Di Cristo *et al.* 2016).

The system of Equations (1)–(5) is solved with a numerical code developed by the authors employed also in Di Cristo *et al.* (2016) and Di Cristo *et al.* (2018b). The numerical method, using 2D unstructured quadrilateral meshes, relies on a mixed cell-centred (CCFV) and node-centred finite-volume discretization. The former is adopted for the variables δ_l, δ_s, $\mathbf{U}_l$ and $\mathbf{U}s$, defined at the grid cell centres, the latter for the bed elevation z_B, with the control volumes constructed around the mesh nodes by the median-dual partition (Barth & Jespersen 1989; Delis *et al.* 2011). The numerical fluxes in the CCFV discretization of Equations (1)–(4) are calculated through the first-order Harten–Lax–Van Leer (HLL) scheme (Harten *et al.* 1983), with second-order reconstruction of the free-surface elevation for subcritical flows, and an appropriate treatment of the bed slope source term is included (Greco *et al.* 2008).

RESULTS AND DISCUSSION

The model discussed in the previous section has been applied to the numerical simulation of the problem sketched in Figure 1. In the first simulation, the floodplain is assumed to be non-erodible, while in the second series of simulations it is assumed constituted by loose sediments. Two scenarios have been considered by changing the sediment diameter from 5×10^{-4} m (denoted as Test 1), representative of fine sand, to 5×10^{-3} m (denoted as Test 2), which is typical of very coarse sand. In the simulations, the following values of the parameters are assumed: $\rho_s = 2{,}650$ kg/m^3, $\mu_d = 25°$, $\mu_s = 38°$, $K_{KMPM} = 8$ and $c_1 = 0.25$. To isolate the effect of bed mobility, in both erodible and non-erodible tests, the same value of the dimensionless Chezy coefficient, $C_{Ch} = 12$, is adopted. A computational mesh with $\Delta x = \Delta y = 5 \times 10^{-3}$ m used, along with $\Delta t = 1/1024$ s. Based on the simulated flow fields, the impact force is evaluated by numerically computing the following integral (Di Cristo *et al.* 2018b):

$$\mathbf{F} = \rho_l \int_\sigma \left(\delta_l \mathbf{U}_l U_{l,n} + g\frac{h^2}{2}\hat{\mathbf{n}}\right) d\sigma + \rho_s \int_\sigma \left(\delta_s \mathbf{U}_s U_{s,n} + \frac{r}{r+1}\frac{g\delta_s^2}{2C_s}\hat{\mathbf{n}}\right) d\sigma \quad (14)$$

where σ denotes the boundary of the obstacle and $\hat{\mathbf{n}}$ is the corresponding normal unit vector. The values of the flow variables in the cells adjacent to the obstacle are used.

Figure 2 shows the time history of the impact force F on the obstacle for the fixed and mobile bed cases, along with the simulation results from the shallow-water model of Aureli *et al.* (2015). As far as the non-erodible case is

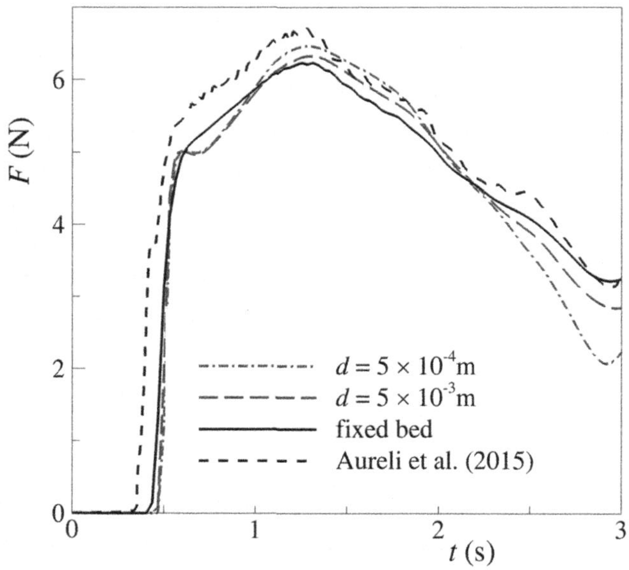

Figure 2 | Time history of the impact forces over the downstream obstacle for fixed and mobile bed cases.

concerned, the results fairly agree with those of Aureli *et al.* (2015), even though the force values predicted herein are slightly smaller. This is attributed to the different resistance formula employed in Aureli *et al.* (2015), i.e., Manning instead of Chezy, and to the value of the dimensionless Chezy coefficient used in the present calculations. Indeed, as shown in Di Cristo *et al.* (2019), increasing the C_{Ch} value up to 25, the agreement between the results of two simulations is strongly improved. After the impact of the wave on the obstacle ($t \sim 0.5$ s), an abrupt increase of the force is observed, which reaches its maximum value ($F \sim 6$ N) at $t \sim 1.25$ s, and then it quasi-monotonically decreases, reaching half of the peak value at $t \sim 3.0$ s, when the flood wave completely surrounds the obstacle. Bottom mobility only marginally influences the load dynamics on the obstacle, especially in the first instants. The time at which the impact and the peak force occur, along with the peak force, only slightly depends on the sediment composing the mobile bottom.

A preliminary discussion of the local characteristics of the flow field has been presented in Di Cristo *et al.* (2018a). That discussion focused mainly on the results at $t = 3$ s, at which the flood wave reattaches behind the obstacle. This time is representative of the final part of the investigated process, as the bottom geometry changes slightly going on. In the previous paper, it has been noted that the fine sediment promotes the reattachment of the flow and a reduced flow height in front of the obstacle is observed, which is consistent with the lower values of the impact force (Figure 2). As far as the bottom topographies are concerned, the case with the fine sediment is characterized by a more pronounced scour hole just downstream the original dam position and at the upstream corners of the obstacle (with depths of about 0.07 m). Finally, it is observed that the interaction with the coarser sediment produces a reduction of the momentum in the region close to the original dam.

In what follows a more detailed analysis is provided, which is mainly focused on the near- and mid-field phase of the evolution of the process ($t \leq 2.0$ s). Namely, the free- and bed-surface elevation of the solid-phase volume (for the unit bottom surface) transported as bed load (δ_s) for the two cases $d = 5 \times 10^{-4}$ m (denoted as Test 1) and $d = 5 \times 10^{-3}$ m (denoted as Test 2) are discussed in Figures 3–7 for different instants. Moreover, for the sake of comparison, the free-surface elevation of the fixed bed (denoted as Test 0) is also shown.

At $t = 0.25$ s, Figure 3(b) and 3(c) indicate that the free-surface elevation for Test 2 and Test 0 shows only marginal differences. On the other hand, in the $d = 5 \times 10^{-4}$ m case (Figure 3(a)), the shape of the wave is characterized by a more pronounced diffusion in the transversal direction. Namely, the longitudinal distance reached by the wave front does not substantially differ from the other two tests, whereas the width of the wave body in the transversal direction is sensibly larger. Indeed, while in Test 1 the bed load is already appreciable, in the $d = 5 \times 10^{-3}$ m case, it is only perceivable (Figure 3(d) and 3(e)). At the same instant, the erosion process is taking place in the $d = 5 \times 10^{-4}$ m case while it is negligible in Test 2 (Figure 3(f) and 3(g)).

For all the tests, the time $t = 0.5$ s is very close to the moment in which the wave impacts against the obstacle (see Figure 2). Soon after, the wave is deflected and splits into two symmetrical bodies flowing laterally to the obstacle (Figure 4(a)–4(c)). The comparison among Figure 4(a)–4(c) indicates that the fixed-bed case is characterized by the highest value of the celerity propagation, as suggested by the time evolution of the impact force shown in Figure 2.

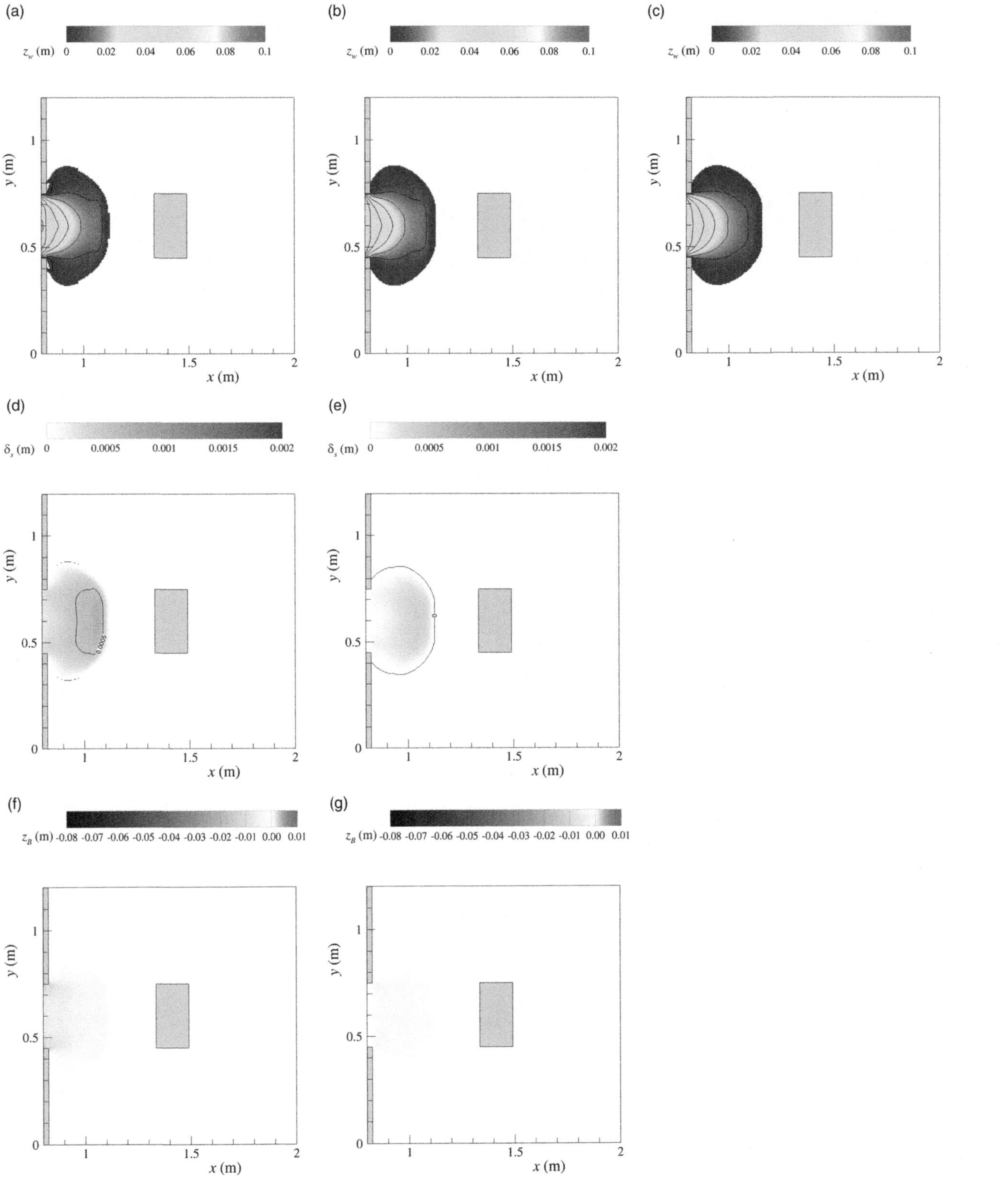

Figure 3 | Results at $t=$ 0.25 s. Free-surface elevation for $d = 5\times10^{-4}$ m (a), $d = 5\times10^{-3}$ m (b) and fixed bed (c). Bedload solid-phase volume (for the unit bottom surface) for $d = 5\times10^{-4}$ m (d) and $d = 5\times10^{-3}$ m (e). Bottom erosion/deposition for $d = 5\times10^{-4}$ m (f) and $d = 5\times10^{-3}$ m (g).

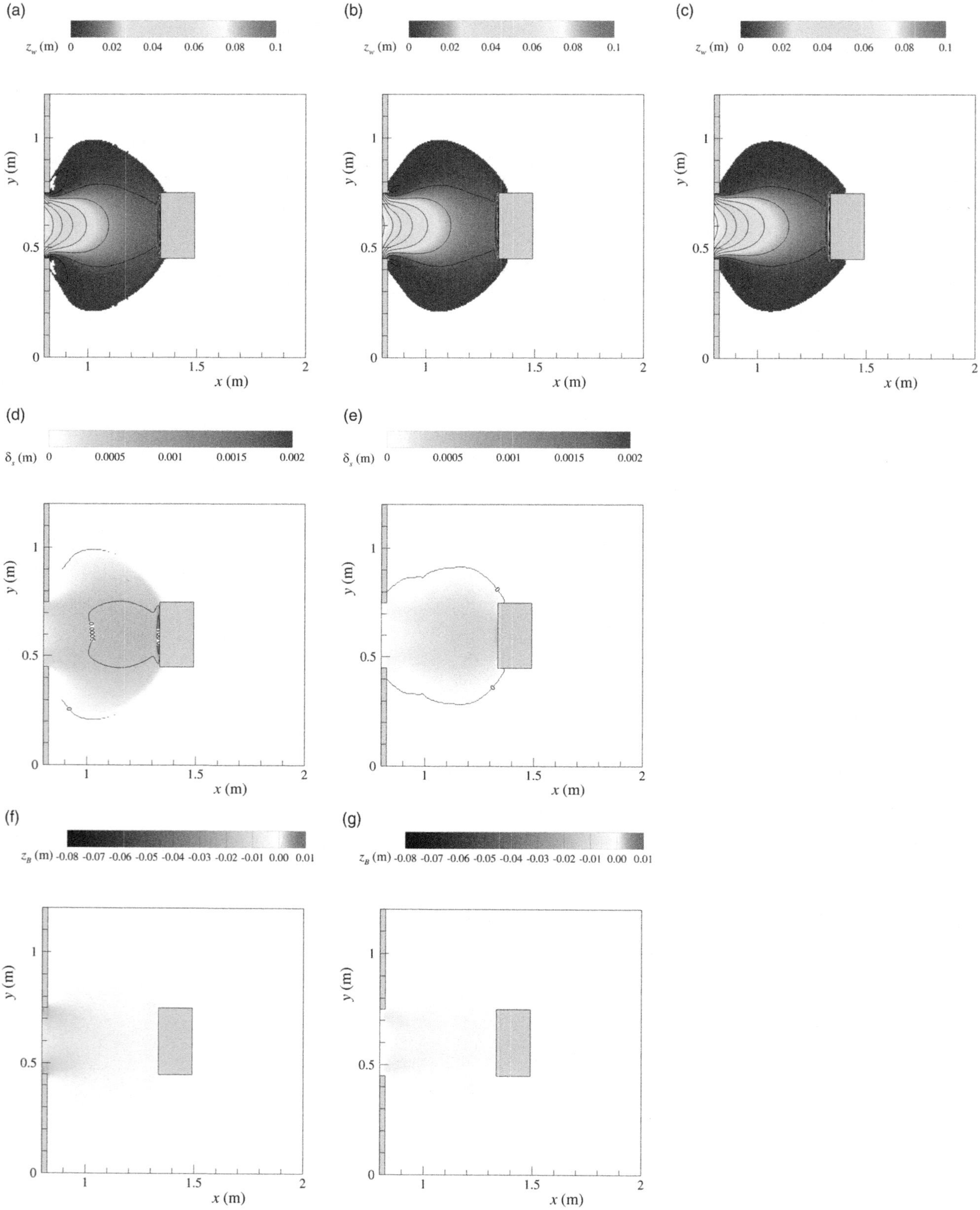

Figure 4 | Results at $t = 0.5$ s. Free-surface elevation for $d = 5 \times 10^{-4}$ m (a), $d = 5 \times 10^{-3}$ m (b) and fixed bed (c). Bedload solid-phase volume (for the unit bottom surface) for $d = 5 \times 10^{-4}$ m (d) and $d = 5 \times 10^{-3}$ m (e). Bottom erosion/deposition for $d = 5 \times 10^{-4}$ m (f) and $d = 5 \times 10^{-3}$ m (g).

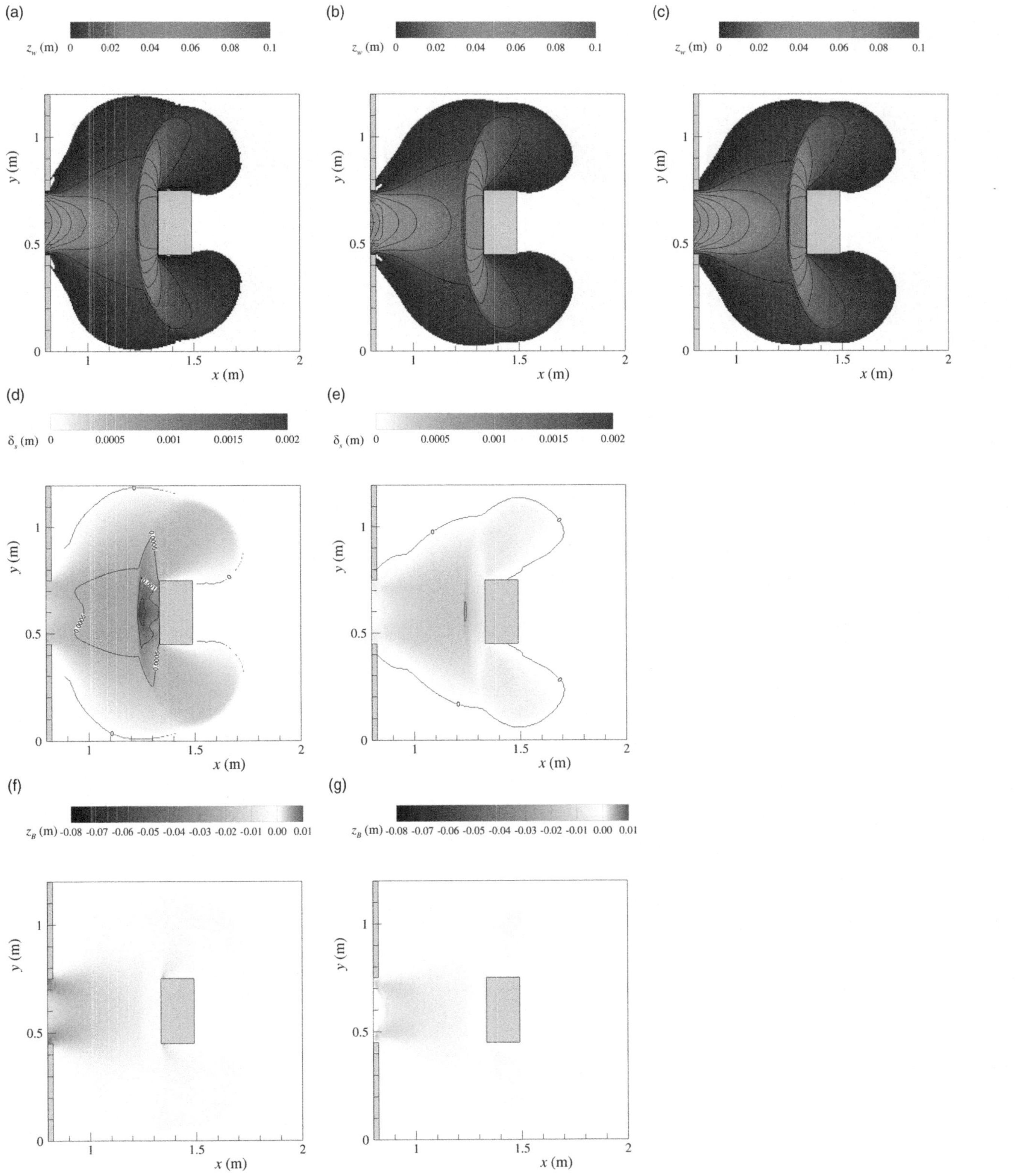

Figure 5 | Results at $t = 1.0$ s. Free-surface elevation for $d = 5 \times 10^{-4}$ m (a), $d = 5 \times 10^{-3}$ m (b) and fixed bed (c). Bedload solid-phase volume (for the unit bottom surface) for $d = 5 \times 10^{-4}$ m (d) and $d = 5 \times 10^{-3}$ m (e). Bottom erosion/deposition for $d = 5 \times 10^{-4}$ m (f) and $d = 5 \times 10^{-3}$ m (g).

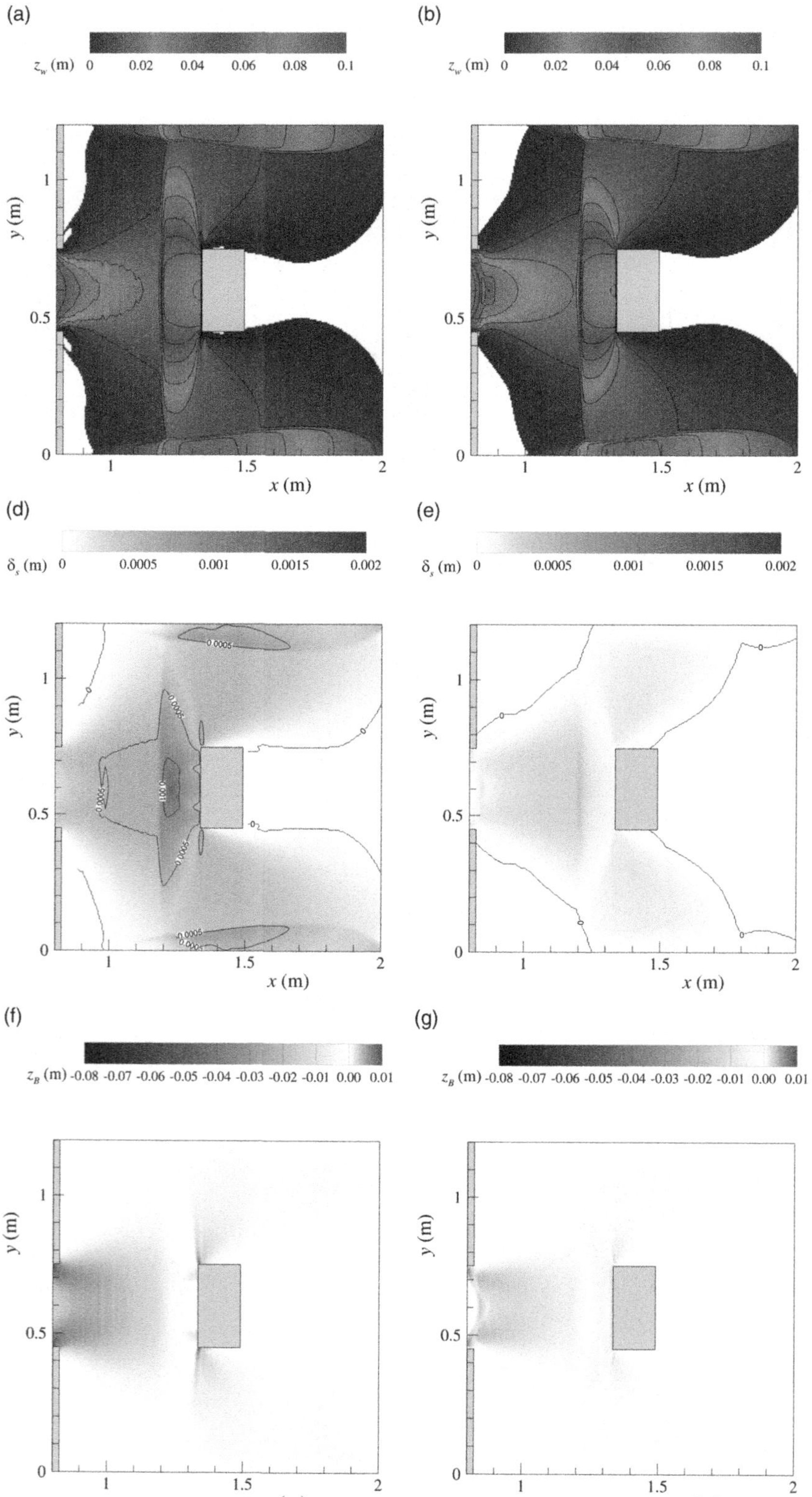

Figure 6 | Results at $t = 1.5$ s. Free-surface elevation for $d = 5 \times 10^{-4}$ m (a), $d = 5 \times 10^{-3}$ m (b) and fixed bed (c). Bedload solid-phase volume (for the unit bottom surface) for $d = 5 \times 10^{-4}$ m (d) and $d = 5 \times 10^{-3}$ m (e). Bottom erosion/deposition for $d = 5 \times 10^{-4}$ m (f) and $d = 5 \times 10^{-3}$ m (g). Please refer to the online version of this paper to see this figure in colour: http://dx.doi.org/10.2166/hydro.2019.014.

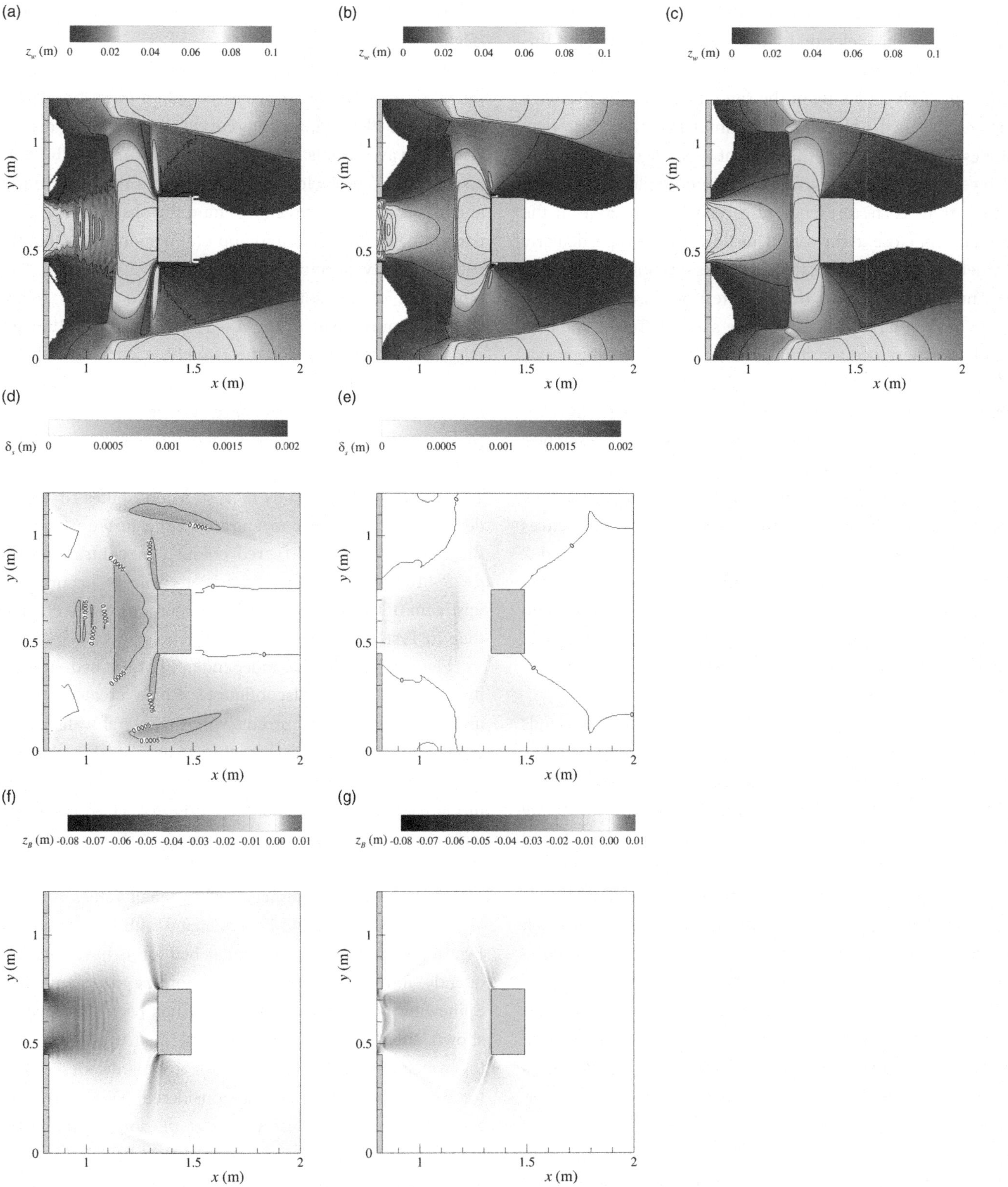

Figure 7 | Results at $t = 2.0$ s. Free-surface elevation for $d = 5 \times 10^{-4}$ m (a), $d = 5 \times 10^{-3}$ m (b) and fixed bed (c). Bedload solid-phase volume (for the unit bottom surface) for $d = 5 \times 10^{-4}$ m (d) and $d = 5 \times 10^{-3}$ m (e). Bottom erosion/deposition for $d = 5 \times 10^{-4}$ m (f) and $d = 5 \times 10^{-3}$ m (g).

Interestingly, the presence of erodible bottom appears to slow down the celerity of the wave, even though this effect is marginally influenced by the sand diameter value. Globally, independently of the sand diameter, the presence of erodible bed only marginally influences the hydrodynamics of the wave at this instant (see Figure 4(a)–4(c)). However, the main peculiarity of the erodible bed cases at this instant is the presence of a recirculation zone in the region very close to the gate. This flow feature is particularly evident in Test 1 (Figure 4(a)), whereas it is not found in Test 0. The δ_s maps (Figure 4(d) and 4(e)) put in evidence that, owing to the large diameter of the sand, in Test 2, the bedload transport occurs mainly in the centre of the wave, where the high velocity values are able to entrain sediments from the bottom (Figure 4(e)). In contrast, in the $d = 5 \times 10^{-4}$ m case (Figure 4(a)), the δ_s footprint strongly resembles one of the liquid phases and therefore almost everywhere the fluid flow is able to entrain sediment from the erodible bottom inducing the bedload movement. The differences in the erosion/deposition process between Tests 1 and 2 are clearly shown in Figure 4(f) and 4(g). The high velocity values very close to the gate produce an intense erosion process, which is more evident for the simulation with fine sand (Figure 4(f)) rather than in the $d = 5 \times 10^{-3}$ m case (Figure 4(g)).

At $t = 1.0$ s (Figure 5), in all tests, the wave surrounds the obstacle, laterally deflecting and overcoming it. However, the deflected wave has not reached the lateral walls yet. Moreover, created by the reflection of the impacting wave against the obstacle, a discontinuous front just upstream the obstacle is visible in all cases. While in Test 2 the shape of the wave still strongly resembles one of the fixed-bed simulation (Figure 5(b) and 5(c)), the small value of sand diameter in Test 1 causes a transversal diffusion stronger than in Test 0. In the $d = 5 \times 10^{-4}$ m case, the intense bedload, witnessed by the high δ_s values just upstream of the obstacle, suggests that a strong momentum transfer from the liquid phase towards the solid one is occurring (Figure 5(d)). The erosion process, which is particularly violent in the zone just downstream of the gate, induces a consistent reduction of the surface bed elevation (Figure 5(f)). Moreover, starting from the upstream edges of the obstacle, two thin eroded limbs are clearly appreciable. In Test 2, the bed load activity is weaker than in Test 1, and it takes place in a smaller portion of the channel (Figure 5(e) and 5(g)).

At $t =$ 1.5 s in all tests (Figure 6(a)–6(c)), the impact of the wave with the lateral walls has taken place because of the lateral deflection, and moreover, two additional discontinuous wave fronts are generated symmetrically with respect to the obstacle. The wave reaches the flume outlet and leaves a large dry region behind the obstacle. The comparison between Figure 6(a) and 6(c) demonstrates that the bed erodibility promotes the lateral wave distortion. The movement of the solid phase is seen to strictly follow the water movement in the $d = 5 \times 10^{-4}$ m case, although, due to the interaction of the two-phase current with the solid walls (obstacle and lateral walls), three localized islands with high values of δ_s have been created (Figure 6(d)). In Test 2, the solid transport is less vigorous than in Test 1, and it occurs only dowstream of the obstacle (Figure 6(e)). The intense bedload leads in Test 1 contributes to a solid deposition in a very concentrated zone just upstream of the obstacle (Figure 6(f), red colour). In contrast, in the $d = 5 \times 10^{-3}$ m case, the deposition process occurs, still upstream of the obstacle, but in a larger part of the channel than in Test 1 (Figure 6(g)).

Figure 7 shows that, independently of the bed erodibility, at $t = 2.0$ s the discontinuous fronts created by the interaction of the two currents with the lateral walls propagate symmetrically towards the centre of the channel, significantly reducing the extension of the dry zone present in the back of the obstacle. Such a modification is particularly evident in the $d = 5 \times 10^{-4}$ m case (Figure 7(a)). In Test 1, the erodibility of the bed determines the presence of superficial waves characterized by small values of wavelength which are seen to be associated with bed forms (see Figure 7(a), 7(d) and 7(f)). Similar bed forms have been frequently found in analogous geomorphic transients (e.g. Spinewine & Zech 2010) and, as it has been theoretically shown, their formation may be described even by a shallow-water approach, provided that a dynamic description for the sediment transport is considered (Di Cristo *et al.* 2006; Vesipa *et al.* 2012; Greco *et al.* 2018). Finally, the zones close to the upstream edge of the obstacle in which a deposition started at $t = 1.0$ s increase their extension and, moreover, a deposition process occurs symmetrically close to the lateral walls. In Test 2, the presence of bed forms is not detected, and only a deposition process is found upstream of the obstacle but in a smoother way than in Test 1. The geometry of this depositional pattern is elongated in the transversal for an extension approximately three times the obstacle width. No appreciable deposition is detected alongside the lateral walls.

The analysis of the present process with a two-phase morphodynamical model provides the opportunity to discuss the effects of the grain size on the dynamics of the solid phase during the impact against the obstacle. To this aim, Figure 8 compares the velocity vector plots referring to the two Test Cases 1 and 2, representing the velocity field of both the liquid and the solid phases at $t = 1.0$ s. At this time, the wave overcomes the obstacle, and the fluid/sediment mixture is deflected in the lateral direction.

The results obtained with the fine sand (Figure 8(a)) show that the velocity vectors of the solid phase have a smaller magnitude of the liquid-phase ones, but they present nearly the same direction. The magnitude of maximum difference between the velocities, i.e. the maximum of the module of the difference between the liquid and the solid velocities, is 0.44 m/s. In contrast, the coarse sand (Figure 8(b)) is characterized by a much more pronounced resistance to adapt to the change in the flow direction induced by the impact against the obstacle, and a sensible disalignment of the velocity vectors of the two phases is detected in the region close to the obstacle corner. In this case, the magnitude of maximum difference between the velocity of the solid and liquid phases is 0.64 m/s. This result and the comparison between Figure 8(a) and 8(b) demonstrate that the hypothesis of collinearity of the motion direction of the two phases, which is assumed in the single phase and the mixture models, may fail in the presence of coarse sand, whereas it is reasonably satisfied if the sediment is sufficiently fine.

CONCLUSIONS

In the present paper, a numerical study of the impact of a dam-break wave propagating over an erodible floodplain against a rigid obstacle has been discussed. The investigated condition has been inspired by a laboratory experiment recently reported in the literature with a non-erodible bed. To cope with the features of the complex morphodynamics characterizing the present test case, a recent shallow-water two-phase model, numerically integrated with a proper method, has been adopted. Results have been analysed in terms of impact force on the obstacle, flow field and bottom deformation. The effect of the bottom erodibility has been discussed by comparing with the corresponding fixed-bed condition. Moreover, the effect of the diameter of the sediment constituting the loose bed on the two-phase

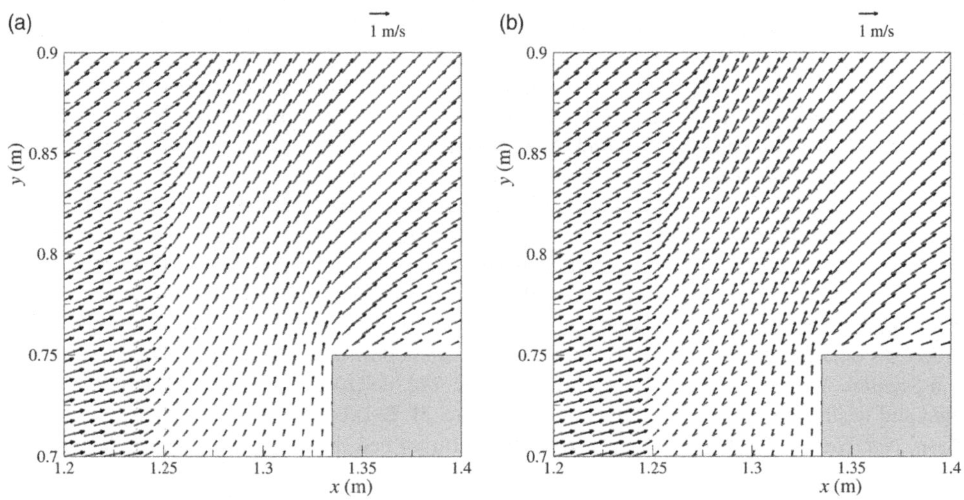

Figure 8 | Velocity vector plots for $d = 5 \times 10^{-4}$ m (a) and $d = 5 \times 10^{-3}$ m (b). Black (red) arrows refer to liquid (solid)-phase velocities. Vector density has been reduced to improve readability. Please refer to the online version of this paper to see this figure in colour: http://dx.doi.org/10.2166/hydro.2019.014.

flow field and on the bottom evolution has been discussed for fine and coarse sand.

Bottom mobility has been found to affect the shape of the wave starting from the very beginning of the geomorphic transient, influencing the propagation of the wet-dry front in the transversal direction, especially for fine sand. As far as the evolution of the erodible bottom is concerned, fine sand has been found to determine more pronounced scour at the original dam location and at the upstream corners of the impacted obstacles compared to the case with coarse sand. On the other hand, the coarse sediment induces the appearance of a larger deposit area immediately upstream of the obstacle, elongated in the transversal direction with a width of about three times the obstacle. Additionally, the analysis of the flow field of solid and liquid phases has shown that under similar circumstances, the hypothesis of collinearity of the motion direction of solid and liquid phases, adopted in the single phase and mixture models, may fail in the presence of coarse sand.

ACKNOWLEDGEMENTS

The work described in the present paper was realized in the framework of the project MISALVA and financed by the Italian Minister of the Environment, Land Protection and Sea (CUP H36C18000970005).

REFERENCES

Aureli, F., Dazzi, S., Maranzoni, A., Mignosa, P. & Vacondio, R. 2015 Experimental and numerical evaluation of the force due to the impact of a dam-break wave on a structure. *Adv. Wat. Res.* **76**, 29–42.

Barth, T. J. & Jespersen, D. C. 1989 *The design and application of upwind schemes on unstructured meshes.* AIAA Paper 89-0366.

Bukreev, V. I. 2009 Force action of discontinuous waves on a vertical wall. *J. Appl. Mech. Tech. Phys.* **50** (2), 278–283.

Canelli, L., Ferrero, A. M., Migliazza, M. & Segalini, A. 2012 Debris flow risk mitigation by means of rigid and flexible barriers – experimental tests and impact analysis. *Nat. Hazard Earth Syst. Sci.* **12**, 1693–1699. doi:10.5194/nhess-12-1693-2012.

Capart, H. & Young, D. 2002 Two-layer shallow water computations of torrential geomorphic flows. In: *Proc. of River Flow 2002.* Swets & Zeitlinger, Lisse, Netherlands, pp. 1003–1012.

Chen, H. Y., Xu, W. L., Deng, J., Xue, Y. & Li, J. 2014 Experimental investigation of pressure load exerted on a downstream dam by dam-break flow. *J. Hydr. Eng.* **140** (2), 199–207. doi:1943-7900.0000743.

Chiou, M. C., Wang, Y. & Hutter, K. 2005 Influence of obstacles on rapid granular flows. *Acta. Mech.* **175**, 195–122. doi:10.1007/s00707-004-0208-9.

Cui, P. & Gray, J. M. N. T. 2013 Gravity driven granular free-surface flow around a circular cylinder. *J. Fluid. Mech.* **720**, 314–337.

Delis, A. I., Nikolos, I. K. & Kazolea, M. 2011 An unstructured node-centered finite volume scheme for shallow water flows with wet/dry fronts over complex topography. *Arch. Comput. Meth. Eng.* **18**, 57–118.

Dewals, B., Rulot, F., Erpicum, S., Archambeau, P. & Pirotton, M. 2011 *Advanced Topics in Sediment Transport Modelling: Non-Alluvial Beds and Hyperconcentrated Flows, Sediment Transport.* Available from: http://www.intechopen.com/books/sediment-transport/advanced-topics-in-sediment-transport-modelling-non-alluvial-beds-and-hyperconcentrated-flows.

Di Cristo, C., Iervolino, M. & Vacca, A. 2006 Linear stability analysis of a 1-D model with dynamical description of bed load transport. *J. Hydr. Res.* **44**, 480–487. doi:10.1080/00221686.2006.9521699.

Di Cristo, C., Greco, M., Iervolino, M., Leopardi, A. & Vacca, A. 2016 Two-dimensional two-phase depth-integrated model for transients over mobile bed. *J. Hydr. Eng.* **142** (2), 04015043.

Di Cristo, C., Greco, M., Iervolino, M. & Vacca, A. 2018a Numerical simulation of a dam-break wave propagating over an erodible floodplain in presence of a structure. In: *HIC 2018 EPiC Series in Engineering* (G. La Loggia, G. Freni, V. Puleo & M. De Marchis, eds), 3, pp. 564–571.

Di Cristo, C., Evangelista, S., Greco, M., Iervolino, M., Leopardi, A. & Vacca, A. 2018b Dam-break waves over an erodible embankment: experiments and simulations. *J. Hydr. Res.* **56** (2), 196–210. doi:10.1080/00221686.2017.1313322.

Di Cristo, C., Greco, M., Iervolino, M. & Vacca, A. 2019 Numerical simulation of mud-flows impacting structures. *J. Mount. Sc.* **16** (2), 364–382. doi:10.1007/s11629-018-5279-5.

El Kadi Abderrezzak, K., Paquier, A. & Mignot, E. 2009 Modelling flash flood propagation in urban areas using a two-dimensional numerical model. *Nat. Haz.* **50**, 433–460.

Faug, T. 2015 Depth-average analytical solution for free-surface granular flow impacting rigid walls down inclines. *Phys. Rev. E* **92**, 062310. doi:10.1103/PhysRevE.9292.062310.

Gao, L., Zhang, L. M. & Chen, H. X. 2017 Two dimensional simulation of debris flow impact pressure on buildings. *Eng. Geol.* **226**, 236–244. doi:10.1016/j.enggeo.2017.06.012.

Greco, M., Iervolino, M. & Leopardi, A. 2008 Discussion of 'divergence form for bed slope source term in shallow water equations' by Alessandro Valiani and Lorenzo Begnudelli. *J. Hydr. Eng.* **134** (5), 676–678. doi:10.1061/(ASCE)0733-9429(2008)134:5(676).

Greco, M., Iervolino, M., Leopardi, A. & Vacca, A. 2012 A two-phase model for fast geomorphic shallow flows. *Int. J. Sed. Res.* **27** (4), 409–420.

Greco, M., Iervolino, M. & Vacca, A. 2018 Analysis of bedform instability with 1-D two-phase morphodynamical models. *Adv. Water Resour.* **120**, 50–64. doi:10.1016/j.advwatres.2017.07.002.

Harten, A., Lax, P. D. & van Leer, B. 1983 On upstream differencing and Godunov-type schemes for hyperbolic conservation laws. *SIAM Rev.* **25** (1), 35–61.

Hoerling, M. P., Hurrell, J. W. & Xu, T. 2001 Tropical origin for recent North Atlantic climate change. *Science* **292**, 90–92.

Hung, O., Morgan, G. C. & Kelerhals, R. 1984 Quantitative analysis of debris torrent hazard for design of remedial measures. *Can. Geotech. J.* **21**, 663–677.

Kattel, P., Kafle, J., Fischer, J. T., Mergili, M., Tuladhar, B. M. & Pudasaini, S. P. 2018 Interaction of two-phase debris with obstacles. *Eng. Geol.* **242**, 197–217. doi:10.1016/j.enggeo.2018.05.023.

Kelman, I. & Spence, R. 2004 An overview of flood actions on buildings. *Eng. Geol.* **73** (2004), 297–309.

Li, J., Cao, Z., Pender, G. & Liu, Q. 2013 A double layer-averaged model for dam-break flows over mobile bed. *J. Hydr. Res.* **51** (5), 518–534.

Lobovsky, L., Botia-Vera, E., Castellana, F., Mas-Soler, J. & Souto-Iglesias, A. 2014 Experimental investigation of dynamic pressure loads during dam break. *J. Fluid Struct.* **48** (2014), 407–434.

Palumbo, A., Soares-Frazão, S., Goutiere, L., Pianese, D. & Zech, Y. 2008 Dam-break flow on mobile bed in a channel with a sudden enlargement. *Proc. River Flow* **2008**, 645–654.

Pontillo, M., Schmocker, L., Greco, M. & Hager, W. H. 2010 1D numerical evaluation of dike erosion due to overtopping. *J. Hydr. Res.* **48** (5), 573–582.

Pudasaini, S. P. 2012 A general two-phase debris flow model. *J. Geophys. Res.* **117**, F03010. doi:10.1029/2011JF002186.

Richardson, J. F. & Zaki, W. N. 1954 Sedimentation and fluidisation: part 1. *Trans. Inst. Chem. Eng.* **32**, 35–53.

Rosatti, G. & Begnudelli, L. 2013 A closure-independent Generalized Roe solver for free-surface, two-phase flows over mobile bed. *J. Comp. Phys.* **255**, 362–383.

Savary, C. & Zech, Y. 2007 Boundary conditions in a two-layer geomorphological model: application to a hydraulic jump over a mobile bed. *J. Hydr. Res.* **45** (3), 316–332.

Scheidl, C., Chiari, M., Kaitna, R., Mullegger, M., Krawtschuk, A., Zimmermann, T. & Proske, D. 2013 Analysing debris-flow impact model, based on small scale modelling approach. *Surv. Geophys.* **34** (1), 121–140.

Shige-eda, M. & Akiyama, J. 2003 Numerical and experimental study on two-dimensional flood flows with and without structures. *J. Hydr. Eng.* **129** (10), 817–821.

Soares-Frazão, S. & Zech, Y. 2007 Experimental study of dam-break flow against an isolated obstacle. *J. Hydr. Res.* **45** (Suppl1), 27–36. doi:10.1080/00221686.2007.9521830.

Soares-Frazao, S. & Zech, Y. 2008 Dam-break flow through an idealised city. *J. Hydr. Res.* **46** (5), 648–658.

Soares-Frazão, S., Canelas, R., Cao, Z., Cea, L., Chaudhry, H. M., Die Moran, A., El Kadi, K., Ferreira, R., Fraga Cadórniga, I., Gonzalez-Ramirez, N., Greco, M., Huang, W., Imran, J., Le Coz, J., Marsooli, R., Paquier, A., Pender, G., Pontillo, M., Puertas, J., Spinewine, B., Swartenbroekx, C., Tsubaki, R., Villaret, C., Wu, W., Yue, Z. & Zech, Y. 2012 Dam-break flows over mobile beds: experiments and benchmark tests for numerical models. *J. Hydr. Res.* **50** (4), 364–375.

Spinewine, B. & Zech, Y. 2010 Small-scale laboratory dam-break waves on movable beds. *J. Hydr. Res.* **45** (Suppl1), 73–86.

Swartenbroekx, C., Zech, Y. & Soares Frazão, S. 2013 Two-dimensional two-layer shallow water model for dam break flows with significant bed load transport. *Int. J. Num. Meth. Fluids* **73** (5), 477–508.

Tai, Y. C., Gray, J. M. N. & Hutter, C. 2001 Flow dense avalanches past obstructions. *Ann. Glaciol.* **32**, 281–284.

Teufelsbauer, H., Wang, Y., Chou, C. & Wu, W. 2009 Flow obstacle-interaction in rapid granular avalanches: DEM simulation and comparison with experiments. *Granular. Matter.* **11** (4), 209–220. 10.1007/s10035-009-0142-6.

Thieken, A. H., Muller, M., Kreibich, H. & Merz, B. 2005 Food damage and influencing factors: new insights from the August 2002 flood in Germany. *Water Resour. Res.* **41** (12), W12430.

Van Rijn, L. C. 1984 Sediment pick-up functions. *J. Hydr. Eng.* **110** (10), 1494–1502.

Vesipa, R., Camporeale, C. & Ridolfi, L. 2012 A shallow-water theory of river bedforms in supercritical conditions. *Phys. Fluids.* **24**, 094104. doi:10.1063/1.4753943.

Wang, J. S., Ni, H. G. & He, Y. S. 2000 Finite-difference TVD scheme for computation of dam break problems. *J. Hydr. Eng.* **126** (4), 253–262. doi:10.1061/(ASCE)0773-9429.

Wu, W. & Wang, S. S.-Y. 2007 One dimensional modeling of dam break flow over movable beds. *J. Hydr. Eng.* **133** (1), 48–58.

First received 14 January 2019; accepted in revised form 18 April 2019. Available online 21 May 2019

Non-hydrostatic modelling of shallow water flow around a circular array of emergent cylinders

Jian Wang, Jingxin Zhang, Dongfang Liang and Lian Gan

ABSTRACT

This paper focuses on the investigation of the wake flow around a circular patch of cylinders with different solid volume fractions (SVF). The wake flow pattern was found to be dependent on the SVF and the flow stability factor S. A non-hydrostatic model was used to simulate the open channel flow through the cylinders. The turbulence was simulated using the SST (shear strain transport) k-ω turbulence model. The model was first validated against experimental data to ensure the accuracy of the numerical simulation. Then the local turbulent flow and the wake flow field structures were investigated by changing the SVF and the water depth. The spatial evolution of the wake flow behind the patch was analysed, and the correlation between the flow pattern and the wake flow stability was discussed. The cylinder-scale turbulence was intensive, even at low SVFs and low water depths. In contrast, the patch-scale turbulence in the wake was suppressed and the unsteady bubble wake was formed when the water depth decreased, which revealed the importance of the bed friction force to the flow pattern. The parameters S and SVF were adjusted to demonstrate their contributions to the formation of the different wake flow patterns.

Key words | bed friction, shallow water flow, steady wake, vegetation flow, wake stability parameter

Jian Wang
School of Naval Architecture, Ocean and Civil Engineering,
Shanghai Jiao Tong University,
Shanghai 200240,
China

Jingxin Zhang (corresponding author)
MOE Key Laboratory of Hydrodynamics,
Shanghai Jiao Tong University,
Shanghai 200240,
China
E-mail: *zhangjingxin@sjtu.edu.cn*

Dongfang Liang
Department of Engineering,
University of Cambridge,
Cambridge CB2 1PZ,
UK

Lian Gan
Department of Engineering,
Durham University,
Durham DH1 3LE,
UK

INTRODUCTION

Vegetation in fluvial systems plays a significant role in the sustainable development of aquatic systems because of its biological and hydrodynamic effects. It improves the water equality by the production of oxygen and uptake of nutrients. The biodiversity in rivers and oceans is promoted by providing food and creating different habitats. Also, by altering hydrodynamic conditions, such as flow velocity, turbulence structure and mass mixing, the exchanges of sediment, metals and other contaminants between terrestrial and aquatic systems can be well mediated. Aquatic plants can change the velocity field across different scales, ranging from individual branches and blades on a single plant to a patch of plants, called canopy or meadow (Nepf 2012; Aberle & Järvelä 2015). Generally, vegetation often exists in large-scale in coastal regions and rivers, and it forms a complex hydrodynamic environment (Yu *et al.* 2014). The drag caused by large-scale vegetation in rivers can stabilize sediment and change the patterns of erosion and deposition. In nearshore regions, vegetation dissipates wave energy, which protects shoreline and riverbed from flood and wave attack. Therefore, it is of great significance to investigate the hydrodynamic influence of vegetation on shallow water flows (Wang *et al.* 2018).

A variety of laboratory experiments on flow through vegetation have been conducted over the last few decades. In these studies, the vegetation elements were usually simplified as rigid bodies, and the vegetation zone was modelled by an array of solid cylinders (Poggi *et al.* 2004; Nezu & Sanjou 2008; Zhan *et al.* 2017). Ball *et al.* (1996) measured the velocity vector fields associated with flow through pile

doi: 10.2166/hydro.2019.018

groups in shallow water using particle tracking velocimetry (PTV). A steady low velocity near wake followed by an unstable far wake was observed downstream of the structure. Chen & Jirka (1995) investigated the wake structure behind a porous plate further. Their observations show that the von Kármán vortex street originates at a distance downstream of the porous obstruction. The region from the obstruction to the point where the vortex street forms is called the steady wake. Nicolle & Eames (2011) found that the vortex street did not appear at all for the solid volume fraction (SVF) less than 0.05. Additionally, as the SVF decreases, the vortex street moves further downstream, resulting in a longer steady wake. In the experiments of Zong & Nepf (2012), the vortex development behind a patch of cylinders was studied, and a monotonous relationship between the dimensionless length of wake and the SVF was generalized.

Both small-scale and large-scale turbulence are generated when the water flows through a patch of cylinders. The small-scale turbulence is generated by the stem wake behind single cylinders, and the large-scale turbulence is produced in the wake behind the patch. Takemura & Tanaka (2007) investigated the flow structures and drag characteristics around a colony-type model, which comprises seven equally spaced emergent cylinders. They concluded that the change in the large-scale and small-scale vortex structures was related to the interaction between the water flow and the solid obstruction. The production of patch-scale turbulence is mainly caused by the separated shear layers (SSLs) formed at the shoulders of the patch (Chang & Constantinescu 2015). The patch-scale turbulence is often considered as horizontal 2D or quasi-2D because of the low ratio of water depth to the patch scale. For shallow water flow, the turbulent flow is considered to be constrained by the rough bottom. Chu & Babarutsi (1988) concluded that the bed-friction generates not only the small-scale turbulence but, at the same time, exerts a stabilizing force on the large-scale turbulent coherent structures in shallow turbulent mixing layers. The vortex street behind a solid obstruction may be suppressed by the strong bed friction, which is characterized by a critical stability parameter, $S_c = 0.20$ (Ingram & Chu 1987). Therefore, the development of patch-scale turbulence of the wake flow can also be impeded by the bed friction.

There is a periodic unsteadiness in the flow when the vortex shedding off bluff bodies occurs. It is one of the toughest tasks for turbulence modelling to accurately predict the unsteady turbulent flows (Durbin 2002; Wegner *et al.* 2004; Xu & Ma 2009). Durbin (2002) argued that Reynolds-averaged Navier–Stokes (RANS) cannot be applied to unsteady flow, unless there is a spectral gap between the unsteadiness and the turbulence. In the experiments of Zong & Nepf (2012), the vortex shedding frequencies were calculated to be around 2 Hz (cylinder-scale) and 0.1 Hz (patch-scale), and the gap between these shedding frequencies and the background turbulence was also observed. Thus, the RANS methods were still commonly used in practical simulations. Brito *et al.* (2016) conducted the RANS simulations of compound open-channel flows with vegetated floodplains. Neary (2003) used a RANS-k-ω model to simulate flow through submerged vegetation. Yu *et al.* (2014) simulated flow through a circular patch of emergent cylinders using a horizontal 2D multi-body model and porous model, respectively. Their achievements agreed well with the experimental data regarding the mean velocities, but failed to predict the magnitude of the turbulent fluctuation. Moreover, the influence of free surface and bed shear were not included in the 2D model. When the patch diameter D is much larger than the water depth H, the bed friction is large enough to suppress the vortex street and cannot be ignored (Zong & Nepf 2012). Under this condition, a robust 3D model should be considered.

In this paper, a 3D non-hydrostatic FVM model (HydroFlow®) based on unstructured grids is developed and used to simulate shallow water flow through a circular array of emergent cylinders. The flow patterns, turbulence intensity and wake structures are analysed in detail referring to different solid volume fractions and water depths.

METHODS

Governing equations

In the present mathematical model, the total pressure is represented as a superposition of the hydrostatic component $p = \rho g\ (\eta - z)$ and the non-hydrostatic component q (Casulli & Stelling 1998). The vertical coordinate z is transformed to the σ coordinate (Phillips 1957) to fit the free surface and uneven bottom (see Figure 1). The modified equations are rewritten as:

$$\frac{\partial \eta}{\partial t}+\frac{\partial Hu}{\partial x}+\frac{\partial Hv}{\partial y}+\frac{\partial Hw_\sigma}{\partial \sigma}=0 \tag{1}$$

$$\frac{\partial Hu}{\partial t}+\frac{\partial Huu}{\partial x}+\frac{\partial Huv}{\partial y}+\frac{\partial Huw_\sigma}{\partial \sigma}=-gH\frac{\partial \eta}{\partial x}-\frac{H}{\rho}\frac{\partial q}{\partial x}+\frac{\partial}{\partial x}\left(v_t\frac{\partial Hu}{\partial x}\right)+\frac{\partial}{\partial y}\left(v_t\frac{\partial Hu}{\partial y}\right)+\frac{1}{H}\frac{\partial}{\partial \sigma}\left(\frac{v_t}{H}\frac{\partial Hu}{\partial \sigma}\right) \tag{2}$$

$$\frac{\partial Hv}{\partial t}+\frac{\partial Hvu}{\partial x}+\frac{\partial Hvv}{\partial y}+\frac{\partial Hvw_\sigma}{\partial \sigma}=-gH\frac{\partial \eta}{\partial y}-\frac{H}{\rho}\frac{\partial q}{\partial y}+\frac{\partial}{\partial x}\left(v_t\frac{\partial Hv}{\partial x}\right)+\frac{\partial}{\partial y}\left(v_t\frac{\partial Hv}{\partial y}\right)+\frac{1}{H}\frac{\partial}{\partial \sigma}\left(\frac{v_t}{H}\frac{\partial Hv}{\partial \sigma}\right) \tag{3}$$

$$\frac{\partial Hw}{\partial t}+\frac{\partial Hwu}{\partial x}+\frac{\partial Hwv}{\partial y}+\frac{\partial Hww_\sigma}{\partial \sigma}=-\frac{1}{\rho}\frac{\partial q}{\partial \sigma}+\frac{\partial}{\partial x}\left(v_t\frac{\partial Hw}{\partial x}\right)+\frac{\partial}{\partial y}\left(v_t\frac{\partial Hw}{\partial y}\right)+\frac{1}{H}\frac{\partial}{\partial \sigma}\left(\frac{v_t}{H}\frac{\partial Hw}{\partial \sigma}\right) \tag{4}$$

where η is the free surface elevation, h is the still water depth and $H=h+\eta$ is the total water depth. u, v, w are the velocities in x, y and z directions, respectively. $\sigma=(z-\eta)/(h+\eta)$ is the vertical coordinate in the computational domain. g is the gravitational acceleration. v_t is the eddy viscosity coefficient. w_σ is the vertical velocity in σ coordinate, and can be derived from:

$$w_\sigma=\frac{w}{H}-\frac{u}{H}\left(\sigma\frac{\partial H}{\partial x}+\frac{\partial \eta}{\partial x}\right)-\frac{v}{H}\left(\sigma\frac{\partial H}{\partial y}+\frac{\partial \eta}{\partial y}\right)-\frac{1}{H}\left(\sigma\frac{\partial H}{\partial t}+\frac{\partial \eta}{\partial t}\right) \tag{5}$$

Turbulence model

The two-equation models have been widely adopted for turbulence closure in shallow water flows (Pu *et al.* 2014; Pu 2015). In the present study, the eddy viscosity coefficient v_t is determined by the Shear-Stress Transport (SST) $k-\omega$ turbulence model (Menter 1993). The transport equations for the turbulent kinetic energy k and the turbulence frequency ω are as follows:

$$\frac{\partial(\rho k)}{\partial t}+\frac{\partial(\rho u_j k)}{\partial x_j}=\tau_{ij}\frac{\partial u_i}{\partial x_j}-\beta^*\rho\omega k+\frac{\partial}{\partial x_j}\left[(\mu+\sigma_k\mu_t)\frac{\partial k}{\partial x_j}\right] \tag{6}$$

$$\frac{\partial(\rho \omega)}{\partial t}+\frac{\partial(\rho u_j \omega)}{\partial x_j}=\frac{\gamma}{v_t}\tau_{ij}\frac{\partial u_i}{\partial x_j}-\beta\rho\omega^2+\frac{\partial}{\partial x_j}\left[(\mu+\sigma_\omega\mu_t)\frac{\partial \omega}{\partial x_j}\right]+2\rho(1-F_1)\sigma_{\omega 2}\frac{1}{\omega}\frac{\partial k}{\partial x_j}\frac{\partial \omega}{\partial x_j} \tag{7}$$

The eddy viscosity coefficient v_t is calculated as:

$$v_t=\frac{\mu_t}{\rho}=\frac{k}{\omega} \tag{8}$$

where ρ is the water density, μ is the dynamic viscosity and τ_{ij} is the Reynolds stress calculated as:

$$\tau_{ij}=\mu_t\left(\frac{\partial u_i}{\partial x_j}+\frac{\partial u_j}{\partial x_i}\right)-\frac{2}{3}\rho k\delta_{ij} \tag{9}$$

where δ_{ij} is the Kronecker delta. F_1 is a blending function, and the original sets of corresponding constants β^*, σ_k, γ, σ_ω and $\sigma_{\omega 2}$ are employed (Menter 1993).

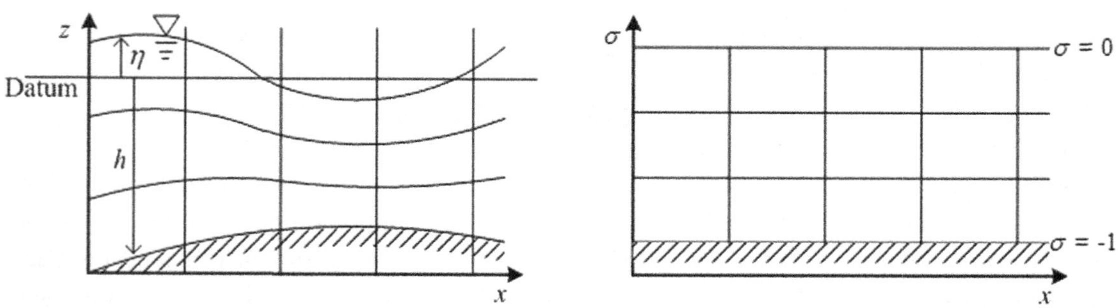

Figure 1 | Sketch of transformed coordinates.

Numerical scheme

A two-step predictor–corrector scheme is used in the numerical method. The flow driven by the hydrostatic pressure is first calculated in the predictor step, and then the flow driven by the non-hydrostatic pressure is updated in the corrector step. In the numerical model, the grid system consists of unstructured meshes in the horizontal plane and multi-layers in the vertical direction. The governing equations are discretized based on the finite volume method (FVM). A second order total variation diminishing scheme (OSHER scheme) is adopted to discretize the convective terms, and the central difference method is used to discretize the diffusion terms. In the corrector step, the Poisson-type equation for the non-hydrostatic pressure is numerically solved by a pre-conditioned BI-CGSTAB approach. The in-house codes (HydroFlow®) are paralleled by the OpenMP library (Zhang *et al.* 2014).

SIMULATION IMPLEMENTATION

Patch geometry

The same array geometry from experiments (Zong & Nepf 2012) was chosen in the present work. Their measurements were conducted in a channel with a length of 13 m and a width of 1.2 m, which is shown in Figure 2. A circular patch of cylinders was put in the middle of the channel in a staggered arrangement. The leading edge of the patch was 3.0 m away from the inlet. The diameter d of a single cylinder was 0.6 cm, and the diameter D of the patch was 22 cm. The density of the cylinders within the patch was defined as the solid volume fraction (SVF), and was determined using $\varphi = n\pi d^2/4$, in which n (cm^{-2}) is the number of cylinders per unit horizontal area.

Study cases and simulation implementation

A high grid resolution was designed in fitting the solid cylinder surface to meet the no slip solid wall boundary condition. A minimum normal scale of 0.1 mm of the adjacent grids to the cylinder wall was chosen, which was about 2 ~ 5 dimensionless wall distance, i.e., $\Delta_n^+ = 2 - 5$. On the bottom, the wall functions (10) were used to model the solid boundary, which required the first grids at the bottom lie in the fully developed turbulent boundary layer, i.e., $y^+ > 30$.

$$\frac{u}{u^*} = \frac{1}{\kappa}\ln(Ey^+) \quad k = \frac{(u^*)^2}{\sqrt{C_\mu}} \quad \omega = \sqrt{C_\mu}\frac{\mu^*}{\kappa y} \tag{10}$$

where u^* is the friction velocity, C_μ is an empirical coefficient, and y^+ is the dimensionless length scale of the

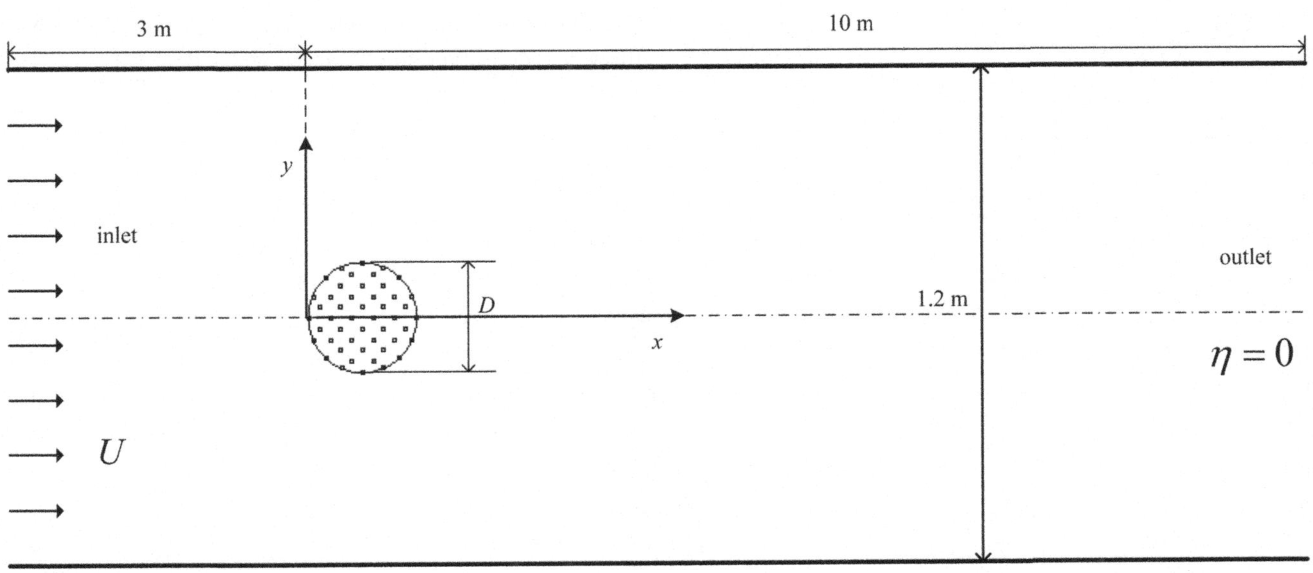

Figure 2 | Definition sketch of the model.

distance to the wall; Von Karman's constant $\kappa = 0.41$ and wall roughness parameter $E = 9.8$.

A steady flow discharge was specified at the inlet of the numerical flume, and a steady water elevation was specified at the downstream outlet. The integration time step was 0.0005 s in all of the study cases. The simulations proceeded until the temporal water elevation probed at a fixed upstream observing point reached nearly steady, i.e., the calculated water elevation fluctuation decreased to ± 0.001 m. The computation was continually carried out during the next 60 s; meanwhile, the flow variables were recorded in a frequency of 20 Hz. The saved database was statistically analysed to identify the wake flow patterns.

Ten cases were numerically investigated in this study, with the SVF φ being 0.03 and 0.10, and the water depth H varying from 0.03 m to 0.266 m (see Table 1). $U = 0.098$ m/s was the free stream velocity at the inlet. The Reynolds number $Re_d = Ud/\nu$ was thus calculated to be 588.

The wake stability parameter is denoted by $S = C_f\, D/H$. The bed friction coefficient C_f was evaluated by the formula proposed by Silberman *et al.* (1963). U_1 is the mean streamwise velocity at the centreline of the steady wake, and U_2 is the velocity outside the wake. L_1 is the length of the steady wake, which was measured from the patch ($x/D = 1$) to the point where the velocity $\bar{u}$ began to increase.

$$\frac{1}{C_f} = -4\log\left(\frac{1.25}{4Re_H\sqrt{C_f}}\right) \qquad (11)$$

RESULTS AND DISCUSSION

Model validations

The simulated mean velocity profiles $\bar{u}$ (at $y = 0$) and $\bar{v}$ (at $y = D/2$) in the longitudinal direction were compared with experiments in Figure 3. To be consistent with the experiments of Zong & Nepf (2012), the velocity was measured at mid-depth. In Figure 3, the outer range of the cylinder patch is outlined from $x/D = 0$ to $x/D = 1.0$. In general, good agreements between the predicted and measured velocities were achieved, which validated the accuracy of the present model.

Mean flow

Mean velocity

Flow around and through an array of cylinders exhibited a great difference to the flow around a solid body. At the centreline, the streamwise velocity decreased gradually from about $2D$ upstream of the patch. Within the patch, the streamwise velocity u decreased downstream because of the drag forces induced by cylinders. The mean velocity profiles (Figures 4(a) and 5(a)) downstream a lower SVF patch ($\varphi = 0.03$) were nearly the same at different water depths. However, the spatial distributions of the mean streamwise velocity behind a dense patch ($\varphi = 0.10$) were distinct with the water depth varying. At large water depths (i.e., P1H5,

Table 1 | Main geometrical and flow variables of the cases

Case	φ	H (m)	Re_H	C_f	S	U_1/U	U_2/U	L_1/D
P1H1	0.10	0.03	2,940	0.00720	0.053	0.16	1.24	4.06
P1H2	0.10	0.04	3,920	0.00687	0.038	0.13	1.26	6.20
P1H3	0.10	0.0665	6,517	0.00606	0.020	0.06	1.28	7.07
P1H4	0.10	0.10	9,800	0.00551	0.012	0.08	1.23	4.81
P1H5	0.10	0.133	13,034	0.00517	0.008	0.05	1.26	4.40
P1H6	0.10	0.200	19,600	0.00473	0.005	0.07	1.28	4.45
P1H7	0.10	0.266	26,068	0.00446	0.003	0.07	1.28	4.40
P2H3	0.03	0.0665	6,517	0.00606	0.020	0.58	1.08	6.10
P2H5	0.03	0.133	13,034	0.00517	0.008	0.57	1.08	9.03
P2H7	0.03	0.266	26,068	0.00446	0.003	0.56	1.10	8.12

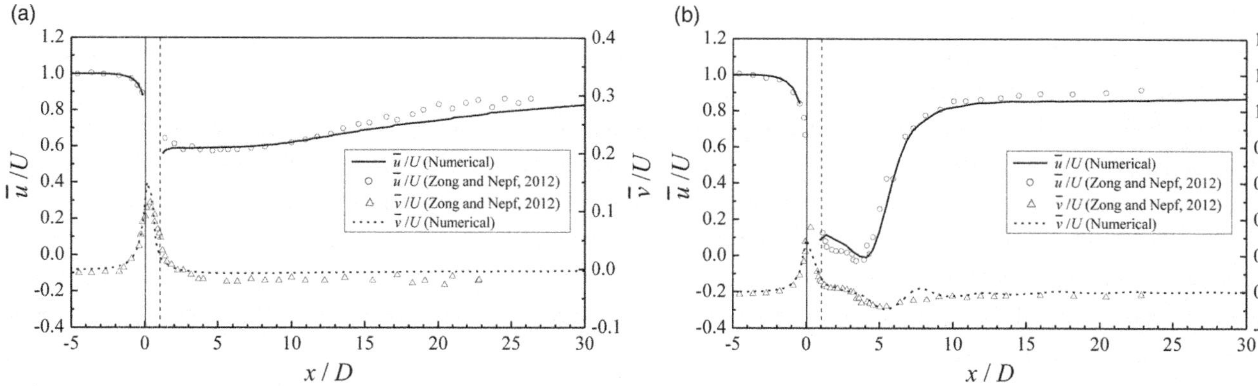

Figure 3 | Longitudinal profiles of mean velocity $\bar{u}/U$ and $\bar{v}/U$ at $y=0$. (a) P2H5. (b) P1H5.

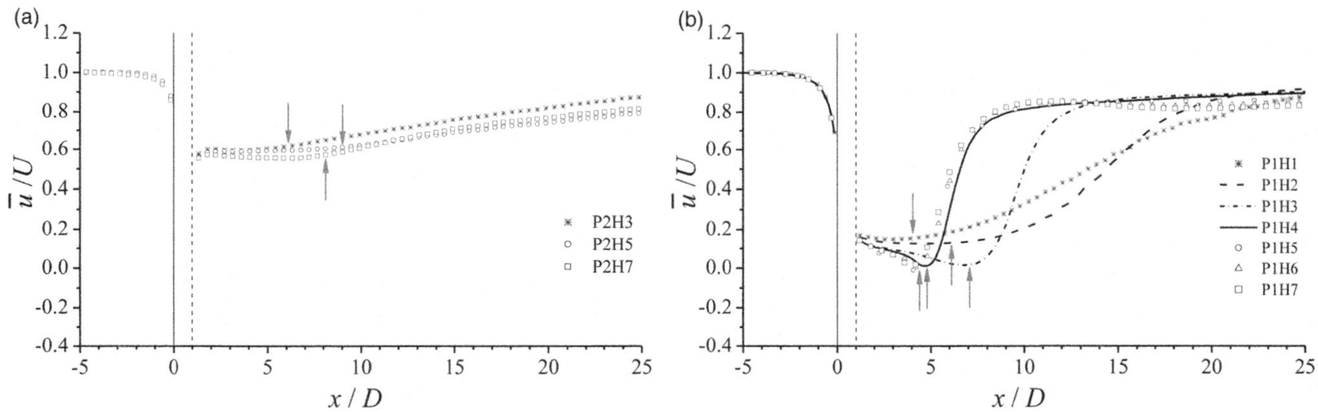

Figure 4 | Longitudinal profiles of mean velocity $\bar{u}/U$ at centreline $y=0$. The grey solid arrows locate the end of the steady wake of different cases. Since the length of the steady wake for P1H5, P1H6 and P1H7 has similar value, it is represented by the same arrow in (b). (a) $\varphi=0.03$ (b) $\varphi=0.10$.

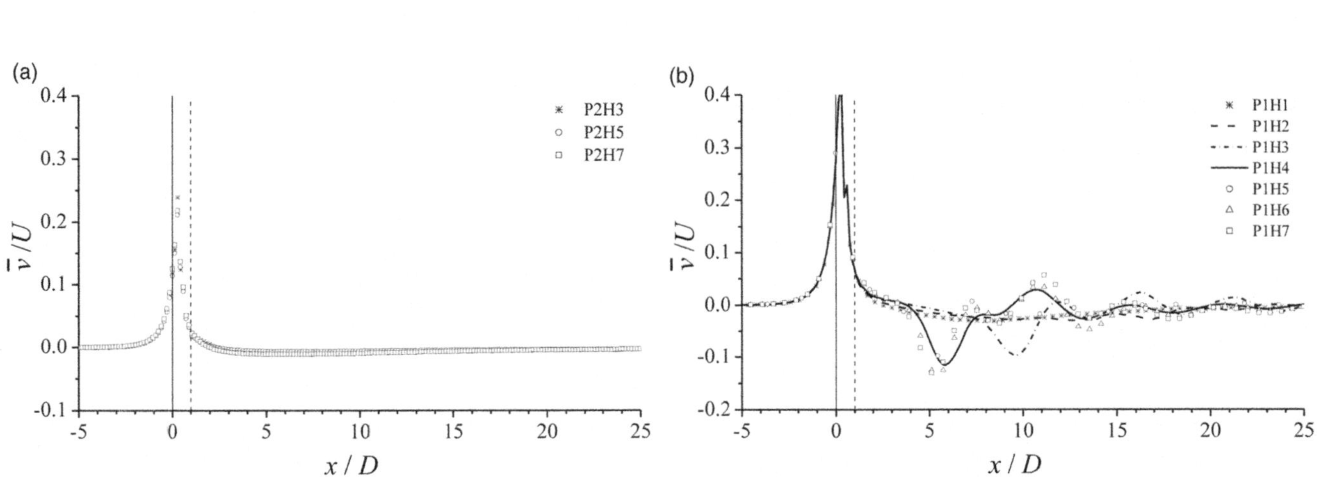

Figure 5 | Longitudinal profiles of mean velocity $\bar{v}/U$ at $y=D/2$. (a) $\varphi=0.03$. (b) $\varphi=0.10$.

P1H6, P1H7), there were only small discrepancies between the mean flow in the wakes. As shown in Figure 4(b), the steady near wake extended further downstream and there was a longer distance for the wake velocity deficit recovery at small water depths. The recovery of the centreline velocity is known so that the cross-wake mixing is mainly driven by the patch-scale vortex street. The analysis revealed that the patch-scale turbulence was weakened with H decreasing. The water depth was critical to the evolution of the wake flow; however, the length of the steady wake was found to be nonmonotonic with the water depth H. It was deduced that the change of wake structures was affected by the strong bed friction force. The flow reversal in the wake, which is indicative of a recirculation zone ($\bar{u} < 0$), was observed at large water depths and high SVF (Figure 4(b)). The recirculation zone moved downstream as H decreased (i.e., P1H3, P1H4, P1H5), but it disappeared at very low water depths (i.e., P1H1, P1H2). Instead of changing the water depth, Zong & Nepf (2012) observed the same trend for the length of the recirculation zone by changing the SVF.

Turbulent kinetic energy

The influence of water depth on the turbulent structures was further illustrated by the profiles of turbulent kinetic energy (TKE) (Figure 6). In the simulations, the TKE comprised two parts, i.e., $\text{TKE} = k_{\text{resolved}} + k_{\text{modelled}}$. The first constituent was obtained using the directly simulated velocity, which was calculated as $k_{\text{resolved}} = {u_{rms}}^2 + {v_{rms}}^2$. The other constituent was obtained from the turbulence Equations (6) and (7), which were modelled by the turbulence model, i.e., $k_{\text{modelled}} = k$. For a RANS-type model, the resolved TKE overwhelms the modelled TKE in the wake, i.e., $x/D > 1$ region, but the modelled TKE overwhelms the resolved TKE within the cylinder patch ($x/D < 1$) (Yu *et al.* 2014). For convenience, the turbulent kinetic energy was normalized by the mean velocity and denoted as the turbulence intensity, $I_{TKE} = \sqrt{\text{TKE}}/U$.

In the wake of a patch, there were two distinct zones of elevated turbulence. The upstream peak of I_{TKE} (denoted as $I_{\text{TKE},\ max}$) within the patch ($0 < x/D < 1$) indicated the small-scale turbulence generated by individual cylinders (Figure 6(a)). $I_{\text{TKE},\ max}$ in Figure 6(b) appeared at a distance away from the patch ($x/D > 1$), which corresponded to the patch-scale turbulence. As shown in Figure 6, the variation of the small-scale turbulence within the patch was relatively weak. That indicated the impact of bed friction on the cylinder-scale turbulence within the patch was negligible. To facilitate analysis, the cylinder-scale turbulence intensity at the end of the patch ($x/D = 1$) was characterized by Istem. Commonly, the turbulence intensity I_{TKE} in the sparse patch ($\varphi = 0.03$) was higher than that in the dense patch ($\varphi = 0.1$), which was because the wake-to-cylinder interactions were suppressed at high SVF. For high SVF, the stem-scale turbulence decayed rapidly through the patch. Behind the patch, the turbulence intensity decreased in the near wake, and then increased approaching the downstream edge of the near wake. When the SSLs met, the turbulence intensity reached maximum and this peak contributed to the formation of large-scale Kármán vortex street (LKV). The profiles of I_{TKE} for large water depths (i.e., P1H5, P1H6 and P1H7) differed slightly. However, when H decreased

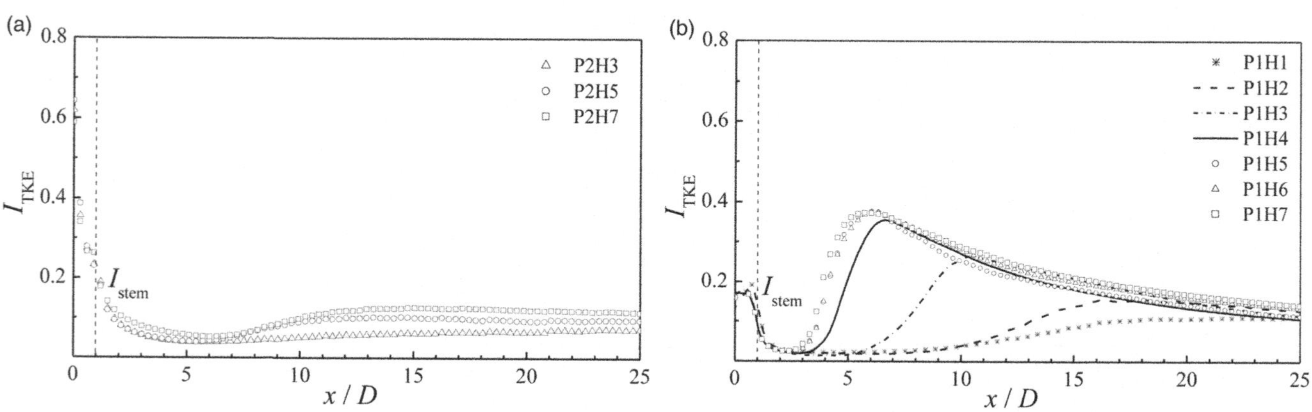

Figure 6 | Longitudinal profiles of turbulent kinetic energy. The cylinder-scale turbulence intensity I_{stem} was evaluated at the end of the patch ($x/D = 1$). (a) $\varphi = 0.03$. (b) $\varphi = 0.10$.

under 0.133 m (i.e., P1H1, P1H2, P1H3 and P1H4), the position of the second $I_{TKE,\ max}$ moved further downstream, which indicated that the onset of patch-scale turbulence was delayed. The strength of the second $I_{TKE,\ max}$ decreased with H decreasing for both of the SVFs. The simulations highlighted the turbulence intensity was weakened as H decreased under one critical value, which resulted in a lower rate of the velocity recovery. Because the movement of SSLs contributed to the TKE behind the patch (Chang & Constantinescu 2015), the spatial evolution of SSLs was suppressed by the bed friction, which was gradually distinguished when the water depth decreased.

Instantaneous vorticity field

It is known that the time-averaged vorticity may smear out important information about the flow structures. Thus, the instantaneous vertical vorticity, ω_z was used to investigate the temporal and spatial evolution of the flow through the cylinder patch. When the upstream flow approached the cylinder patch, some water was forced around the patch and the residual water flowed through the patch. Because of the drag force on individual cylinders, there was a velocity deficit between the flow through the patch and the flow around the patch. Hence, two separated shear layers (SSLs) formed at each shoulder of the patch. These shear layers were separated by the flow through the patch (similar to a bleed flow) and extended a distance downstream of the patch until the LKV formed induced by the instability of the shear layer flow. For a high SVF and a large water depth (Figure 7(a)–7(c)), the LKV behind the steady wake was clearly simulated. The LKV was also observed for a small water depth (Figure 7(d) and 7(e)), but the onset point was delayed compared with those cases shown in Figure 7(a)–7(c). When the water depth varied, the instantaneous vorticity revealed different spatial patterns. In Figure 7(f) and 7(g), although the swirling vortices were visible, the vortex shedding process hardly occurred. This wake pattern was similar to the near wake behind a solid cylinder, which was labelled as unsteady bubble (UB) wake by Chen & Jirka (1995). The bed friction is much stronger as the water depth decreases. The strong bed friction dissipated the flow energy and suppressed the spatial evolution of the separated shear layers originating at the patch shoulders. As a result, the formation of LKV was limited and the wake pattern changed. In contrast, the vorticity field for a sparse patch at different water depths did not change greatly. Since the patch-scale turbulence for a sparse patch was relatively weak, the vortex street was hardly generated.

Correlation of patch-scale turbulence with bed friction

To reveal the effect of the bed friction force on the patch-scale turbulence, the profiles of turbulence intensity in the vertical direction are displayed in Figure 8. In particular, the turbulent intensity was measured at the position where SSLs met (position of $I_{TKE,\ max}$ in Figure 6). Figure 8(a) shows the turbulence intensity at low SVF, in which the cases without cylinders are taken as comparisons. The maximum turbulence intensities of the cases with bare bed were found to be lower than the cylinder-scale turbulence intensities $I_{stem.}$ This explained that the turbulence within the cylinder array was dominated by wake generation, and it agreed with the observations of Nepf *et al.* (1997).

For low SVF, the bleed flow is strong, which weakens the formation and evolution of SSLs, and the turbulence intensity in the patch-scale wake is weak. Commonly, the turbulence intensity in the wake zone is weaker than cases with the bare bed. In free surface flows without lateral shear, the sizes of the turbulence structures are mainly governed by the balance of two forces. One force is the two-dimensional up-cascading of turbulent kinetic energy towards larger length scales at the free surface, and the other force comes from the interaction between large structures and the rough bottom (Uijttewaal & Tukker 1998). Since the bed friction is the only force dominating the turbulence evolution in cases of bare bed, the turbulence intensities monotonously decrease from the near-bed region to the water surface. The curve for P2H3 in Figure 8(a) also exhibited this trend, and it can explain why the overall level of TKE in the patch-scale wake was lower than the other cases in Figure 6(a). Because the bed shear effect is more and more strong as the water depth H decreases, the turbulence intensity was gradually weaker (shown in Figure 8(a) and 8(b)).

By contrast, the turbulence intensity behind a dense patch was much stronger. The suppression of patch-scale

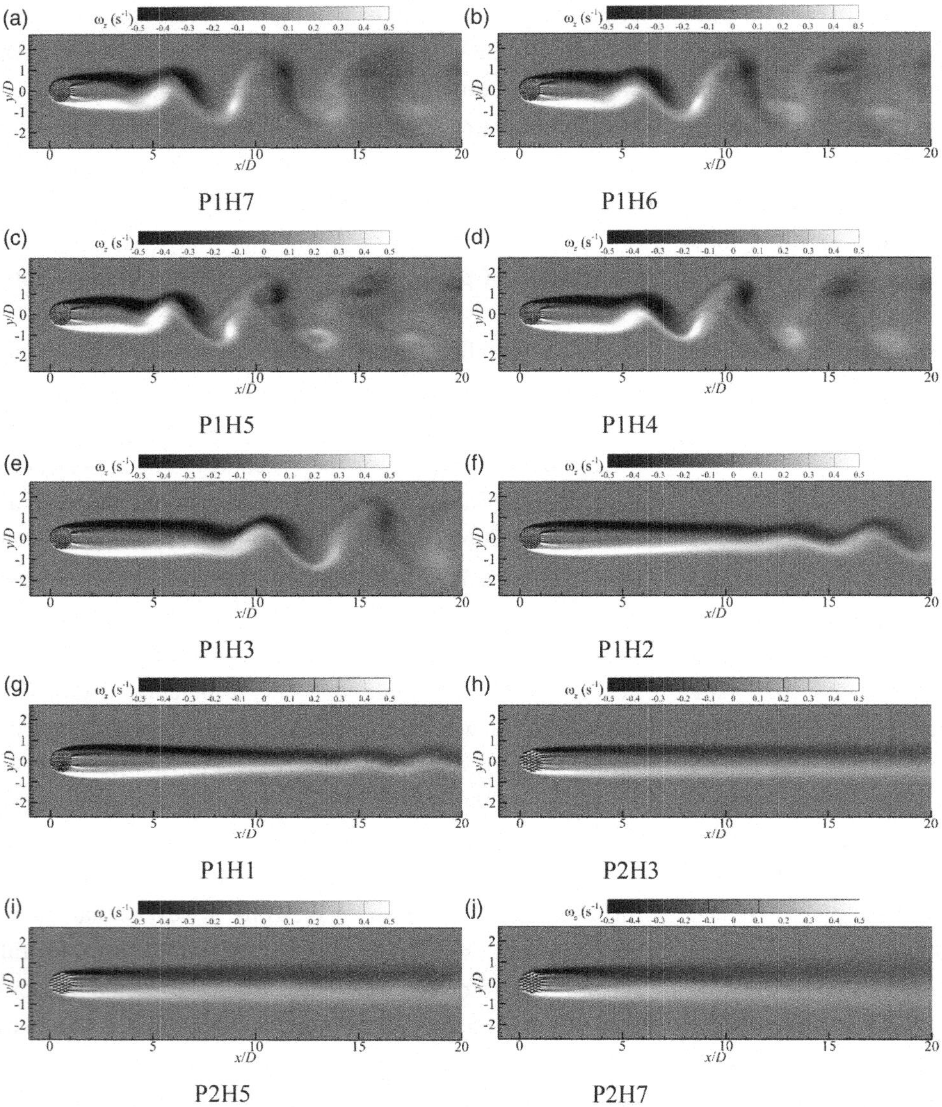

Figure 7 | Contour map of instantaneous vertical vorticity. (a) P1H7. (b) P1H6. (c) P1H5 (d) P1H4. (e) P1H3. (f) P1H2. (g) P1H1. (h) P2H3. (i) P2H5. (j) P2H7.

turbulence was not distinctively observed until the water depth reduced to under 0.04 m (see P1H2 in Figure 8(b)). As the turbulence generated by SSLs was highly suppressed by the bed friction at small water depths, the large-scale Kármán vortex street was gradually stabilized by the enhanced bed friction force. Additionally, the values of I_{TKE} measured at middle-depth $z/H = 0.5$ for cases P1H5, P1H6 and P1H7 were almost the same. The tiny differences between the results of these three cases were also observed in the mean flow and instantaneous vorticity field.

The spatial evolution of the steady wake

In the steady wake of the cylinder patch, the SSLs were formed and enhanced downstream along the two sides of the wake region until they developed to the centreline of the wake region. Beyond the interaction position, the patch-scale vortex street formed. The length of the steady wake region, L_1, was therefore determined by the scale of the patch (D) and the growth rate of the SSLs. As shown in Figure 9, the characteristic width of the shear layer is δ,

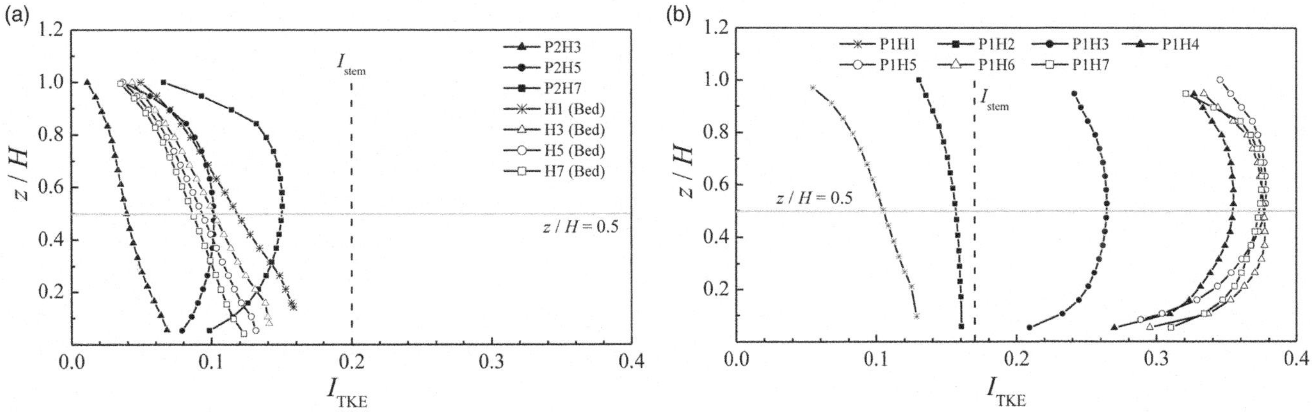

Figure 8 | Profiles of turbulent kinetic energy in the vertical direction, measured at the position of I_{max} along centreline $y = 0$. The grey solid line $z/H = 0.5$ is the position of middle water depth, and the vertical dashed line represents the turbulence intensity at the end of the patch ($x/D = 1$), I_{stem}. The profiles of turbulence intensity without cylinders were plotted and labelled as 'Bed'. (a) $\varphi = 0.03$. (b) $\varphi = 0.10$.

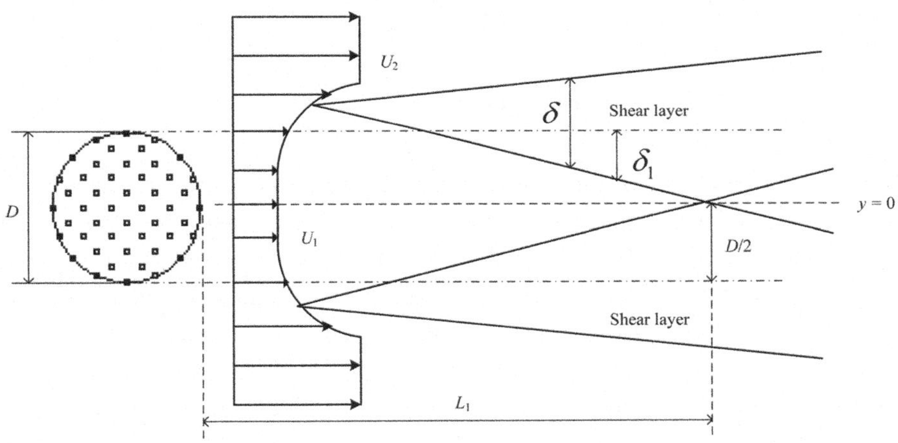

Figure 9 | Sketch of the plane shear layer, and the definition of δ_1. U_1 is the low velocity within the steady wake, and U_2 is the velocity outside the wake.

and a new width, δ_1, instead of δ, was used to calculate the growth rate of δ_1, i.e., $d\delta_1/dx$.

δ_1 was defined as the distance from the shoulder of the array, i.e., $y/D = 0.5$, to the position where the mean velocity began to increase from U_1 to U_2. The values of δ_1 were obtained by measuring the lateral distribution of the mean streamwise velocity. Because of the porous obstruction blocking the flow, the velocity U_2 is always larger than the velocity U_1. The shear layer flows generated by the deficit between U_1 and U_2 developed laterally downstream until the two SSLs merged at the centreline of the wake zone. When the two SSLs merged at the centreline, the steady wake region was closed and the width δ_1 reached maximum, i.e., $\delta_1 = D/2$. Beyond the steady wake region, the velocity at the centreline increased because of the strong exchange of the flow momentum. As a result of the interaction of SSLs, the flow speed recovered gradually to the upstream free-stream velocity U.

Figure 10 shows the dependence of $d\delta_1/dx$ on the stability parameter S. The shear layer growth rate for low SVF ($\varphi = 0.03$) was less than that for high SVF ($\varphi = 0.10$), which explained a longer steady wake region downstream of a sparse cylinder patch. The shear layer growth rate decreased as S increased, which indicated the transverse spreading rate of the shear layer decreased when the bed friction became the primary dynamic force on the turbulent flow (Chu & Babarutsi 1988).

The length of the steady wake was extensively measured. Drawing on the descriptions of planar shear layer growth

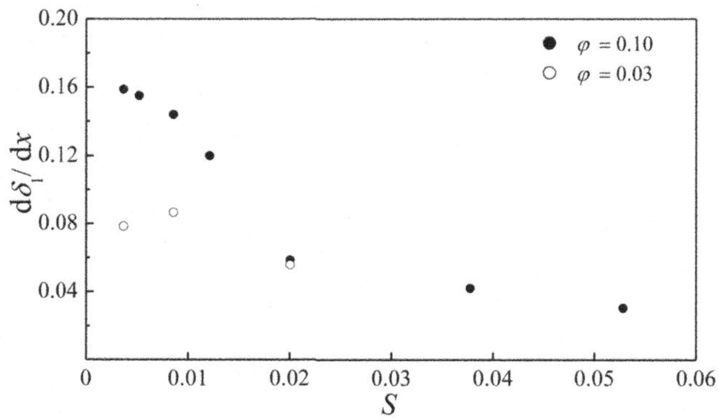

Figure 10 | The growth rate of the shear layer against the parameter *S*.

(Champagne *et al.* 1976) and assuming that the shear layer grows linearly, Zong & Nepf (2012) introduced a relationship to predict the length scale of the steady wake:

$$L_1 = \frac{D/2}{S_{\delta 1}} \frac{\overline{U}}{\Delta U} \quad (12)$$

where $\Delta U = U_2 - U_1$ is the velocity difference, $\overline{U} = (U_1 + U_2)/2$ is the averaged velocity. $S_{\delta 1}$ is an empirical parameter and was estimated as 0.10 in the experiments of Zong & Nepf (2012). Given the values of U_1 and U_2, the predicted dimensionless length scale L_1/D by Equation (12) was plotted against the stability parameter S in Figure 11. As a comparison, the estimated L_1/D from the mean velocity profiles are also shown. The predicted results agreed well with simulated results when S was less than 0.02 regardless of the patch density. The relationship (12) failed to predict L_1 for both patches when S exceeds 0.02. One possible reason was that the parameter $S_{\delta 1}$ used by Zong & Nepf (2012) was constant and obtained

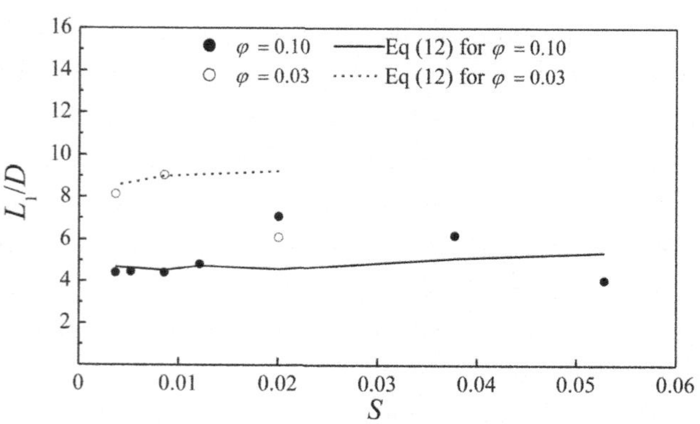

Figure 11 | Dimensionless length of the steady wake, L_1/D.

from only one water depth, which was deduced to be critical to the bed friction force. Another reason was deduced that the assumption of the linear shear layer growth rate was invalid. Chu & Babarutsi (1988) also observed that the width of the mixing layer no longer increased linearly with distance when the water depth was 2.5 cm. Particularly, the steady wake length L_1 reduced clearly as the bed friction force exceeded a critical value (e.g., $S > 0.03$ for $\varphi = 0.10$ and $S > 0.01$ for $\varphi = 0.03$). A conjectured explanation is the appearance of a recirculating bubble in the 'UB' wake (see Figure 7). For a single cylinder, the wake is attached to the body, and the bubble length becomes stretched out as S grows (Chen & Jirka 1995). Likewise, the reversal flow within the recirculating bubble pushed part of the fluid backwards into the cylinder patch.

CONCLUSIONS

In this paper, shallow water flows past a circular array of emergent cylinders have been numerically investigated. Ten study cases with different solid volume fractions of the patch and water depths were simulated, and the investigations focused on the correlations of the wake flow pattern with the SVF, water depth, bottom friction and the stability dynamic factor. The topic can help researchers investigate the influence of vegetation on patterns of deposition and erosion in shallow water flows.

The wake structure behind a circular array of cylinders was not only determined by the SVF, but also dependent on the bed friction. Two typical types of wake, the vortex street (VS) wake and the unsteady bubble (UB) wake were both observed in the simulations. For a solid cylinder, the VS–UB transition was characterized by a wake stability parameter, $S_c \approx 0.2$ (Chen & Jirka 1995). However, Chen & Jirka (1995) also observed the unsteady bubble wake behind a porous plate for $S = 0.073$. This means that the critical wake stability parameter, S_c, for a porous patch may be different from a solid body. Based on the present study cases, the critical S was preliminarily found to be 0.02 for $\varphi = 0.10$.

The bed friction played a different role in the cylinder-scale and patch-scale turbulent structures. Within the patch, even for the case of the sparse patch, the production of cylinder-scale turbulence was higher than the production by the bed shear. However, the patch-scale turbulence behind the patch can be greatly influenced by the bed friction. The peak of patch-scale turbulence occurred at the position where SSLs met and the shear layer instability took place. The von Kármán vortex street may also form at this position, but it depended on the level of the patch-scale turbulence intensity. The VS was suppressed by the bed friction force, which happened when $S > 0.02$ for $\varphi = 0.10$.

The lateral growth of SSLs was slower when the bed friction increased or the SVF decreased. The length of the steady wake, L_1, was related to the growth rate of SSLs and the patch diameter D. Therefore, the L_1 obtained from the simulations were compared with an empirical relationship. It was noted that the empirical relationship failed to predict L_1 at $S > 0.03$ with $\varphi = 0.10$ and $S > 0.01$ with $\varphi = 0.03$. The assumption of linear shear layer growth and a constant empirical parameter $S_{\delta 1}$ possibly contributed to the failure of the empirical formula.

A critical parameter, S_c, was estimated to be 0.02 for high SVF ($\varphi = 0.10$), and the S_c decreased as φ decreased. In order to extend a robust correlation between S_c and φ, a series of study cases with different SVFs must be investigated.

ACKNOWLEDGEMENTS

This work was sponsored by the National Natural Science Foundation of China (No. 11572196) and Shanghai Science and Technology Committee (No. 17230741200).

REFERENCES

Aberle, J. & Järvelä, J. 2015 Hydrodynamics of vegetated channels. In: *Rivers – Physical, Fluvial and Environmental Processes* (P. Rowiński & A. Radecki-Pawlik eds). Springer International Publishing, Cham, Switzerland.

Ball, D. J., Allison, N. & Stansby, P. K. 1996 Modeling shallow water flow around pile groups. *Ice Proceedings Water Maritime & Energy* **118**, 226–236.

Brito, M., Fernandes, J. & Leal, J. B. 2016 Porous media approach for RANS simulation of compound open-channel flows with submerged vegetated floodplains. *Environmental Fluid Mechanics* **16**, 1247–1266.

Casulli, V. & Stelling, G. S. 1998 Numerical simulation of 3D quasi-hydrostatic, free-surface flows. *Journal of Hydraulic Engineering-ASCE* **124**, 678–686.

Champagne, F., Pao, Y. & Wygnanski, I. 1976 On the two-dimensional mixing region. *Journal of Fluid Mechanics* **74**, 209–250.

Chang, K. & Constantinescu, G. 2015 Numerical investigation of flow and turbulence structure through and around a circular array of rigid cylinders. *Journal of Fluid Mechanics* **776**, 161–199.

Chen, D. Y. & Jirka, G. H. 1995 Experimental-study of plane turbulent wakes in a shallow-water layer. *Fluid Dynamics Research* **16**, 11–41.

Chu, V. H. & Babarutsi, S. 1988 Confinement and bed-friction effects in shallow turbulent mixing layers. *Journal of Hydraulic Engineering* **114**, 1257–1274.

Durbin, P. A. 2002 A perspective on recent developments in RANS modeling. *Engineering Turbulence Modelling and Experiments* **5**, 3–16.

Ingram, R. G. & Chu, V. H. 1987 Flow around islands in Rupert Bay – an investigation of the bottom friction effect. *Journal of Geophysical Research-Oceans* **92**, 14521–14533.

Menter, F. 1993 Zonal two equation kw turbulence models for aerodynamic flows. In: *23rd Fluid Dynamics, Plasmadynamics, and Lasers Conference*, 6–8 July, Nashville, TN, USA.

Neary, V. S. 2003 Numerical solution of fully developed flow with vegetative resistance. *Journal of Engineering Mechanics-ASCE* **129**, 558–563.

Nepf, H. M. 2012 Hydrodynamics of vegetated channels. *Journal of Hydraulic Research* **50**, 262–279.

Nepf, H., Sullivan, J. & Zavistoski, R. 1997 A model for diffusion within emergent vegetation. *Limnology and Oceanography* **42**, 1735–1745.

Nezu, I. & Sanjou, M. 2008 Turburence structure and coherent motion in vegetated canopy open-channel flows. *Journal of Hydro-Environment Research* **2**, 62–90.

Nicolle, A. & Eames, I. 2011 Numerical study of flow through and around a circular array of cylinders. *Journal of Fluid Mechanics* **679**, 1–31.

Phillips, N. A. 1957 A coordinate system having some special advantages for numerical forecasting. *Journal of Meteorology* **14**, 184–185.

Poggi, D., Porporato, A., Ridolfi, L., Alberston, J. D. & Katul, G. G. 2004 The effect of vegetation density on canopy sub-layer turbulence. *Boundary-Layer Meteorology* **111**, 565–587.

Pu, J. H. 2015 Turbulence modelling of shallow water flows using Kolmogorov approach. *Computers & Fluids* **115**, 66–74.

Pu, J. H., Shao, S. D. & Huang, Y. F. 2014 Numerical and experimental turbulence studies on shallow open channel flows. *Journal of Hydro-Environment Research* **8**, 9–19.

Silberman, E., Carter, R., Einstein, H., Hinds, J. & Powell, R. 1963 Friction factors in open channels. *Journal of the Hydraulics Division ASCE* **89**, 97–143.

Takemura, T. & Tanaka, N. 2007 Flow structures and drag characteristics of a colony-type emergent roughness model mounted on a flat plate in uniform flow. *Fluid Dynamics Research* **39**, 694–710.

Uijttewaal, W. & Tukker, J. 1998 Development of quasi two-dimensional structures in a shallow free-surface mixing layer. *Experiments in Fluids* **24**, 192–200.

Wang, J., Li, L., Zhang, J., Liang, D. & Yang, Q. 2018 A 3D hydrodynamic model for shallow water flow through a circular patch of emergent cylinders. In: *HIC 2018, 13th International Conference on Hydroinformatics*, 1–6 July, Palermo, Italy, pp. 2268–2275.

Wegner, B., Maltsev, A., Schneider, C., Sadiki, A., Dreizler, A. & Janicka, J. 2004 Assessment of unsteady RANS in predicting swirl flow instability based on LES and experiments. *International Journal of Heat and Fluid Flow* **25**, 528–536.

Xu, J. & Ma, H. 2009 Applications of URANS on predicting unsteady turbulent separated flows. *Acta Mechanica Sinica* **25**, 319–324.

Yu, L. H., Zhan, J. M. & Li, Y. S. 2014 Numerical simulation of flow through circular array of cylinders using multi-body and porous models. *Coastal Engineering Journal* **56**, 1–26.

Zhan, J.-M., Hu, W.-Q., Cal, W.-H., Gong, Y.-J. & Li, C.-W. 2017 Numerical simulation of flow through circular array of cylinders using porous media approach with non-constant local inertial resistance coefficient. *Journal of Hydrodynamics, Ser. B* **29**, 168–171.

Zhang, J. X., Sukhodolov, A. N. & Hua, L. 2014 Non-hydrostatic versus hydrostatic modelings of free surface flows. *Journal of Hydrodynamics* **26**, 512–522.

Zong, L. J. & Nepf, H. 2012 Vortex development behind a finite porous obstruction in a channel. *Journal of Fluid Mechanics* **691**, 368–391.

First received 15 January 2019; accepted in revised form 18 April 2019. Available online 22 May 2019

On the application of the depth-averaged random walk method to solute transport simulations

Fan Yang, Dongfang Liang, Xuefei Wu and Yang Xiao

ABSTRACT

Most numerical studies on solute mixing rely on mesh-based methods, and complicated schemes have been developed to enhance numerical stability and reduce artificial diffusion. This paper systematically studies the depth-averaged random walk scheme, which is a meshfree method with the merits of being highly robust and free of numerical diffusion. First, the model is used to solve instantaneous release problems in uniform flows. Extensive parametric studies are carried out to investigate the influences of the number of particles and the size of time steps. The predictions are shown to be independent of time steps but are sensitive to the particle numbers. Second, the model is applied to the solute transport problem along an estuary subject to extensive wetting and drying during tidal oscillations. Finally, the model is used to investigate the wind-induced chaotic mixing in a shallow basin. The effect of diffusion on the chaotic mixing is investigated. This study proposes a generic sampling method to interpret the output of the random walk method and highlights the importance of accurately taking diffusion into account in analysing the transport phenomena. The sampling technique also offers a guideline for estimating the total number of particles needed in the application.

Key words | advective diffusion, numerical diffusion, pollutant transport, random walk method, shallow water flows

Fan Yang
Dongfang Liang (corresponding author)
Department of Engineering,
University of Cambridge,
Trumpington Street, Cambridge CB2 1PZ,
UK
E-mail: *d.liang@eng.cam.ac.uk*

Xuefei Wu
School of Environment and Resource,
Southwest University of Science and Technology,
Sichuan 621010,
China

Yang Xiao
State Key Laboratory of Hydrology-Water Resources and Hydraulic Engineering,
Hohai University,
Nanjing 210098,
China

INTRODUCTION

Pollutant transport in waters has been extensively studied since it is relevant to a wide range of environmental problems and may cause significant economic losses (Rajar 1997). Among the natural waters, many can be regarded to be shallow, such as estuaries, lakes, reservoirs and near-shore waters. The horizontal scale of these flows is much larger than their depth. Therefore, it can be assumed that the solute is well-mixed vertically over the water column, allowing a depth-integrated approach to model the solute transport process. This paper focuses on the solute transport in shallow waters.

With the development of computing techniques, many modern numerical methods have been established to solve such problems (Lin & Falconer 1997; Gupta *et al.* 2004; Begnudelli & Sanders 2006; Benkhaldoun *et al.* 2007). Previously, most research has been based on the Eulerian approach to solve the advection-diffusion equation using finite-difference, finite-volume or finite-element method (Mingham *et al.* 2001; Liang *et al.* 2006; Yang *et al.* 2018). However, these grid-based methods have proven to be problematic in addressing steep concentration gradients and tend to produce numerical diffusion (Wu & Liang 2018). By contrast, Lagrangian methods use a group of discrete particles to represent the pollutant cloud, and these particles are tracked independently in flows. The Lagrangian approach can usually yield a good estimation regardless of the concentration gradients (Zhang & Chen 2007; Wu *et al.* 2018), and the computation can be easily parallelised. Nowadays, thanks to the rapid progress in electronic technology, the high computational cost of the Lagrangian approach has

doi: 10.2166/hydro.2019.015

become increasingly affordable. Its merits include inherently conserving mass and being free from artificial diffusion.

The computational cost of the Lagrangian approach depends on two factors: the number of particles to represent the pollutant cloud and the size of the computational time step. The choice of these two factors is crucial to the efficiency of random walk modelling. Traditionally, it is necessary to manually tune the number of particles and the size of time steps. Such a trial-and-error procedure is time-consuming and relies on the modeller's experience. This paper first uses an ideal case study to investigate the sensitivity of numerical solutions to the time step and particle number in depth-averaged random walk modelling. Then, the model is used to solve a solute oscillation problem along a hypothetical tidal estuary. The results of a TVD–MacCormack model are used as a reference. Finally, the paper describes the application of the random walk scheme to simulate the wind-driven chaotic mixing in a shallow circular lake. The results are qualitatively compared with those in previous research, and the importance of diffusion is highlighted.

DEPTH-AVERAGED ADVECTION-DIFFUSION EQUATION

The governing equation of the conservative solute transport is the advection-diffusion equation (Gresho & Sani 1998). Advection is the transport of a substance by a fluid flow as the result of the fluid's bulk motion. In this process, it is assumed that particles of the solute exactly follow the flow. Diffusion is the process whereby solute transports from higher concentration to lower concentration due to the small-scale random movements. It is assumed that the diffusive substances do not affect the motion of flows. In other words, the flow field is independent of the existence of the diffusive materials. With these assumptions, the conservation of solute mass is presented as:

$$\frac{\partial(sh)}{\partial t}+\frac{\partial(ush)}{\partial x}+\frac{\partial(vsh)}{\partial y}=\frac{\partial}{\partial x}\left(dD_{xx}\frac{\partial s}{\partial x}+dD_{xy}\frac{\partial s}{\partial y}\right)+\frac{\partial}{\partial y}\left(dD_{yx}\frac{\partial s}{\partial x}+dD_{yy}\frac{\partial s}{\partial y}\right)+q_s \tag{1}$$

where t is time; s is the depth-averaged concentration of the solute; h is the water depth; u and v represent the velocities along x and y-axis, respectively; q_s is the sources term of the governing equation, representing the increase ($q_s > 0$) or decrease ($q_s < 0$) in the total amount of the solute; D_{xx}, D_{xy}, D_{yx} and D_{yy} represent the dispersion-diffusion tensor of depth-averaged mixing in Cartesian coordinates. The relationship between the streamwise-transverse system and Cartesian coordinate can be expressed as:

$$D_{xx}=D_s\cos^2\theta+D_t\sin^2\theta \tag{2}$$

$$D_{xy}=D_{yx}=(D_s-D_t)\sin\theta\cos\theta \tag{3}$$

$$D_{yy}=D_s\sin^2\theta+D_t\cos^2\theta \tag{4}$$

where

$$D_s=\varepsilon_s hu_*,\ \ D_t=\varepsilon_t hu_* \tag{5}$$

D_s and D_t are the streamwise and transverse diffusion coefficients, respectively; ε_s and ε_t are two dimensionless constants. $\theta = arctan(v/u)$ is the angle between the direction of the x-axis and the direction of the local flow. The shear velocity u_* is expressed as Equation (6) using *Chézy* coefficient:

$$u_*=\frac{\sqrt{g}}{Ch\acute{e}zy}\cdot\sqrt{u^2+v^2} \tag{6}$$

DEPTH-AVERAGED RANDOM-WALK METHOD

Equation reformulation

In this paper, the depth-averaged advection-diffusion equation is recast into a new form that follows a standard advection-diffusion equation to design the particle-tracking algorithm. A new concentration variable, $S = sh$, is introduced. Then, the new form of the equation can be written as:

$$\frac{\partial S}{\partial t}+\frac{\partial(US)}{\partial x}+\frac{\partial(VS)}{\partial y}=\frac{\partial^2(D_{xx}S)}{\partial x^2}+2\frac{\partial^2(D_{xy}S)}{\partial x\partial y}+\frac{\partial^2(D_{yy}S)}{\partial y^2} \tag{7}$$

$$U=u+\frac{\partial D_{xx}}{\partial x}+\frac{\partial D_{xy}}{\partial y}+\frac{D_{xx}}{h}\frac{\partial h}{\partial x}+\frac{D_{xy}}{h}\frac{\partial h}{\partial y} \tag{8}$$

$$V=v+\frac{\partial D_{yy}}{\partial y}+\frac{\partial D_{xy}}{\partial x}+\frac{D_{yy}}{h}\frac{\partial h}{\partial y}+\frac{D_{xy}}{h}\frac{\partial h}{\partial x} \tag{9}$$

The source term q_s in the previous equation is neglected in the new form, as all the following case studies do not include any source or sink. According to the stochastic process theory, S can considered as a probability density function. U and V in Equations (8) and (9) represent the modified advective velocity components. The more detailed explanations of this scheme can be found in Wu & Liang (2018). This depth-averaged random walk model is then implemented by performing the advection and diffusion transport processes in each time step, as explained in the following sections.

Advective transport

In the advective transport process, the particles exactly follow the flow. However, most of the flow fields are solved by traditional grid-based methods. The advective velocities in Equations (8) and (9) must be evaluated at the position of each particle. In each grid cell that contains a particle, the velocity components u and v are interpolated to the second-order accuracy. The new particle position after the advective transport process can be expressed as Equations (10) and (11) using the second-order accurate iterative scheme:

$$x^a = x^{old} + \bar{U}\Delta t \tag{10}$$

$$y^a = y^{old} + \bar{V}\Delta t \tag{11}$$

where $\bar{U}$ and $\bar{V}$ are the flow velocity used in calculating the particles' advective displacement in each time step. To increase the order of accuracy, they are taken to be the averaged velocity within each time step of movement.

$$\bar{U} = \frac{1}{2}(U(x^{old}, y^{old}, t) + U(x^a, y^a, t + \Delta t)) \tag{12}$$

$$\bar{V} = \frac{1}{2}(V(x^{old}, y^{old}, t) + V(x^a, y^a, t + \Delta t)) \tag{13}$$

Diffusive transport

After the advection process, the particles undergo diffusive transport in a time step. The random streamwise and transverse velocity components are calculated as:

$$U_s^d = r_s\sqrt{\frac{2D_s}{\Delta t}} \tag{14}$$

$$V_t^d = r_t\sqrt{\frac{2D_t}{\Delta t}} \tag{15}$$

Here, the random numbers r_s and r_t are independent and follow a normal distribution with a mean of zero and a standard deviation of unity. The subscripts s and t represent the streamwise and transverse directions, respectively. The superscript d represents the diffusion-related velocity components. In the Cartesian coordinate system, the diffusion-related velocity components are expressed as:

$$U_x^d = U_s^d \cos\theta - V_t^d \sin\theta \tag{16}$$

$$V_y^d = U_s^d \sin\theta + V_t^d \cos\theta \tag{17}$$

Finally, particles' new coordinates at the end of one time step can be evaluated as:

$$x^{new} = x^a + U_x^d \Delta t \tag{18}$$

$$y^{new} = y^a + V_y^d \Delta t \tag{19}$$

Boundary condition

Particles follow a 'random walk' trajectory with the flow inside the computational domain. The boundary treatment is designed to prevent them from crossing solid boundaries. In this study, the fully reflective boundary condition is applied to particles that penetrate the solid boundaries after implementing Equations (18) and (19). When the time step is not too large, such a treatment then reflects the particles back into the computational domain.

SOLUTE TRANSPORT IN UNIFORM FLOWS

The instantaneous release problem in uniform flows is tested first using the random walk model. When the uniform flow

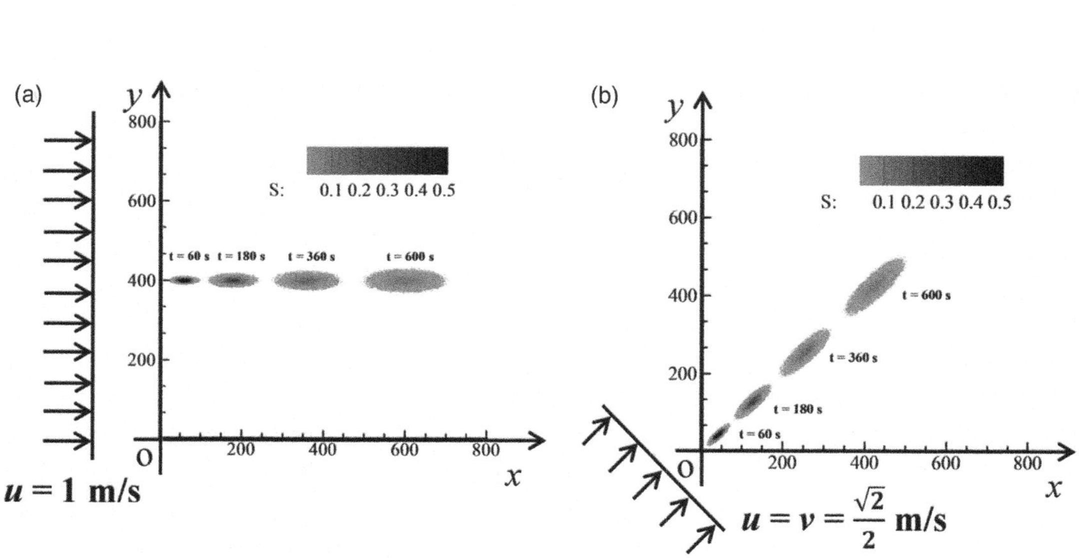

Figure 1 | The instantaneous release problem in a uniform flow: (a) *x*-direction flow and (b) diagonal-direction flow.

only follows the x-axis ($v=0$, $D_{xy}=0$, $D_{xx}=D_s$, $D_{yy}=D_t$), the analytical solution of this ideal test case can be expressed as:

$$s(x,y,t)=\frac{\frac{M}{h}}{4\pi t\sqrt{D_sD_t}}e^{-\frac{(x-x_0-ut)^2}{4D_st}-\frac{(y-y_0)^2}{4D_tt}} \tag{20}$$

In this case study, the total amount of solute material $M=233.06$ kg is released suddenly at the initial location (x_O, y_O). As shown in Figure 1, the water depth is set to be $h=1$ m, and the *Chézy* coefficient is $40\,\mathrm{m}^{1/2}/\mathrm{s}$ over the whole study area. The streamwise dispersion and transverse diffusion coefficients are typical values of 13.0 and 1.2, respectively, in straight open channel flows (Falconer 1991). Two flow conditions are considered in this section. The first one is a uniform flow with $u=1$ m/s along the x-axis ($\theta=0$). The solute material is initially located at $(x_O, y_O)=(0,\ 400\,\mathrm{m})$. The second one is a uniform flow aligned diagonal direction ($\theta=45°$). The velocity is set to be $u=v=1/\sqrt{2}$ m/s. For computations in both scenarios, the particle numbers P is 2.33×10^6 and the time step is 1 s.

The development of concentration contours for the flows along the x-axis direction and diagonal direction are presented in Figure 1(a) and 1(b), respectively. It is notable that the major axis of the elliptical patches is along the flow direction. The cloud of the solute experiences rapid elongation along its path. The reason is that streamwise dispersion is over ten times larger than the transverse diffusion. The results for the same test case obtained by a grid-based method, TVD–MacCormack, can be found in Liang *et al.* (2010).

The influence of time steps

The advantage of the random walk model includes high accuracy and small numerical diffusion. However, the disadvantage that is associated with this Lagrangian approach is its high computational cost. As seen in the scheme description, the amount of the computation depends on two factors: the size of the computational time step and the number of particles to present the pollutant cloud. Therefore, the choices of these two factors are crucial to the efficiency of the random walk model. The following part of this section is to discuss the influence of the two parameters on the random walk model when applied to the instantaneous release problem in uniform flows.

A great deal of research on time steps has been undertaken for the Euler scheme. The size of Δt is restricted by the Courant–Friedrichs–Lewy condition. Usually, the smaller the time step is, the more stable the simulation will be. However, the behaviour of the random walk model is very different. Figure 2 presents the longitudinal concentration profile at 600 s after the solute release. Regardless of the time step of 600 s or 0.01 s, the same concentration profile is predicted by the model. Variations of the peak concentration are not affected by the size of time steps as well, as seen in Figure 3.

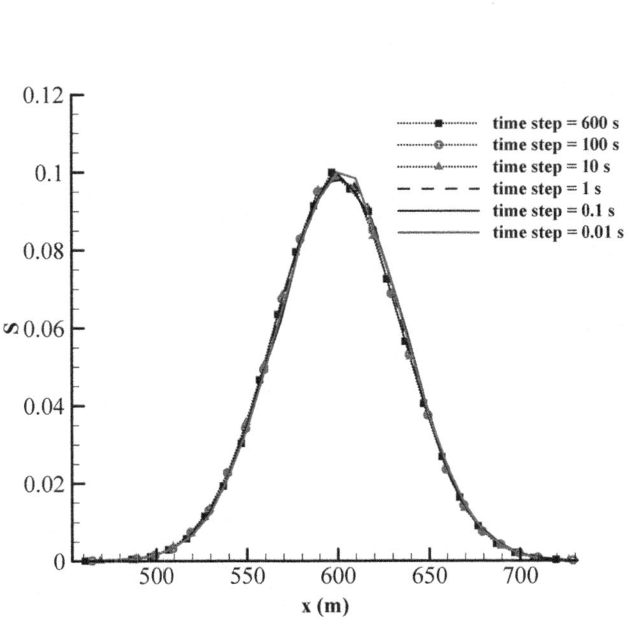

Figure 2 | Concentration distributions at 600 s with changes in time steps (x-direction flow).

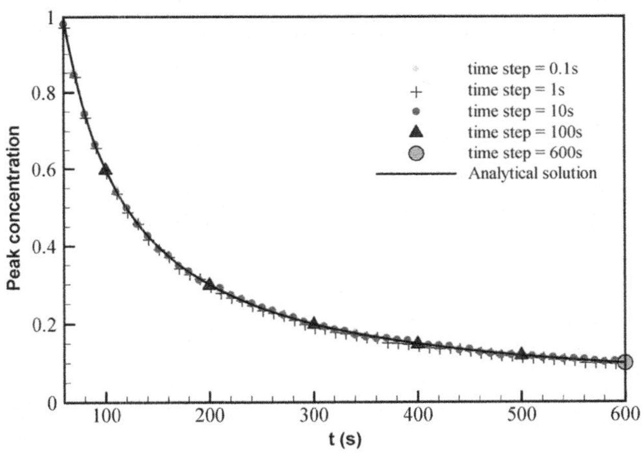

Figure 3 | Development of peak concentrations with changes in time steps (x-direction flow).

This independency of time steps can be proved through the time marching procedure of the calculation. After n time steps, the position of the particles at t is expressed as:

$$x_1(t) = x_0 + \bar{U}n\Delta t + \sqrt{2D_s\Delta t} \cdot \sum_{i=1}^{n} r_i \tag{21}$$

$$y_1(t) = y_0 + \bar{V}n\Delta t + \sqrt{2D_t\Delta t} \cdot \sum_{j=1}^{n} r_j \tag{22}$$

If the time step changes to be $m\Delta t$, n/m times of iterations are needed to reach the same time level. Then, the new position of the particles at t time is expressed as:

$$x_2(t) = x_0 + \bar{U}n\Delta t + \sqrt{2D_s\Delta t} \cdot \sqrt{m} \sum_{i=1}^{n/m} r_i \tag{23}$$

$$y_2(t) = y_0 + \bar{V}n\Delta t + \sqrt{2D_t\Delta t} \cdot \sqrt{m} \sum_{j=1}^{n/m} r_j \tag{24}$$

In the present model, the random numbers r_i follow a normal distribution with a mean of zero and a standard deviation of unity, as shown in Equation (25). According to the properties of a normal distribution (Bryc 2012), $\sum_{i=1}^{n} r_i$ and $\sqrt{m}\sum_{j=1}^{n/m} r_j$ have the same expectation and deviation:

$$r_i \sim N(\mu, \quad \sigma^2), \text{ where } \mu = 0, \sigma^2 = 1 \tag{25}$$

$$\sum_{i=1}^{n} r_i \sim N(n\mu, n\sigma^2) \tag{26}$$

$$\sqrt{m} \cdot \sum_{j=1}^{\frac{n}{m}} r_j \sim N\left(\sqrt{m}\frac{n}{m}\mu, \frac{n}{m}\left(\sqrt{m}\sigma\right)^2\right) \tag{27}$$

Therefore, the results will obey the distribution of $N(0, n)$ no matter what size of the time step is used. It can be concluded that the change of the time step does not affect the accuracy of the computation for this case of uniform flow with constant water depth.

The influence of particle numbers

Figure 4 shows a qualitative illustration of the solute transport process predicted by the random walk model using different numbers of particles. In general, the contour turns out to be more reasonable with the particle number increases. Larger particle numbers significantly improve the visualisation of the elliptical cloud for solute distribution.

To get a quantitative analysis, the predicted results are compared with analytical solutions. Figure 5 shows variations of peak concentration with different particle numbers used in the simulation. The size of sampling bins is set to be

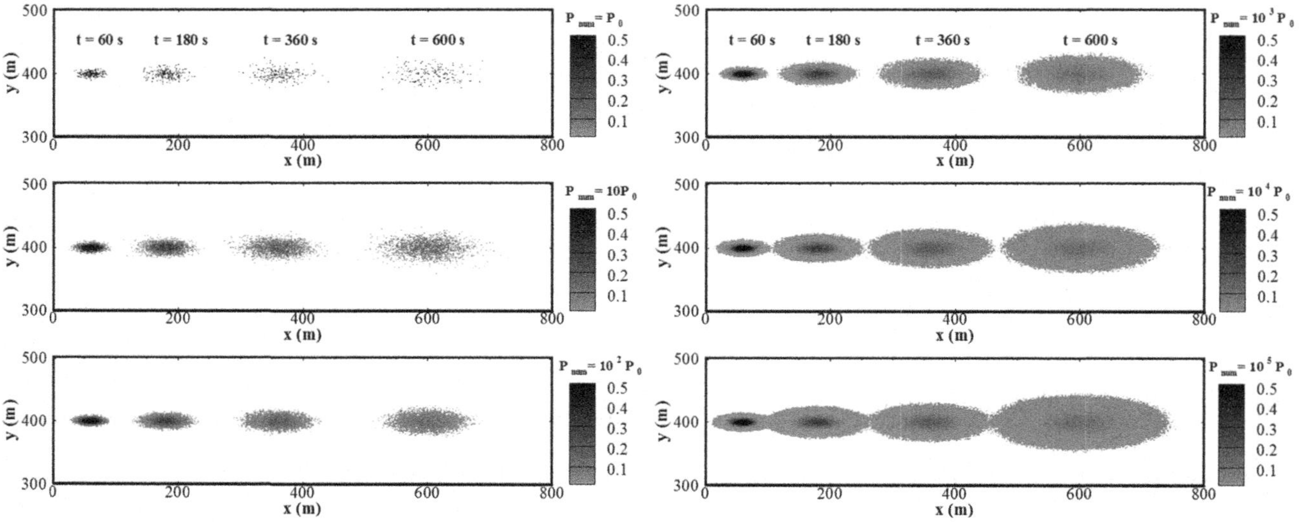

Figure 4 | Evolution of the solute cloud in *x*-direction uniform flows ($P_0 = 233$).

a circle with a radius of 1 m. It is apparent that small particle numbers tend to produce numerical oscillations. On the contrary, by introducing a larger number of particles in the model, the prediction further approaches the analytical solution. For example, when 2.33×10^7 particles are introduced in the model, the peak concentration is identical to the analytical solution. The relative errors with different particle numbers and different sampling bin sizes are compared in Figure 6(a). In the legend, *r* represents the radius of the bin. It is clear that, with the increase of particle numbers involved in the simulation, the error decreases significantly. The error for insufficiently small particle numbers can jump beyond 18%, which is unacceptable in most simulations.

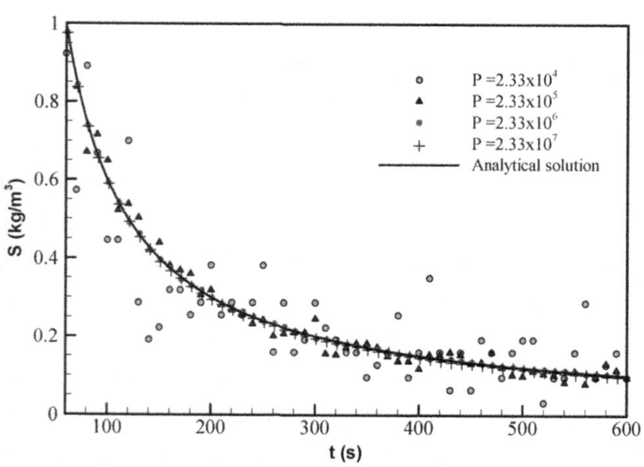

Figure 5 | Variation of peak concentration for *x*-direction flows.

When the number of particles in each sampling bin approaches 200 to 300, the relative error is reduced to less than 5%. This means that the particle should be guaranteed

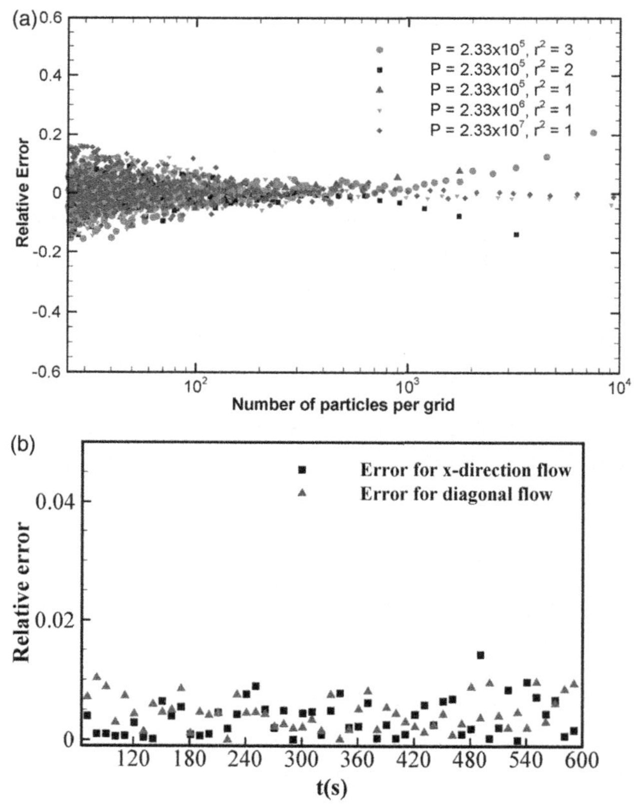

Figure 6 | Relative errors of peak concentration: (a) with the changes in particle numbers in each sampling bin and (b) with changes in time.

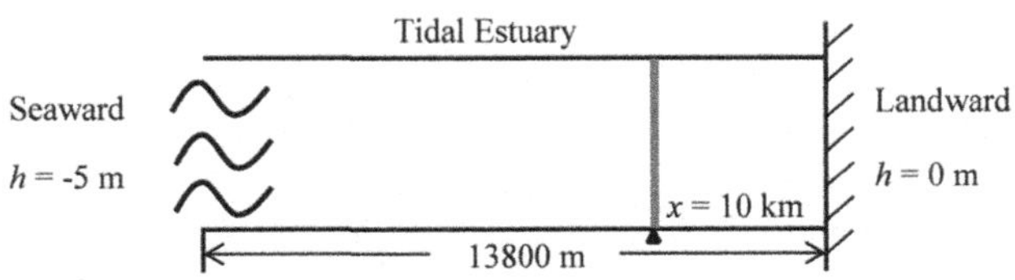

Figure 7 | One-dimensional hypothetical tidal estuary.

to be higher than hundreds per bin to give a reasonable estimation of the solute concentration. In addition, it is worth noting that the error for both x-direction and diagonal flows is always very similar, as shown in Figure 6(b). This means the random-walk model is not affected by the flow direction, unlike most of the mesh-based models.

SOLUTE OSCILLATION ALONG A ONE-DIMENSIONAL TIDAL ESTUARY

This section considers the situation where the solute is transported forward and backward in a hypothetical tidal estuary. As shown in Figure 7, the boundary condition for the left end is regarded as a sinusoidal tide, while the right side of this estuary is considered as a vertical wall. The total length of the estuary is 13,800 m, and the bed level changes from −5 m at the end of the seaward side to 0 m at the right end. In this case, the tidal flow has an average water level of 0 m and amplitude of 2 m, rising from the average water level at the beginning of the simulation. Before the pollutant transport simulation, the predictions of the flow field were obtained by using TVD–MacCormack scheme to solve the SWEs (shallow water equations). Detailed information can be found in Liang *et al.* (2010). The discretised velocity field is then reconstructed over the whole domain using linear interpolation.

The initial concentration is set to be 100 units at the cell located at $x = 10$ km, while the concentration at the other positions is set to zero. An illustration of the particles' position and their distribution at different times is given by Figure 8. It is well presented that particles oscillate with tidal flows and gradually disperse in the longitudinal direction. The instantaneously released solute moves with the rising tide to the closed wall under the influence of tidal currents, while the receding tide carries the particles back to the open boundary. At the same time, the spread of the solute grows because of the effect of dispersion.

The aforementioned sampling algorithm is required to convert the scatter of particles, as illustrated in Figure 8, to the concentration profile, as illustrated in Figure 9. This sampling algorithm is crucial in the interpretation of the results of the random walk method and in the evaluation of the concentrations. Traditionally, this concentration profile, i.e., histogram, uses sampling bins of fixed length. However, this fixed length may not suit all the positions. In this work, the length of the bin is chosen automatically by including 200 particles within each bin. This dynamic determination of the bin size is found to be able to avoid spurious fluctuations and achieve good comparisons with analytical and previous results. A series of smooth concentration profiles along the estuary at both high and low water levels are shown in Figure 9. When the solute cloud moves close to the landward position, the tidal flow generally decelerates because of the wall boundary. It is notable

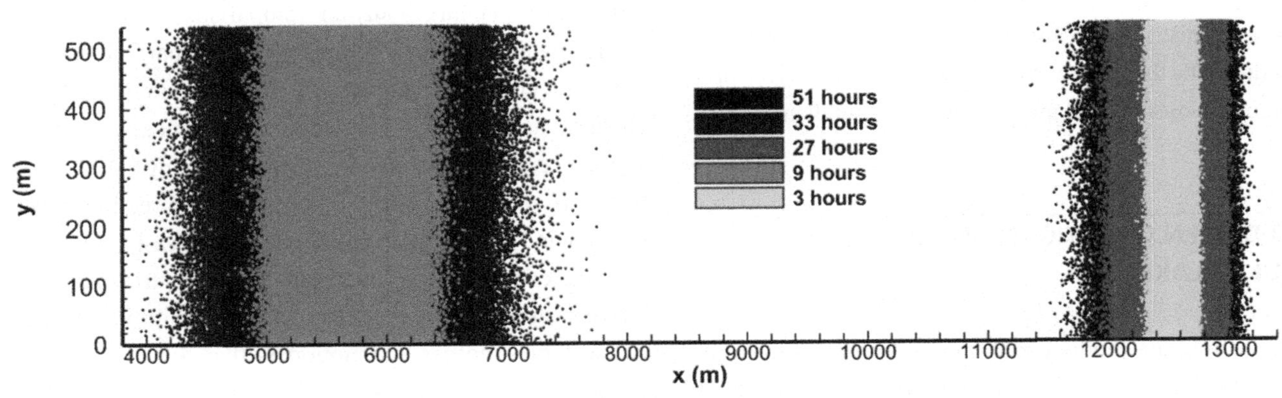

Figure 8 | Distribution of solute particles along the one-dimensional hypothetical estuary.

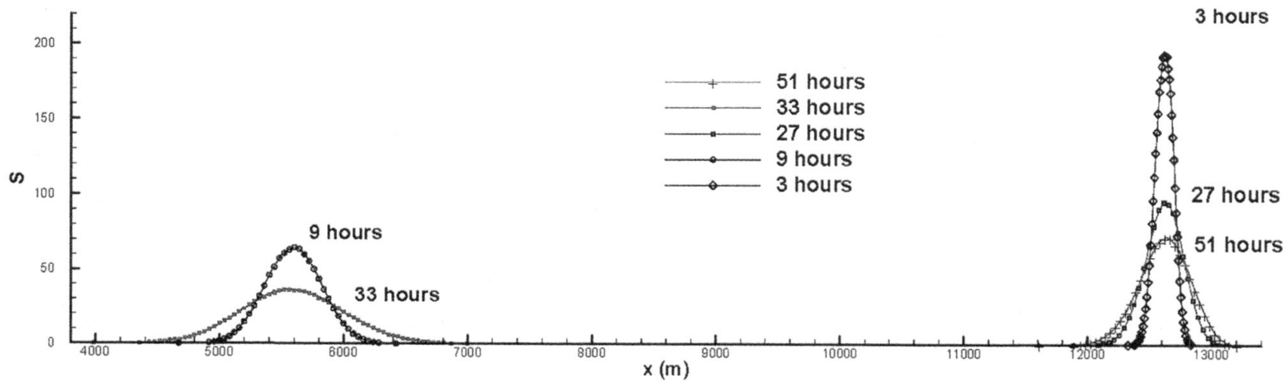

Figure 9 | Distribution of the concentration along the one-dimensional hypothetical estuary.

that the peak concentration at 27 hours is even higher than that at 9 hours, as the water depth is smaller near the right-side wall which amplifies the value of the concentration to conserve mass of the solute.

The numerical solution of the TVD–MacCormack model is used as a reference for this case study. For such a grid-based method, solutions are sensitive to the grid size. The finer the grid is, the more accurate the simulation will be. The numerical diffusion for the Eulerian model is most obvious at the early stage of the simulation because of the relatively sharp concentration gradient. Take Figure 10(a) as an example, which presents the concentration profile at 3 hours. Several different cell sizes, from 3 m to 270 m, are used in the mesh-based simulations. When the mesh size is increased to 270 m, the concentration distribution is far too flattened because of large numerical diffusion. The peak concentration for the grid size of 3 m is nearly three times higher than that for the grid size of 90 m. However, very small grid size leads to very large computational cost. On the contrary, the random walk model does not involve numerical diffusion. A narrower distribution and a higher peak concentration are predicted by the present model, and it is similar to the prediction of the mesh-based method with very small grid size.

WIND-DRIVEN CHAOTIC MIXING IN A SHALLOW CIRCULAR LAKE

In this section, the random walk model is applied to simulate the particle motion due to wind-driven mixing in a shallow lake. As shown in Figure 11, the property of the circular lake is suggested by Kranenburg (1992). The velocity field is described by a stream function in the polar coordinate system:

$$\psi = Br\left(1 - \frac{r}{R}\right)\sin(\varphi - \theta), \text{ where } B = \frac{\ln Z}{4\kappa}u_*H,\ Z = \frac{H}{z_0} \tag{28}$$

The von Kármán constant κ is set to be 0.4; the mean depth H is 0.5 m and the radius of the lake R_0 is 120 m; z_0 is a roughness height of the bed, set to 2.8 mm. The water depth h is described by a function of radial distance r from the basin centre as follows:

$$h = H\left(\frac{1}{2} + \sqrt{\frac{1}{2} - \frac{1}{2}\frac{r}{R_0}}\right) \tag{29}$$

These physical parameters are set to be the same as those in Kranenburg (1992) for comparison purposes.

Kranenburg (1992) found that the particle motion becomes chaotic when surface wind stress periodically changes its direction. Therefore, a sequence of periodic storm events is designed for this case study. During the first and second halves of a period, the direction of the storm wind alternates between the north-east and north-west directions. It is assumed that the wind stress suddenly changes its direction every half period, while its intensity is constant. Another assumption is that the velocity field instantaneously adapts to wind conditions. The resulting flow field is governed by a dimensionless storm duration

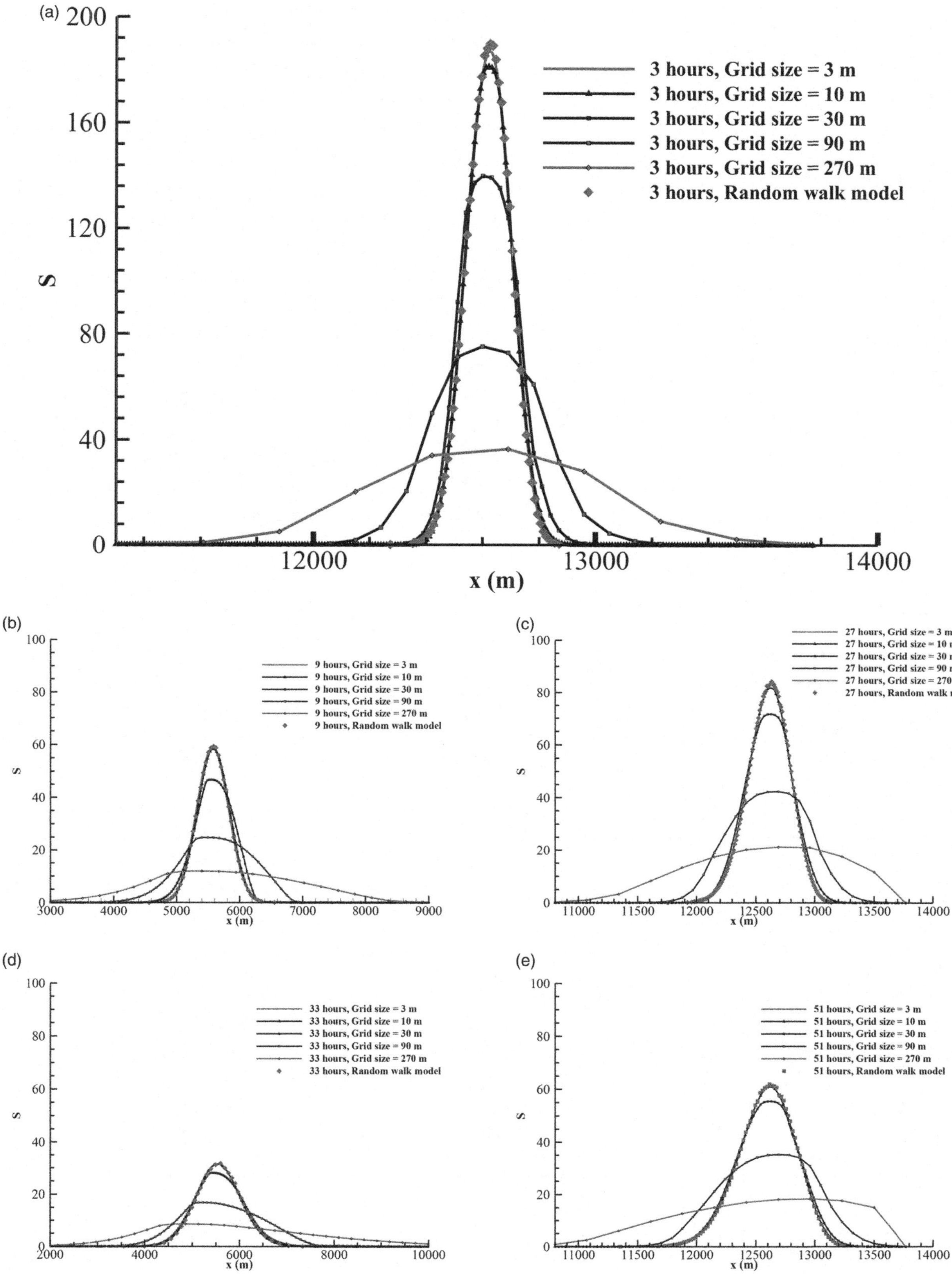

Figure 10 | Distribution of the concentration along the estuary with the changes in grid size: (a) 3 hours; (b) 9 hours; (c) 27 hours; (d) 33 hours; (e) 51 hours.

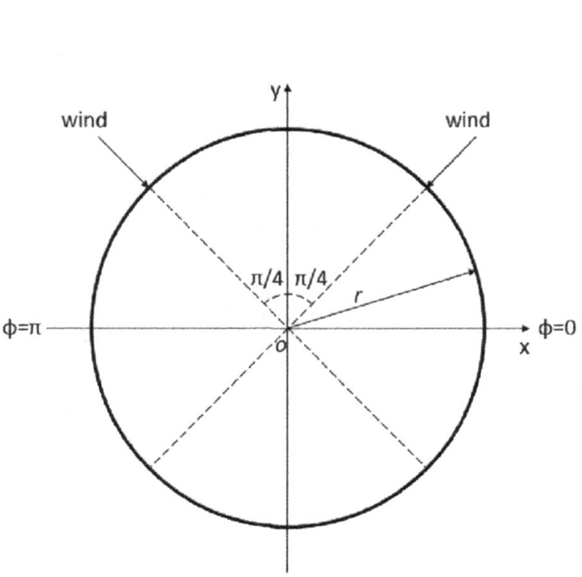

Figure 11 | Aerial view of the model lake with a periodic sequence of storm events.

parameter μ as follows:

$$\mu = \frac{\ln Z}{8\kappa R_0} u_* t_s \tag{30}$$

where t_s is the storm duration, i.e., half of the period T. For all the cases considered, Poincaré sections are used to illustrate the mixing properties, which is the superposition of particle trajectories at the end of each cycle.

For the purpose of comparison with previous studies, the diffusive process is first ignored so as to test the behaviour of the advection part of the random walk model. These particles are tracked for 500 periods. For small μ, as shown in Figure 12(a), the superposed particles with the same phase form elliptical patterns that occupy the left and right halves of the lake. The area of the chaotic region increases with the increase of μ, while the sizes of the noticeable elliptical patterns decrease. When μ is beyond 0.70, the regular patterns are no longer obvious, and only chaotic distribution is noticed (for $\mu = 0.84$). These findings are consistent with previous results of Kranenburg (1992).

Then, both advection and diffusion processes are considered as tracer particles are also spread by turbulence diffusion and shear dispersion in reality. Figure 13 depicts

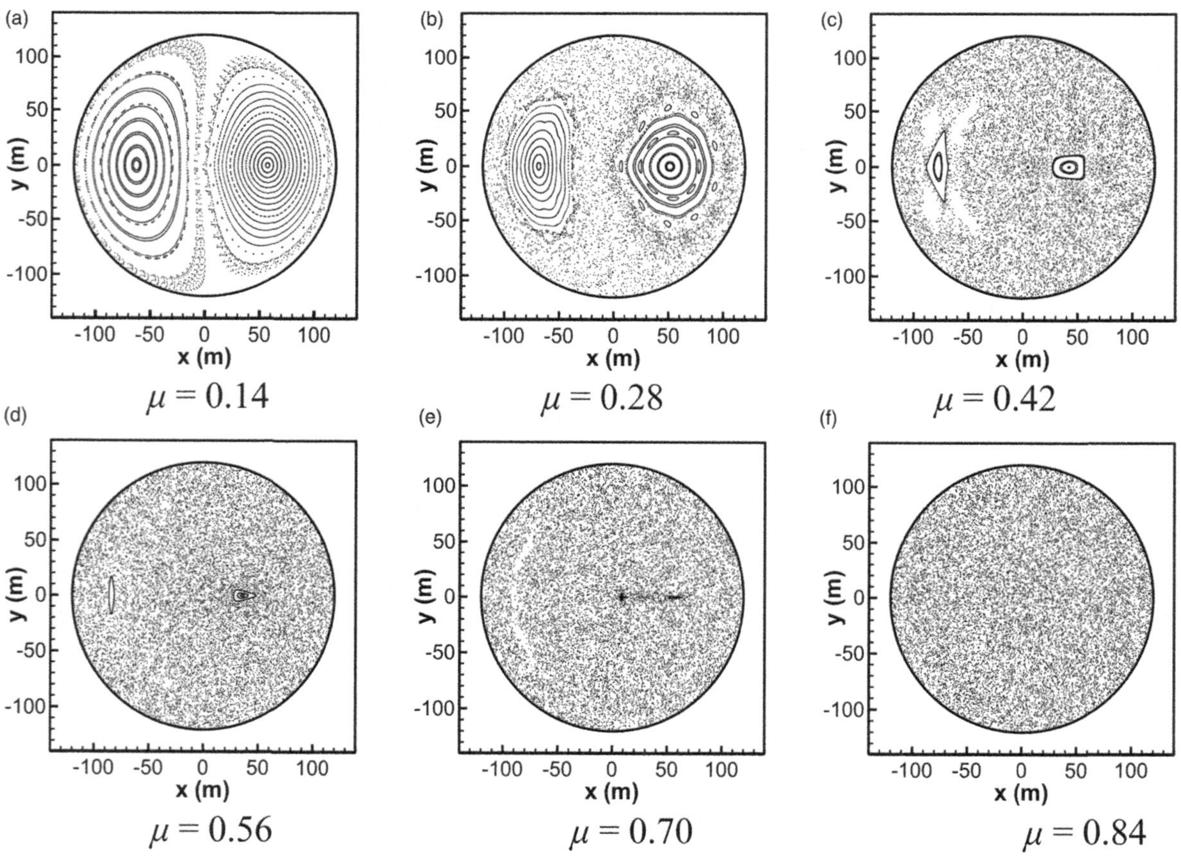

Figure 12 | Poincaré sections for wind periodically blowing from the north-east and north-west with different μ (without diffusive process).

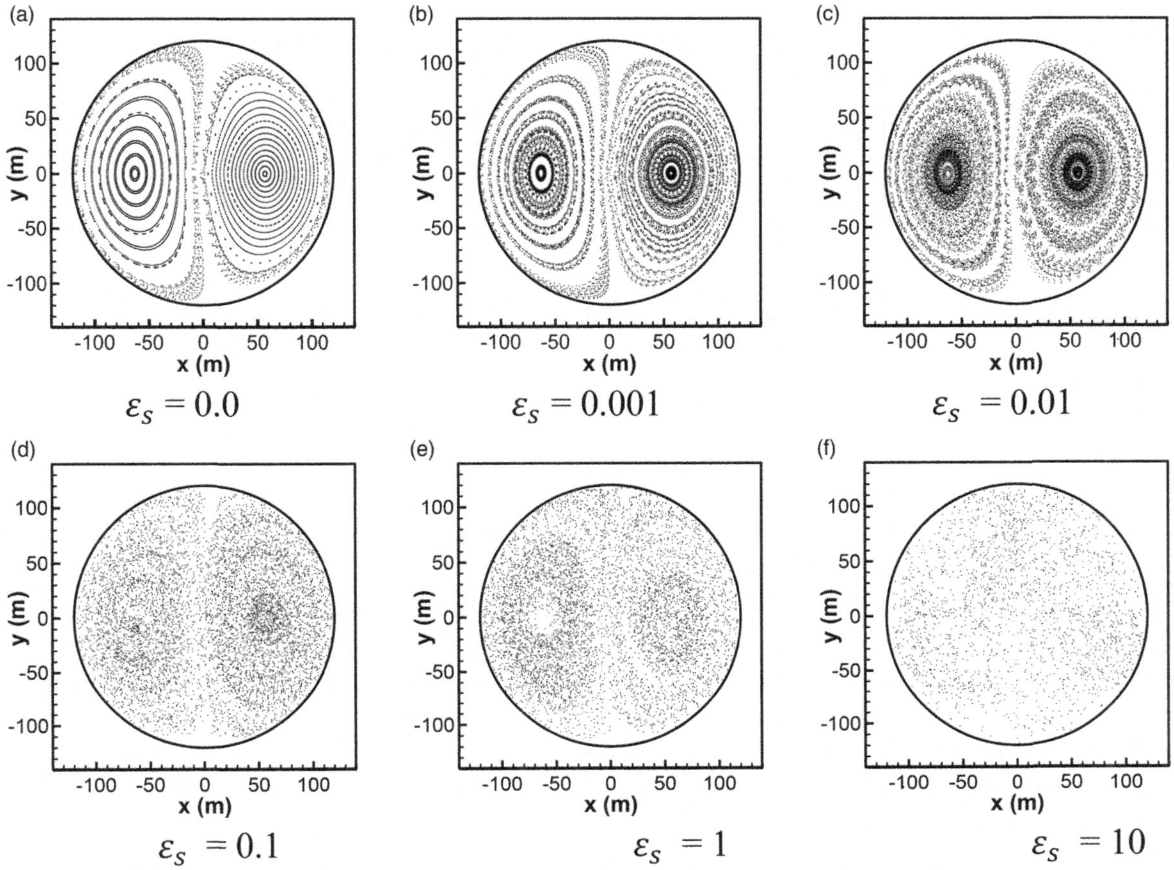

Figure 13 | Poincaré sections for both advection and diffusion processes with different ε_s ($\mu = 0.14$).

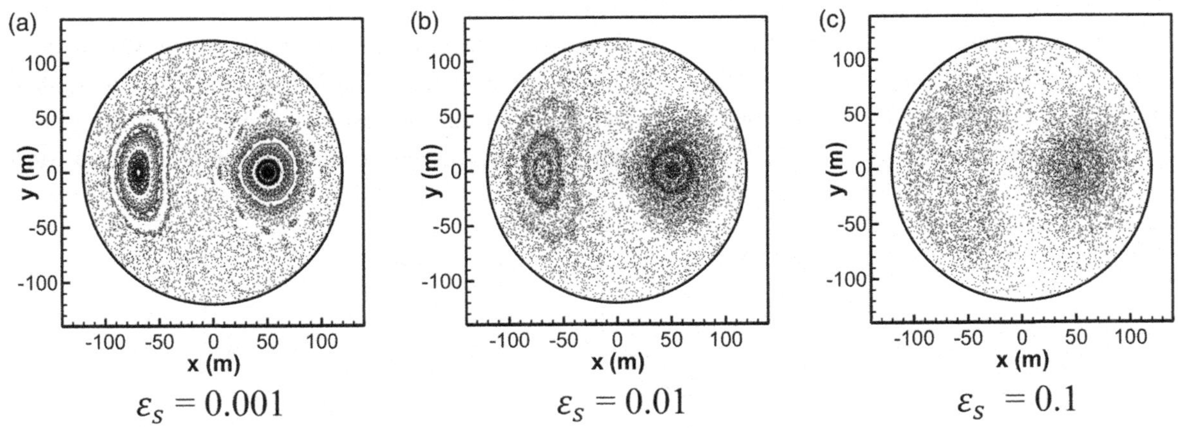

Figure 14 | Poincaré sections for both advection and diffusion processes ($\mu = 0.28$).

the impact of streamwise dispersion ε_s (the transverse diffusion ε_t is set to 0 in this case study) on this chaotic mixing phenomenon. For ε_s equal to 0.001, it is still clear that the circular lake is divided into two halves and patterns of the elliptic movement are still obvious. With the increase of the streamwise dispersion coefficient, this pattern becomes more and more blurred until the KAM (Kolmogorov–Arnold–Moser) (Kranenburg 1992) curves no longer exist

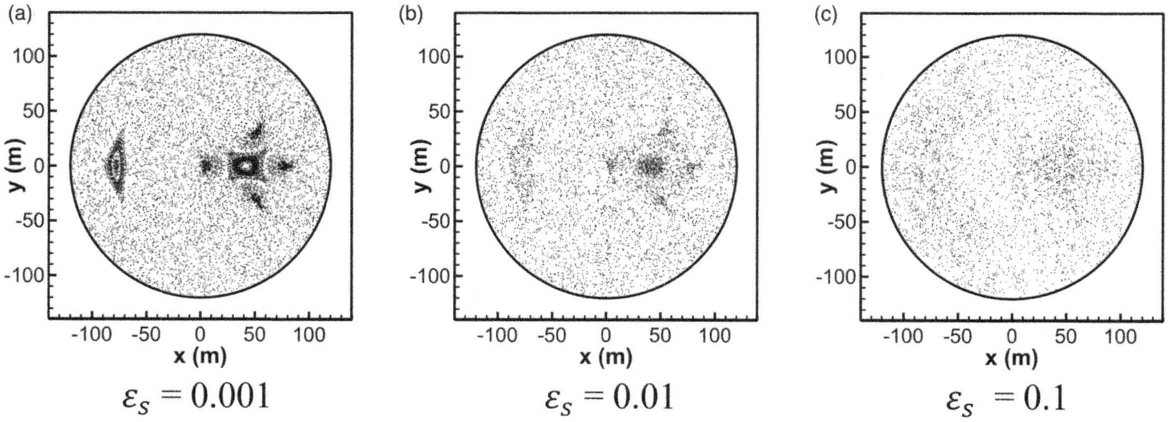

Figure 15 | Poincaré sections for both advection and diffusion processes ($\mu = 0.42$).

($\varepsilon_s > 0.1$). It is worth mentioning that the growth of the streamwise dispersion will not change the mean position of points, but it will remove the large local concentration gradients by increasing the random motion.

For $\mu = 0.28$ and 0.42, the entire calculation domain is chaotic when the dispersion ε_s is as small as 0.1, as shown in Figures 14 and 15. This means that the increase in the duration of the periodic storm will emphasise the influence of the diffusion coefficient, making it easier to achieve a chaotic state. Figure 16 shows the advection and diffusion behaviour of a line of 10,000 particles after 32 periods of the storm event. The line is initially positioned along the x-axis. The whorl and tendril structures coexist when μ is 0.28. With the increase of the dispersion parameter, these fine structures both disappear and particles are spead anywhere in the lake, although it can still be seen that the particle distributions in the left and right halves are symmetrical.

CONCLUSIONS

The traditional random walk model has been extended to solve the depth-integrated advection-diffusion equation. First, this model is verified by solving an instantaneous release problem in a uniform flow. The analytical solution is used as a reference. The results reveal several advantages of this model, including high accuracy, stability and simplicity. Extensive parametric studies have been carried out to investigate the sensitivity of the predictions to the computational parameters. It has been found that simulations are independent of the size of the time step. The particle number significantly influences the performance of the random walk model. Too few particles degrade the visual inspection and make it impossible to quantitatively examine the solute distribution. In uniform-flow applications, a relatively large time step will reduce the computation expense without compromising accuracy. Particle

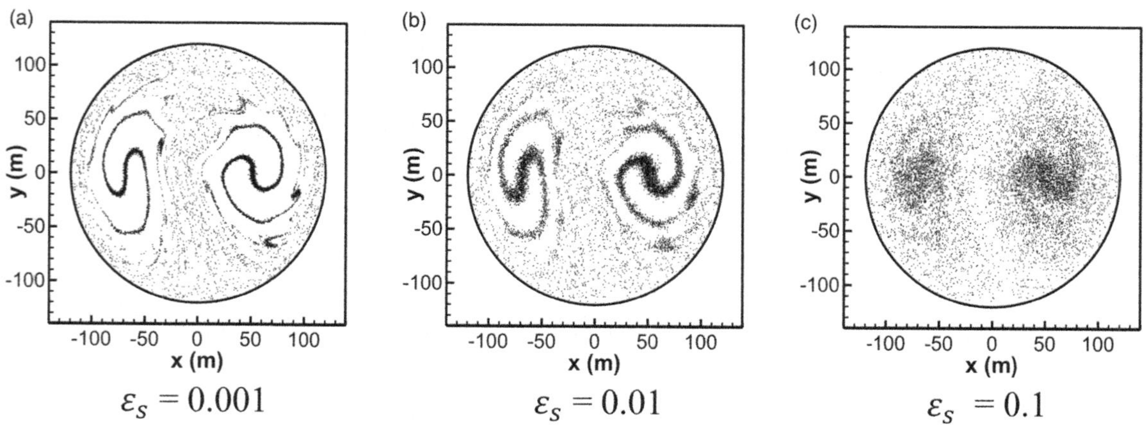

Figure 16 | Snapshots of the particle distribution in Kranenburg's model lake ($t = 32\,T$; $\mu = 0.28$).

numbers should be chosen to ensure at least 200–300 particles in each sampling bin. Subsequently, investigations are carried out regarding the oscillation of a pollutant cloud in a tidal estuary. The sampling method is optimised to convert the particle distributions into concentration profiles. The random walk simulations display high accuracy, which can only be achieved by the mesh-based simulations with extremely fine resolutions. The mesh-based methods are shown to be highly sensitive to grid resolution. Finally, the model is used to simulate the chaotic mixing process in a circular lake. The results for the case of pure advection are consistent with the findings reported by previous researchers. Because of the presence of turbulent diffusion and bottom friction in any real-world lake, the advection is always accompanied by transverse diffusion and longitudinal dispersion. This study shows that diffusion plays an important role in the mixing. The well-structured chaotic mixing pattern produced by pure advection cannot be observed in reality.

In summary, this study demonstrates that the random-walk model is highly stable and free of artificial diffusion in solving the solute transport problems in aquatic environments. In particular, this study proposes a generic sampling technique to convert the scatter of discrete particles to the distribution of solute concentration. In order to obtain a good estimate of the solute concentration, this study proposes a guideline for estimating the total number of particles needed in the simulation.

ACKNOWLEDGEMENTS

We are grateful for financial support provided by the Royal Academy of Engineering UK-China Urban Flooding Research Impact Programme (No. UUFRIP\100051), the 111 Project (No. B17015) and the China Scholarship Council (No. 201708060090).

REFERENCES

Begnudelli, L. & Sanders, B. F. 2006 Unstructured grid finite-volume algorithm for shallow-water flow and scalar transport with wetting and drying. *Journal of Hydraulic Engineering* **132** (4), 371–384.

Benkhaldoun, F., Elmahi, I. & Seaı, M. 2007 Well-balanced finite volume schemes for pollutant transport by shallow water equations on unstructured meshes. *Journal of Computational Physics* **226** (1), 180–203.

Bryc, W. 2012 *The Normal Distribution: Characterizations with Applications*. Springer Science & Business Media, New York, USA.

Falconer, R. A. 1991 Review of modelling flow and pollutant transport processes in hydraulic basins. In: *Water Pollution: Modelling, Measuring and Prediction* (C. A. Wrobel & C. A. Brebbia eds). Springer, Dordrecht, The Netherlands, pp. 3–23.

Gresho, P. M. & Sani, R. L. 1998 *Incompressible Flow and the Finite Element Method*, Vol. 1. Advection-diffusion and isothermal laminar flow. Wiley, New York, USA.

Gupta, I., Dhage, S., Chandorkar, A. A. & Srivastav, A. 2004 Numerical modeling for Thane creek. *Environmental Modelling & Software* **19** (6), 571–579.

Kranenburg, C. 1992 Wind-driven chaotic advection in a shallow model lake. *Journal of Hydraulic Research* **30** (1), 29–46.

Liang, D., Falconer, R. A. & Lin, B. 2006 Comparison between TVD-MacCormack and ADI-type solvers of the shallow water equations. *Advances in Water Resources* **29** (12), 1833–1845.

Liang, D., Wang, X., Falconer, R. A. & Bockelmann-Evans, B. N. 2010 Solving the depth-integrated solute transport equation with a TVD-MacCormack scheme. *Environmental Modelling & Software* **25** (12), 1619–1629.

Lin, B. & Falconer, R. A. 1997 Tidal flow and transport modeling using ULTIMATE QUICKEST scheme. *Journal of Hydraulic Engineering* **123** (4), 303–314.

Mingham, C. G., Causon, D. M. & Ingram, D. M. 2001 A TVD MacCormack scheme for transcritical flow. *Proceedings of the Institution of Civil Engineers-Water and Maritime Engineering* **148** (3), 167–175.

Rajar, R. 1997 The role of mathematical models, physical models and field measurements in water pollution problems. *Transactions on Ecology and the Environment* **14**, 545–555.

Wu, X. F. & Liang, D. 2018 Study of pollutant transport in depth-averaged flows using random walk method. *Journal of Hydrodynamics* **2018**, 1–13.

Wu, X. F., Yang, F. & Liang, D. F. 2018 Study of pollutant transport in environmental flows using depth-averaged random walk method. *EPiC Series in Engineering* **3**, 2342–2350.

Yang, F., Liang, D. & Xiao, Y. 2018 Influence of Boussinesq coefficient on depth-averaged modelling of rapid flows. *Journal of Hydrology* **559**, 909–919.

Zhang, Z. & Chen, Q. 2007 Comparison of the Eulerian and Lagrangian methods for predicting particle transport in enclosed spaces. *Atmospheric Environment* **41** (25), 5236–5248.

First received 14 January 2019; accepted in revised form 23 April 2019. Available online 24 May 2019

Optimal water quality sensor positioning in urban drainage systems for illicit intrusion identification

Mariacrocetta Sambito, Cristiana Di Cristo, Gabriele Freni IWA and Angelo Leopardi

ABSTRACT

In the last decade, the growth of the micro-industry in urban areas has produced an increase in the frequency of xenobiotic polluting discharges in drainage systems. Wastewater treatment plants are usually characterized by low removal efficiencies in respect of such pollutants, which may have an acute or cumulative impact on environmental and public health. To facilitate the early isolation of illicit intrusions, this study aims to develop an approach for positioning water quality sensors based on the Bayesian decision network (BDN). The analysis is focused on soluble conservative pollutants, such as metals. The proposed methodology incorporates several sources of information, including network topology, flows and non-formal 'grey' information about the possible locations of contamination sources. The methodology is tested using two sewer systems with increasing complexity: a literature scheme from the Storm Water Management Model (SWMM) manual and a real combined sewer in Italy. In both cases, the approach identifies the optimal sensor location gaining advantage from additional information, which reduces the computational effort needed to obtain the solution. In the real case, the application of the method yielded a better solution with regards to the real position of the implemented sensor network.

Key words | optimization methods, sensor location, uncertainty analysis, urban drainage, wastewater quality

Mariacrocetta Sambito
Gabriele Freni IWA (corresponding author)
School of Engineering and Architecture,
University of Enna 'Kore', Cittadella Universitaria,
94100 Enna,
Italy
E-mail: *gabriele.freni@unikore.it*

Cristiana Di Cristo
Department of Civil, Architectural and Environmental Engineering,
University of Naples Federico II,
80125, Napoli,
Italy

Angelo Leopardi
Department of Civil and Mechanical Engineering,
University of Cassino and Southern Lazio,
Cassino 03043,
Italy

INTRODUCTION

Wastewater quality and monitoring needs

In both water distribution systems and sewer systems, the monitoring of water quality is very important for preserving resources and public health. Monitoring physical, chemical and biological parameters increases the possibility of early detection of water quality deterioration and individuation of pollution sources. The quality of wastewater impacts the proper functioning of a sewer system and a wastewater treatment plant (WWTP) and the receiving water body in the case of combined sewer overflow (CSO) activation (Even *et al.* 2004). CSOs, which contain untreated domestic and industrial waste, toxic materials and debris, impact the physicochemical, biological, hydraulic and aesthetic status of receiving water bodies. For example, overflows can cause oxygen depletion, increased turbidity and higher concentrations of micropollutants, heavy metals and pathogenic and faecal organisms in surface waters (Passerat *et al.* 2011). Xenobiotic substances, unlike organic substances, are only slightly affected by biological degradation processes. Metals often show a remarkable tendency for bio-accumulation and are unaffected by wastewater treatment and/or dangerous for common plant technologies.

Since the adoption of the Water Framework Directive 2000/60/EC, Member States in EU countries must apply local measures to address the pollution that affects their surface waters; thus, decreasing the occurrence of overflows and improving discharged water quality are important

doi: 10.2166/hydro.2019.036

parts of pollution-reducing strategies. Models for the characterization of the wastewater quality have been extensively investigated to assess the pollution load that is overflowed and/or transferred to WWTP. Boenne *et al.* (2014) showed that the contribution of WWTP overflows due to just two events of a few hours can produce an increase up to 22% of the measured nutrient load in a river. Jiang *et al.* (2018) asserted that the identification of polluting sources after river spill is critical to improving decision-making about the emergency response.

For these reasons, the implementation of a monitoring network is crucial for an efficient contamination prevention strategy in urban drainage systems, which involves the identification and elimination of illicit polluting discharges.

Monitoring systems and polluting sources identification

The problem of polluting source identification was primarily investigated for pressurized distribution networks as a drinking water contamination event determines an immediate alarm for public health (Di Cristo & Leopardi 2008; Lifshitz & Ostfeld 2019). An abnormal polluting discharge into a sewer system usually has a lower impact on the general public even if it has a relevant impact in the environment.

The collection and the analysis of real data are indispensable to control wastewater quality and to identify the origin of the pollution. Montserrat *et al.* (2015) developed a methodology to evaluate the performance of combined sewer systems (CSSs) using low-cost monitoring to reduce the number and the impact of overflows. Using the measurements of various quality parameters, Srinivas *et al.* (2018) developed an approach for identifying the major sources of pollution in rivers using advanced hierarchical clusters and multivariate statistical analysis.

The development of specific sensors (Qin *et al.* 2012) facilitated online and real-time measurement of wastewater quality. Boenne *et al.* (2014) employed online high-frequency continuous measurements to assess the impact of wastewater quality downstream from a treatment plant on the receiving water body. They demonstrated that continuous *in situ* monitoring can furnish important information about pollution sources, even if the cost of the monitoring setup is higher than that of traditional sampling. Troutman *et al.* (2017) presented a data-driven identification/learning toolchain to manage a large number of measurements for dynamic modelling and prediction of a combined sewer functioning. They identify a near-optimal time record for which measurements must be available to ensure an acceptable forecasting performance.

Regarding sewers, illicit discharges can easily enter systems via intentional or accidental damping or spills as the networks are geographically dispersed and have multiple access points. For this reason, many countries have implemented regulations and projects to support actions for illicit discharge individuation in sewer systems (Irvine *et al.* 2011). Recently, Banik *et al.* (2017a) proposed a methodology for identifying an illicit intrusion in a sanitary or combined sewer system using online pollutant concentration measurements.

Optimization of sensor location

The approaches proposed for illicit discharge individuation require the deployment of sensors. However, installing and operating measurement devices in sewers is expensive and limited by many constraints. The installation and maintenance costs can be reduced by optimizing the position of the sensors while simultaneously obtaining a reliable and inexpensive monitoring infrastructure. The sensor placement problem has been extensively investigated to design contamination warning systems in drinking water distribution networks (Rathi & Gupta 2016) and to monitor rivers (Lee *et al.* 2014).

Few studies address the sampling design in sewer systems. Kim *et al.* (2013) aimed their study at developing a decision-support model for identifying the location of the pathogenic intrusion in a real gravity sewer system as a means of facilitating rapid isolation and efficient containment using artificial neural networks (ANNs). The results showed that ANNs identified the location of the injection sites with 57% accuracy. Increasing the number of available sensors within the basin significantly improved the accuracy of the simulation results (from 57% to 100%). Recently, Banik *et al.* (2017b) developed and compared different multi- and single-objective optimization procedures to optimally locate sensors to detect illicit intrusion in sewer systems with objective functions expressed by the parameters entropy, detection time and reliability. The results show that the obtained sensor displacement in all cases is more efficient than the existing displacement established without any optimization. In Banik *et al.* (2017c), the optimal placement of wastewater monitoring sensors that was formulated as a single objective optimization problem is solved using greedy algorithms. The results indicate the robustness of the methodology with respect to the detection of contaminants, the excellent performance of the single fitness function and the better efficiency of the greedy algorithm with respect to the genetic algorithm.

Many of the literature sensor location procedures assume that any node of a network can be a source with equal probability. Weickgenammt *et al.* (2010) presented an importance-based sampling method for selecting dangerous scenarios for a sensor location problem in a water distribution system. In this study, contaminant inputs in nodes located in highly populated areas are considered to be 'more important scenarios'. Tinelli *et al.* (2017) proposed a procedure based on practical considerations of network topology and operations for sampling the most representative contamination events in the sensor location problem of water distribution systems. The results indicate that the optimal sensor placement does not vary when only the selected sampled events are considered.

Yazdi (2018) proposed a methodology based on entropy theory and employed the differential evolution algorithm for identifying the best monitoring locations for detecting wastewater quality changes in sewers. The results indicate that the method improves the level of information with a limited number of sensors.

Vonach *et al.*(2018) presented a heuristic method for measurement site selection in a sewer system to obtain an efficient calibration of the hydrodynamic model.

Bayesian decision networks

When optimal solutions are needed in an uncertain system, Bayesian approaches can be useful to make decisions (in this case for identifying the best sensor network) and assimilate information from the system in an upgradable and updatable way (in this case, gaining information from numerical simulations and data from the real system).

Bayesianism is the philosophy that asserts that to understand human problems while constrained by ignorance and uncertainty, the probability calculus is the single most important tool for representing appropriate strengths of belief. The probability calculus enables us to represent the interdependencies that other systems require and enables the representation of any dependencies.

A Bayesian decision network (BDN), or Bayesian network, is an acyclic graphical structure that enables us to represent an uncertain domain and the conditional dependencies between independent variables and dependent variables in a probabilistic way. Bayesian networks are ideal for considering an event that occurred and predicting the likelihood that any one of several possible known causes was the contributing factor. The nodes represent a set of random variables from the domain ($X = X1,\ldots,Xn$), while a set of directed arcs connects the pairs of nodes ($Xi \rightarrow Xj$) for representing the direct dependencies among the variables. At least three distinct forms of uncertainty, with which an intelligent system that operates in the real world shall need to cope, exist: ignorance the limits of our knowledge, which cause us to be uncertain about many things; physical randomness or indeterminism and vagueness.

The BDN is a very robust and particularly useful method for assessing risk and uncertainty that provides a complete framework for analysing all cause and effect relationships (Korb & Nicholson 2010).

Few applications of BDN, which are related to urban water systems, are included in recent technical literature. These applications are aimed at guiding technical choices in uncertain domains to incorporate different sources of information into the decision. Phan *et al.* (2016) presented some applications of BDN to water resources with respect to spatial factors, water domains and the consideration of climate change impacts to guide management decisions.

Kabir *et al.* (2015) applied BDN to identify water system main failures considering the vulnerability and sensitivity of the system to failure and the global risk. In this case, collected data about failure were progressively incorporated in the Bayesian method to improve the selection.

Freni & Sambito (2017) proposed a probabilistic approach to the positioning of water quality sensors in urban drainage networks for identifying an illicit intrusion, which shows the progressive increase in the identification probability obtained by the Bayesian approach. In this work, the implementation of the pre-conditioning approach proposed by Banik *et al.* (2015), essentially depending on the network topology, produced an improvement in term of computational efforts.

Aim of the research

This study presents a methodology for solving a sensor location problem and individuating the source of an illicit intrusion in a sewer system. It is aimed at solving a sensor location problem, in which the positioning of fixed-type sensors is assumed. The analysis is performed with the hypothesis that each node of the network has the same probability of being the polluting source. Successively, the hypothesis that some nodes can be more frequently polluted than others is introduced. The different probabilities of a node of the contamination source are derived by the knowledge of the system topology, flows and possible polluting activities based on the grey information about the served area in terms of commerce and industry data. In this application, the capacity for incorporating all available information, to individuate the more risky scenarios, represents the main original aspect of the proposed methodology. Contrary to previous studies, the proposed approach considers the inclusion in the network of several sensors, taking into account the interaction and correlation among their responses. The preliminary analysis presented in Sambito *et al.* (2018) is completed.

The methodology is applied to two different networks with increasing size and complexity: the literature network Example 8 of the Storm Water Management Model (SWMM) application manual (Gironás *et al.* 2009), in which the contamination was analysed in wet weather conditions, and the real test case represented by the sewer system of Massa Lubrense (Italy), which was analysed in dry weather conditions. The paper is organized as follows. First, the sensor location problem formulation is presented, and the Bayesian approach that is employed to solve the problem is described. Second, the test cases are discussed, and the results are presented. Last, some conclusions are formed.

MATERIALS AND METHODS

In the proposed methodology, the sensor location problem is solved using a Bayesian approach. The new information from the analysis enables the operator to gain insight into the system once new contamination events are detected and identified. In this way, the approach is suitable for solving problems, in which data are initially collected and the operator plans to improve the monitoring strategy.

To solve the sensor location problem, two main components are required: a calibrated model for hydraulic and water quality simulations in sewer systems and a Bayesian solver for likelihood estimation and probability updating.

Numerical simulation model

The EPA SWMM (5.022 version) was employed to simulate the urban drainage network and the propagation of contaminants in a sewer system. This model enables the user to select different mathematical models to describe the runoff formation and propagation in sewer systems (Gironás *et al.* 2009). The complete 1D Saint-Venant equations were applied to simulate the flow propagation into a sewer system by adopting an iterative explicit mathematical solver.

Water quality routing within conduit links assumes that the conduit behaves as a continuously stirred tank reactor (CSTR). Although a plug flow reactor assumption may be more realistic, the differences will be small if the travel time through the conduit is on the same order as the routing time step. The concentration of a constituent that exits the conduit at the end of a time step is obtained by integrating the conservation of mass equation and using average values for quantities that may change over the time step, such as flow rate and conduit volume. The quality of the water that exits the node is the mixture concentration of all water that enters the node. Water quality modelling within storage unit nodes and manholes follows the approach used for conduits.

The considered contaminant is assumed to be a soluble conservative xenobiotic, such as some heavy metals or soluble ionic compounds. This hypothesis was introduced as the intrusion of a conservative pollutant represents a more dangerous scenario.

The proposed sensor location approach has to be tested as polluting events occur. Considering that the position, magnitude and duration of contamination can be uncertain, each model application is given by a random simulation in which the contamination parameters are randomly set up in terms of the contaminant mass, contamination duration and contamination node. The contaminant mass is randomly set between 0.01 kg and 0.5 kg; the contamination duration is randomly set between 0.25 hour and 3 hours. As mentioned in the Introduction, in Freni & Sambito (2017), all network nodes were considered to have the same probability to host the contamination event, while different probabilities were established in this analysis based on the information about the system and the served area.

Each sensor network configuration was investigated by 1,000 random contamination events, and its efficiency was evaluated by the uncertainty and isolation likelihood D, i.e., the probability of the sensor network to detect the presence and the origin of the contamination. The isolation likelihood is evaluated as the ratio between the number of events in which the network sensor was able to locate the contamination node and the total number of tested contamination events. The uncertainty is summarized by the probability that the sensor network is able to detect the contamination but unable to locate the source node. In the analysis, the likelihood and the reliability/uncertainty functions were slightly adapted from those presented in Preis & Ostfeld (2008) to comply with sewer networks instead of water distribution networks. According to Preis & Ostfeld (2008), the isolation likelihood F_1 and detection reliability or redundancy F_2 are expressed by the following equations:

$$F_1 = \frac{1}{S}\sum_{i=1}^{S} d_r \quad (1)$$

$$F_2 = \frac{1}{\sum_{i=1}^{S} d_r}\sum_{i=1}^{S} R_r \quad (2)$$

where S is the total number of analysed contamination events; d_r is 1 if the contamination was identified by the sensor network and is 0 otherwise; and R_r is 1 if the contamination was detected by at least two sensors and is equal to 0 otherwise. The indicator F_1 (Equation (1)) provides information on the ability of the sensors' network to locate the contamination source, while F_2 (Equation (2)) indicates the reliability of the sensor network (more than one sensor) in detecting an event. If the contamination is not confirmed by more than one sensor in the system, false positives may be present.

Bayesian network approach for sensor location

As discussed in the Introduction, the BDN is used to guide decisions in an uncertain domain progressively incorporating information in the process. In this framework, in the present study:

- the upstream (independent) nodes of the BDN are related to contamination factors (position, magnitude, duration and starting time) and external factors, such as network characteristics, dry weather flows and wet weather flows;
- the intermediate nodes are related to the distribution of the contaminant concentrations and are connected to the upstream nodes by probabilistic arches depending on the model results, which are subjected to uncertainty;
- the downstream nodes are related to the likelihood of sewer manholes to be a suitable location for sensors alone and in combination with others and are connected to the intermediate nodes by probabilistic arches that express the detectability of the contamination and the reliability of the sensors.

Bayesian approaches start with the formulation of prior knowledge in terms of the probability of events to be representative of the truth (in this case, the probability that a sensor or group of sensors to be correctly located to identify the source of contamination). The system is solicited and investigated to obtain a series of events that can confirm or deny prior assumptions (in this case, a set of simulated events of contamination in which sensors were in place; each event was based on the probability distribution of the contamination in nodes). The number of events considered to be sufficient to verify and update prior knowledge (population of the update) is a parameter in the Bayesian approach. After this number is attained, the posterior probability is calculated by incorporating new information from the series of events in prior knowledge by the application of Bayes' theorem. The number of updates for which additional information does not significantly implement previous knowledge represents the efficiency of the approach in terms of rapid convergence to a stable sensor configuration. Once this asymptotical condition is satisfied, the sensor network can be evaluated in terms of the isolation likelihood and uncertainty. The maximum number of updates, robustness and uncertainty of the approach depend on the complexity of the analysed problem.

This study investigated the effect of applying the pre-screening procedure by Banik *et al.* (2015) in reducing the number of nodes that should be considered as a possible source. In Freni & Sambito (2017), all nodes had an equal chance to be the origin of the contamination. In this study, the probability that the nodes are the source is assumed to be different using information about the system. This information is used to implement a pre-conditioning approach to assigning a prior sensor probability distribution to the BDN approach. Two different pre-conditioning strategies are implemented. The prior probability in the BDN analysis is assumed to be proportional to the wastewater volumes, or alternatively, proportional to the contaminant mass passed through each node.

If the pre-conditioning approach is efficient, the BDN approach convergence to a stable sensor configuration may be faster (and reduce computational efforts), and the isolation likelihood may increase. After the prior sensor probability distribution is defined, the application of the BDN approach is affected by the population of events (in this case, the number of simulated contamination events) that is employed for each Bayesian update. A large number of events requires greater computational effort, but the information is used to update the probability distributions only if it is verified several times. A small number is required to achieve faster updates and possibly faster convergence to an asymptotic solution but introduces the risk that unreliable information may be used to update probability distributions. In this study, the tests are performed considering the following approaches (Prior A, B, C, D) to compare the results:

- Prior A: no pre-screening procedure and no prior knowledge (each node has an equal initial probability to be the location of a sensor).
- Prior B: no prior knowledge and pre-screening procedure based on network topology.
- Prior C: pre-screening procedure and prior knowledge based on water fluxes.
- Prior D: pre-screening procedure and prior knowledge based on the mass of contaminant that potentially passed through each node.

The results of the Bayesian approach are also compared with the results obtained by Banik *et al.* (2017b, 2017c) using the NSGA-II and the greedy algorithms. In the real case study, the efficiency of sensor disposition obtained through the Bayesian approach has been compared with the one of the real monitoring network.

CASE STUDIES

The literature example, i.e., network Example 8 presented in the EPA SWMM reference manual, involves a combined sewer network that serves an area of 0.12 km^2 and consists of 31 nodes, 29 pipes and a pump (Figure 1). The network is characterized by two outfalls: the WWTP and the overflow. The nodes downstream of commercial/industrial activities are assumed to have a greater probability of being subject to an illegal spill of contaminants into the sewers. The probability distribution function of contamination, as reported in Figure 2, is not considered to be uniform. Three nodes (J4, J10 and JI18) are hypothetically considered to host industrial activities, and a double probability of illicit contamination is estimated with respect to the other nodes. This hypothesis does not modify the general applicability of the method in the cases in which industrial nodes may

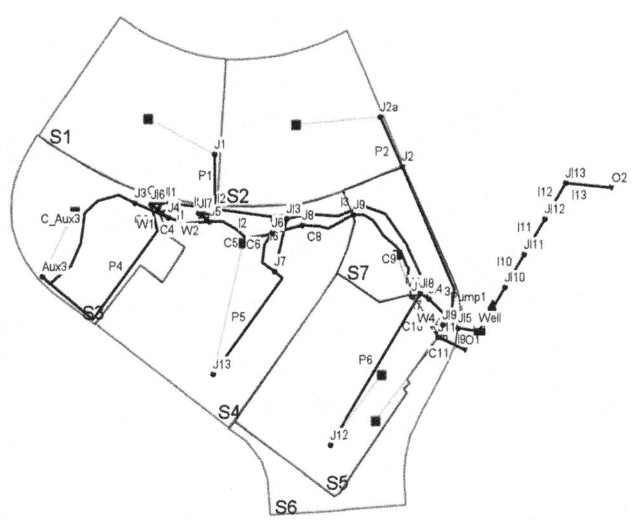

Figure 1 | Example 8 network scheme: Jx and JIx (where x represents a sequential number) represent junctions (only Aux3 has a different name as dry weather and wet weather flows are split); Cx and Ix represent conduits of the combined system and dry weather main to the WWTP; and Ox are outflows to the river and the WWTP.

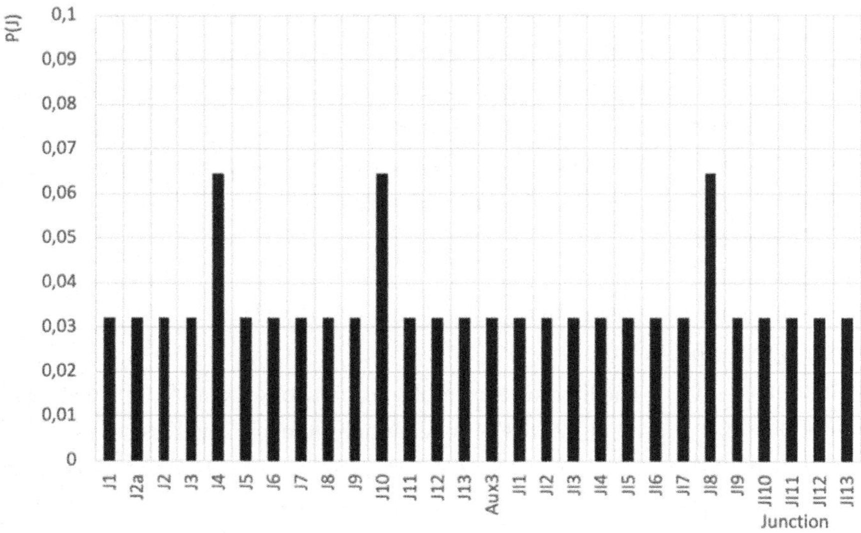

Figure 2 | Probability density function of contamination in the nodes: three nodes have a double probability of contamination.

change with regards to number, contamination probability or location.

The real case study involves the sewer of Massa Lubrense (Figure 3), which is a town located near Naples, Italy. The sewer is a combined sewer system that covers a predominantly hilly area of 19.71 km^2 with a mean altitude of 121 m with respect to sea level. The system is divided into 12 subcatchments that serve 7,452 users who correspond to a population of 14,087 (2011).

The length of the network is 72 km, and the network consists of 1,909 circular conduits that connect 1,902 junctions, 14 pumps, 14 storage units and 1 treatment plant. The pipes have different cross sections and materials but approximately 80% of them are circular with diameters

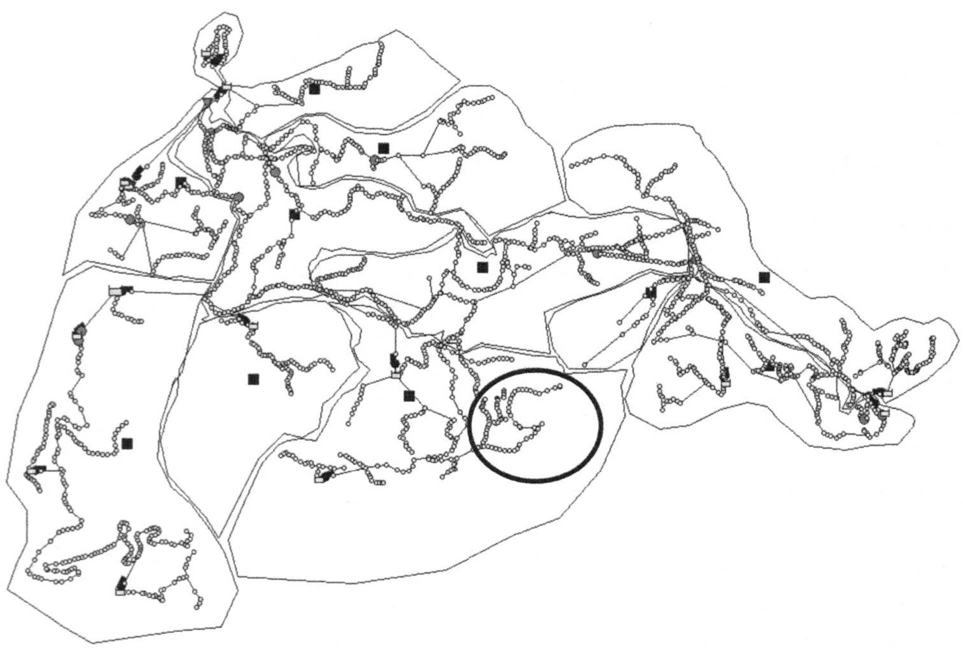

Figure 3 | Massa Lubrense network; the area with a double probability that it is the source of contamination is indicated by a circle. The dots represent the current location of 12 existing sensors that are used to calibrate the model.

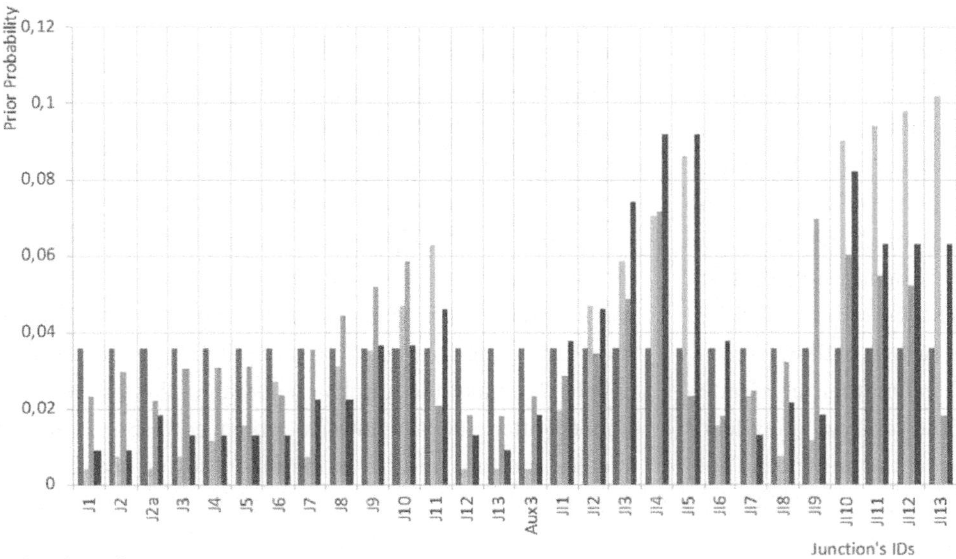

Figure 4 | Prior sensor location distribution in Example 8 network for the different approaches and one sensor.

that vary between 80 mm and 1,000 mm. Due to the variable altimetry of the area, 15 pumping stations exist, 15 outflows are located along the pipes and 10 outflows convey wastewater to the sea, while 5 outflows direct the flow into another pipe. The wastewater is carried to the treatment plant, which represents the final output. All geometric data, which are available on the website htttp://www.progettosimona.it, have been included in the SWMM input file. The daily average values of the dry weather flows in the 1,866 input nodes are estimated considering the population connected to each flow. The input file for the SWMM model has been calibrated using discharge measurements. The system has already installed 12 monitoring stations; their displacements were determined on the basis of practical considerations without any sensor location analysis. Similar to the literature example, the analysis was performed with the hypothesis that the circled area in Figure 3 has a double probability that it will be subjected to more contamination than any other node in the network: 45 nodes have a contamination probability of approximately 0.1%, and the other 1,857 nodes have a contamination probability of approximately 0.05%.

ANALYSIS OF RESULTS

The Bayesian analysis is initially performed by assuming that any node in the network has the same probability that it will be the location of a sensor (Prior A). Prior B was performed by applying the methodology proposed in Banik *et al.* (2015). In a second step, other prior knowledge scenarios are applied, in which a different contamination probability is assigned to the nodes. In Prior C and Prior D, a preliminary analysis is performed by Monte Carlo simulations, in which a single source of contamination is located in one of the nodes of the network according to the contamination probability that was previously assigned. In 50 random simulations, the average water and contaminant fluxes through nodes are calculated to assign the prior probability of sensor location to the BDN analysis.

Example 8 network

The Bayesian analysis is performed for the Example 8 network considering the possible implementation of one, two or three sensors. As denoted in Banik *et al.* (2017b), with more than three sensors, the correlation among the measurements increases with a small increase in the information content.

Figure 4 shows the prior sensor location distribution for the BDN analysis in the case of a single sensor configuration for the Example 8 network for all performed tests. Relevant differences between non-informative distribution (Prior A) and the other distributions are observed, which highlights the importance of a prior discrimination of the possible solutions in the application of the Bayesian approach. Sensors placed in the upper nodes of the network have a smaller probability of detecting contamination as the majority of contamination episodes occur in nodes located downstream of the sensors. Applying the pre-screening procedure based on the topological approach (Prior B), the prior sensor distribution probability is significantly different with respect to that of Prior A. The procedure is limited when several possible paths to the system outflow are present. In these cases, the flow dividers and their efficiencies in the separation of wet weather flows and polluting loads affect the ability of the sensors to detect the presence of contamination. Nodes JI10, JI11, JI12 and JI13 are located on the connection between the network and the WWTP and they are topologically located in the most downstream part of the network. The presence of dividers conveys only a small part of water volumes to these nodes, which renders them less relevant according to the Prior C approach. Comparing Prior C and Prior D, Prior C tends to overestimate the importance of nodes that receive combined sewer overflows (such as J12 and J13) and are characterized by larger volumes but usually small contaminant concentrations.

As indicated in Table 1, in other tests, the magnitude of the event population of each Bayesian update is modified to understand its impact on the analysis, especially with respect to the minimum number of simulations, and highlight the most efficient configuration of the sensors. The modification also explains how the value of this BDN parameter can affect the selection of the most relevant nodes for sensor placement and the number of model simulations that are needed to obtain a stable configuration.

As an example, Figure 5 shows a comparison of the use of Prior A and Prior D for the placement of one sensor by adopting an event magnitude of ten for each update. Figure 6 shows the same comparison of adopting 25 events for each update. To make the graphs comparable, the analysis was stopped after 100 simulations, which corresponds to ten updates and four updates in the first case and second case, respectively.

Figure 5 shows some interesting results:

- The analyses do not differ in terms of selection of the best sensor location, which demonstrates that prior knowledge does not affect the final decision but affects the computational resources and data that are needed to attain the final stable distribution of likelihood.
- Using Prior D, after only two updates (20 simulations), the nodes with the highest likelihood of being selected (JI10) as a sensor location, are identified, and the selection does not change until the end of the analysis.
- Using Prior A, six updates (60 simulations) are necessary to achieve the same results obtained for Prior D after 20 simulations. Prior B and Prior C (not reported) performed equally well and obtained the same selection after four updates (40 simulations).

The comparison between Figures 5 and 6 shows that, even if the number of simulations is identical, the aggregation in the Bayesian update process produces some differences:

- The final selection is not affected by either of the adopted prior distributions. In both cases, three updates (75 simulations) are necessary to identify a stable candidate node for the best sensor location.
- Posterior distributions are affected by a large variability after four updates, which requires a larger number of simulations to obtain a robust solution.

Table 1 reports the results of other tests that were performed considering a maximum of three sensors and varying the event magnitude for each update from 10 to

Table 1 | Results in terms of Bayesian probability after 1,000 simulations

Example 8 network

No. of sensors	No. of updates	Event pop. of updates	Highest sensor posterior probability	Prior probability of each sensor combination
1	10	100	0.1315	0.034
2	10	100	0.0076	0.0025
3	10	100	0.00068	0.00027
1	40	25	0.16	0.034
2	40	25	0.0085	0.0025
3	40	25	0.00075	0.00027
1	100	10	0.178	0.034
2	100	10	0.0085	0.0025
3	100	10	0.00077	0.00027

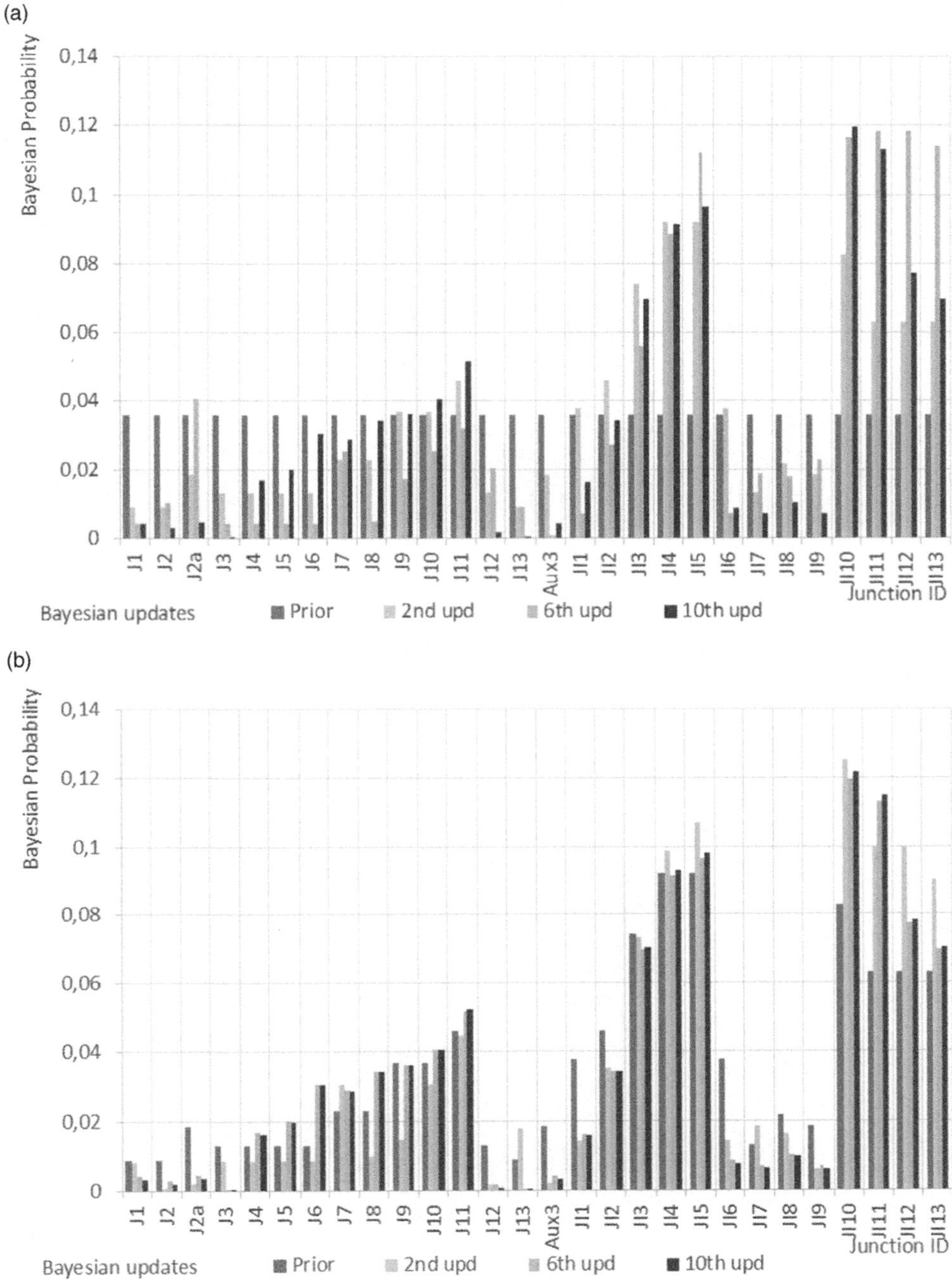

Figure 5 | Prior and posterior distribution after two, six and ten updates. Event population for each update = 10: (a) Prior A and (b) Prior D.

100. The results indicate that, for the Example 8 network, the BDN configuration that enables a more efficient selection of (in terms of Bayesian probability) sensors' location refers to the use of Bayesian updates with small populations (ten events). With ten events after ten updates (100 simulations) to a maximum of 100 updates (1,000 simulations), the improvement is observed (not reported) to be limited, and the posterior distribution attains a stable configuration with very small variations independently from prior knowledge. This behaviour confirms that the Bayesian methods do not retain the memory of the initial assumptions if the number of performed updates is sufficiently large.

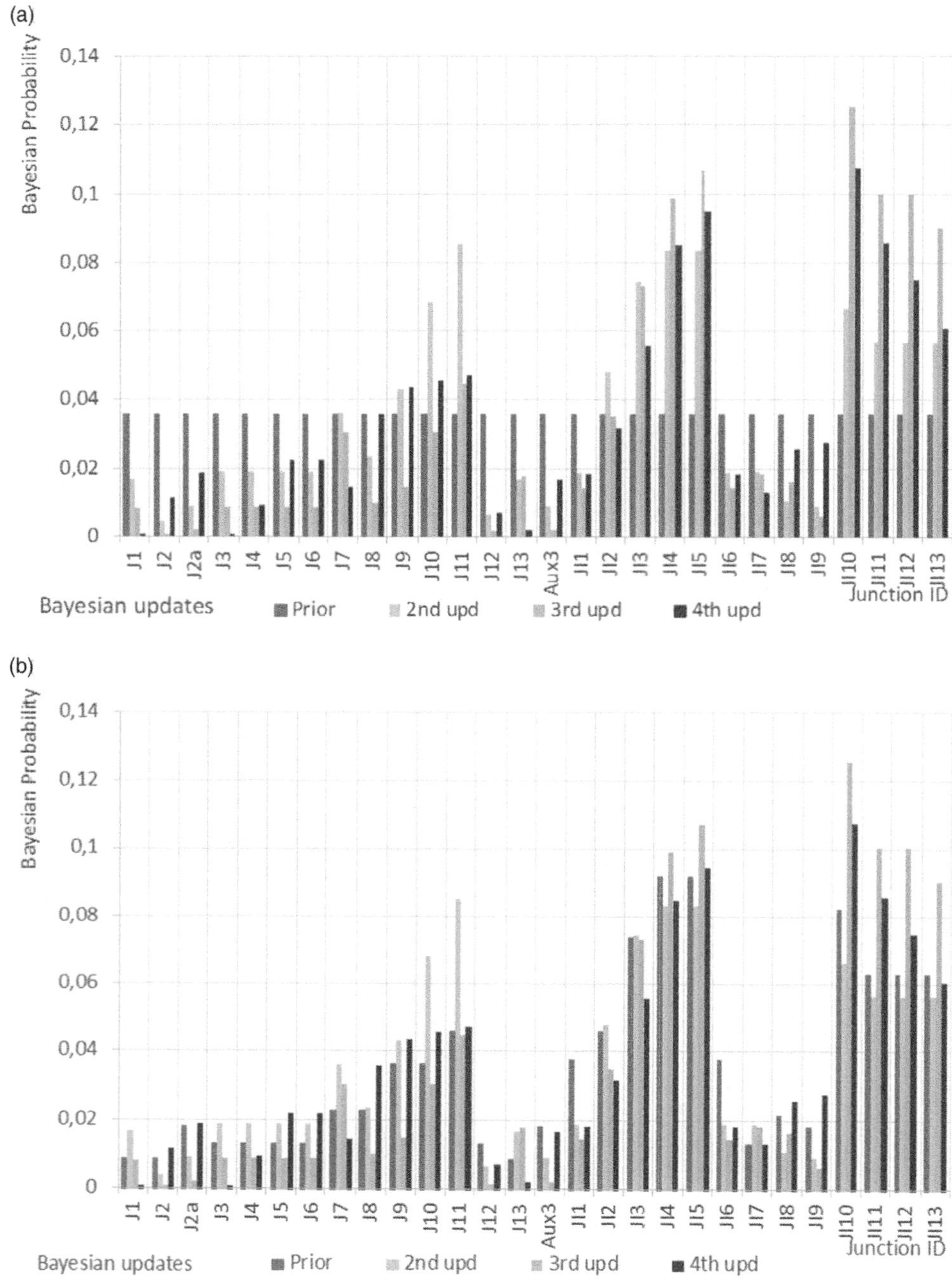

Figure 6 | Prior and posterior distribution after two, three and four updates. Event population for each update = 25: (a) Prior A and (b) Prior D.

All applied approaches, independently from selected prior knowledge and the population of updates, attained the same final selection for the analysed network:

- 1 sensor: node JI10,
- 2 sensors: nodes J11 and JI11,
- 3 sensors: nodes J11, JI5 and JI10.

Starting from this result, the probability that the sensors positioned in these junctions will be able to identify the contamination source (isolation likelihood F_1) is 47% with one sensor, 78% with two sensors and 84% with three sensors. With one sensor, 24% of the contamination events are undetected and 29% are detected; however, the origin of the contaminant is not discovered. With two sensors, only

11% of the events are not detected and an additional 11% of the events are detected without identifying the source. With three sensors, only 6% of the events are undetected even if 10% of the events are detected without identifying the source. Sensor network reliability F_2 is relatively low with two and three sensors reaching 17% and 33%, respectively, which shows that the most relevant part of the contamination events was detected by a single sensor.

The use of a non-uniform informative a priori sensor location distribution reduces the simulation time. The use of distributions based on contaminant mass perform better than other methods, but the simple topological approach (obtained without the use of any additional simulation) can reduce the computational effort by one-third.

Considering the positioning of three sensors, two of the three selected nodes coincide with the ones individuated by two different procedures in Banik *et al.* (2017b). This comparison confirms the validity and robustness of the presented methodology.

Massa Lubrense network

The BDN approach is successively applied to the real network of Massa Lubrense (Italy) considering three sensor configurations that involve 6, 12 (actual number of implemented monitoring stations for model calibration) and 18 sensors.

The analysis is performed to analyse the impact of the BDN update population and the ability of the best possible sensor locations. To reduce computational time, the analysis was performed starting from the informative distribution Prior D based on 100 random contamination simulations.

Table 2 reports the results in terms of the Bayesian probability in the various considered BDN configurations. In this case, the best sensor configuration and the efficiency of the sensor network in identifying the polluting source depend on the number of events and the number of procedure updates. The complexity of the analysed network considers that the best strategy is to increase the number of Bayesian updates, which reduces the population of each update: the use of 40 updates with 25 simulations each provides better results than the use of 10 updates with 100 simulations each. This finding can be explained by the complexity of the system and the number of possible sensor combinations that should be considered.

Table 2 | Results in terms of the Bayesian probability in various BDN configurations (Massa Lubrense)

Massa Lubrense network

No. sensors	No. of Bayes updates	Event pop. of updates	Highest sensor posterior probability	Prior probability of each sensor combination
6	10	100	0.003	5.26×10^{-4}
12	10	100	2.45×10^{-8}	1.53×10^{-17}
18	10	100	6.44×10^{-16}	2.21×10^{-31}
6	40	25	0.046	5.26×10^{-4}
12	40	25	3.23×10^{-4}	1.53×10^{-17}
18	40	25	4.32×10^{-10}	2.21×10^{-31}
6	100	10	0.026	5.26×10^{-4}
12	100	10	6.65×10^{-6}	1.53×10^{-17}
18	100	10	2.15×10^{-13}	2.21×10^{-31}

Positively, the different analyses converge on the same set of sensor locations, which confirms the robustness of the proposed approach. Analysing the three configurations (6, 12 and 18 sensors), the probability that the sensors intercept the contamination (isolation likelihood F_1) is 58% with 6 sensors, 81% with 12 sensors and 92% with 18 sensors.

Detection reliability F_2 is substantially higher in the real case due to the higher number of sensors: 68% of the detected events are reported by two or more sensors in the first configuration (6 sensors) and the probability increases to 77% and 86% with 12 sensors and 18 sensors, respectively. A significant number of events remained undetected: 27% with 6 sensors, 6% with 12 sensors and 4% with 18 sensors.

Figure 7 shows the location of 12 sensors according to the proposed methodology (triangles). Comparing the obtained locations with the actual configuration of the monitoring network (large circles), some considerations can be obtained:

- The optimal configuration of sensors in the downstream part of the network (upper part of the figure) is consistent with the existing configuration with three sensors that are actually located in their optimal positions and two sensors that are only a few nodes from their optimal position.
- In the western part of the network, the methodology provided the same sensor locations of the real network,

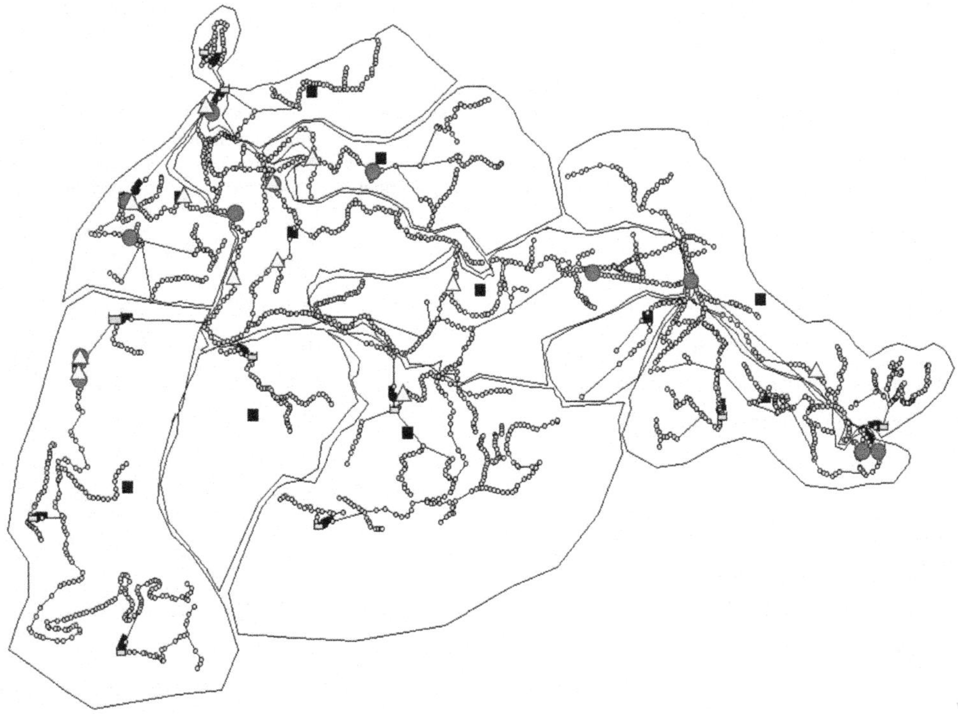

Figure 7 | Massa Lubrense network with the indication of the best location for 12 sensors (triangles) according to the methodology and compared with the actual position of the monitoring stations (large circles).

while a difference was observed in the central and eastern parts of the system; these differences can be primarily connected to the fact that the real sensor location was determined by simple geometric considerations for the network without integrating an analysis by numerical models and optimization methods.

CONCLUSIONS

The contamination of surface waters represents one of the most important aspects of urban management for environmental and sanitary implications and social-economic issues that may arise. The goal of this study is to develop a decision-support approach for identifying the location of the water quality sensors and illicit intrusions in sewers. The study focuses on soluble conservative pollutants, such as heavy metals.

The analysis was based on a Bayesian approach to introduce data assimilation and identification probability in the procedure. The system was simulated by SWMM for different random contamination scenarios. Progressive updates were performed considering contamination events and the need to evaluate the probability that the contamination source will be identified with the sensors located in a specific node. The proposed methodology showed a progressive increase in the identification probability obtained by the Bayesian update. Appropriate pre-screening and/or pre-conditioning approaches are relevant in terms of early identification of the most suitable sensor locations.

The following conclusions were obtained from the analyses:

- The value of the Bayesian probability of identifying the contamination source tends to increase as the number of Bayesian updates increase. However, the number of simulations in each update should be based on the complexity of the problem to be solved to ensure that more complex problems need a larger population for each update.
- As the number of updates increases, with the same number of sensors, the maximum value of the Bayesian

probability increases. However, the marginal increments are progressively lower.

- The use of a pre-screening procedure and/or the inclusion of knowledge about the system characteristics enables convergence towards an optimal solution in less time and with smaller computational resources. Even if this fact may be negligible for a small network, the impact on larger networks is relevant, which enables the same optimal configuration and half of the required number of simulations.
- The sensor locations obtained with the proposed methodology are able to identify a large number of contamination events: maximum of 84% events in Example 8 case with three sensors and maximum of 92% events in Massa Lubrense with 18 sensors.
- Even if a high number of sensors is required to obtain a high probability of contamination source identification, the BDN approach enables a progressive implementation of the sensor network depending on the water manager budget limitations. The comparison with the real configuration of the monitoring network in Massa Lubrense demonstrates that the methodology is also informative, which helps the water manager to understand the dynamics of the network and highlights the areas that require better monitoring.

The analysis shows the potential impact of the proposed methodology. However, further developments are needed to take into account non-conservative pollutants, such as biological contaminants that can have a large impact on the environment and public health. The methodology should be upgraded and tested to take into account the presence of multiple contamination sources and the possibility of deploying Lagrangian sensors carried by the flow. Additionally, the comparison with fault isolation approaches (Blanke *et al.* 2015) may be interesting from the perspective of transferring fault mode and effect analysis (FMEA)-type approaches to the water industry.

REFERENCES

Banik, B. K., Di Cristo, C. & Leopardi, A. 2015 A pre-screening procedure for pollution source identification in sewer systems. *Proocedia Engineering* **119** (1), 360–369.

Banik, B. K., Di Cristo, C., Leopardi, A. & de Marinis, G. 2017a Illicit intrusion characterization in sewer systems. *Urban Water Journal* **14** (4), 416–426.

Banik, B. K., Alfonso, L., Di Cristo, C., Leopardi, A. & Mynett, A. 2017b Evaluation of different formulations to optimally locate sensors in sewer systems. *Journal of Water Resources Planning and Management* **143**, 7.

Banik, B. K., Alfonso, L., Di Cristo, C. & Leopardi, A. 2017c Greedy algorithms for sensor location in sewer systems. *Water* **9**, 856. doi: 10.3390/w9110856.

Blanke, M., Kinnaert, M., Lunze, J. & Staroswiecki, M. 2015 *Diagnosis and Fault-Tolerant Control*, 3rd edn. Springer-Verlag, Berlin, Heidelberg. New York. https://doi.org/10.1007/978-3-662-47943-8

Boenne, W., Desmet, N., Van Looy, S. & Seuntjens, P. 2014 Use of online water quality monitoring for assessing the effects of WWTP overflows in rivers. *Environmental Science: Processes Impacts* **16**, 1510–1518.

Di Cristo, C. & Leopardi, A. 2008 Pollution source identification of accidental contamination in water distribution networks. *Journal of Water Resources Planning and Management* **134** (2), 197–202.

Even, S., Poulin, M., Mouchel, J.-M., Seidl, M. & Servais, P. 2004 Modelling oxygen deficits in the Seine River downstream of combined sewer overflows. *Ecological Modelling* **173** (2–3), 177–196.

Freni, G. & Sambito, M. 2017 Probabilistic approach to the positioning of water quality sensors in urban drainage networks. In: *International Conference of Urban Drainage*, Prague.

Gironás, J., Roesner, L. A., Davis, J., Rossman, L. A. & Supply, W. 2009 *Storm Water Management Model Applications Manual.* National Risk Management Research Laboratory, Office of Research and Development, US Environmental Protection Agency, Cincinnati, OH, USA.

Irvine, K., Rossi, M. C., Vermette, S., Bakert, J. & Kleinfelder, K. 2011 Illicit discharge detection and elimination: low cost options for source identification and trackdown in stormwater systems. *Urban Water Journal* **8** (6), 379–395.

Jiang, J., Shi, B., Huang, A., Wang, N. & Yuan, Y. 2018 Inverse uncertainty characteristics of pollution source identification for river chemical spill incidents by stochastic analysis. *Frontiers of Environmental Science and Engineering* **12** (5), 6.

Kabir, G., Tesfamariam, S., Francisque, A. & Sadiq, R. 2015 Evaluating risk of water mains failure using a Bayesian belief network model. *European Journal of Operational Research* **240**, 220–234.

Kim, M., Choi, C. Y. & Gerba, C. P. 2013 Development and evaluation of a decision supporting model for identifying the source location of microbial intrusions in real gravity sewer systems. *Water Research* **47**, 4630–4638.

Korb, K. B. & Nicholson, A. E. 2010 *Bayesian Artificial Intelligence*, 2nd edn. CRC Press, Boca Raton, FL, USA.

Lee, C., Paik, K., Yoo, D. G. & Kim, J. H. 2014 Efficient method for optimal placing of water quality monitoring stations for an

ungauged basin. *Journal of Environmental Management* **132**, 24–31.
Lifshitz, R. & Ostfeld, A. 2019 Clustering for real time response to water distribution system contamination event intrusion. *Journal of Water Resource Planning and Management* **145**, 2.
Montserrat, A., Bosch, L., Kiser, M. A., Poch, M. & Corominas, L. 2015 Using data from monitoring combined sewer overflows to assess, improve, and maintain combined sewer systems. *Science of the Total Environment* **505**, 1053–1061.
Passerat, J., Ouattara, N. K., Mouchel, J.-M. & Rocher, V. 2011 Impact of an intense combined sewer overflow event on the microbiological water quality of the Seine River. *Water Research* **45**, 893–903.
Phan, T. D., Smart, J. C. R., Capon, S. J., Hadwen, W. L. & Sahin, O. 2016 Applications of Bayesian belief networks in water resource management: a systematic review. *Environmental Modelling & Software* **85**, 98–111.
Preis, A. & Ostfeld, A. 2008 Multiobjective contaminant sensor network design for water distribution systems. *Journal of Water Resources Planning and Management* **134**, 366–377.
Qin, X., Gao, F. & Chen, G. 2012 Wastewater quality monitoring system using sensor fusion and machine learning techniques. *Water Research* **46** (4), 1133–1144.
Rathi, S. & Gupta, R. 2016 A simple sensor placement approach for regular monitoring and contamination detection in water distribution networks. *KSCE Journal of Civil Engineering* **20** (2), 597–608.
Sambito, M., Di Cristo, C., Freni, G., Leopardi, A. & Quintiliani, C. 2018 Pre-conditioning approach to Bayesian Decision Networks for water quality sensors positioning in urban drainage systems. In *13th Hydroinformatics International Conference*, 1–6 July 2018, Palermo, Italy.
Srinivas, R., Singh, A. P., Gupta, A. A. & Kumar, P. 2018 Holistic approach for quantification and identification of pollutant sources of a river basin by analyzing the open drains using an advanced multivariate clustering. *Environmental Monitoring and Assessment* **190**, 720.
Tinelli, S., Creaco, E. & Ciaponi, C. 2017 Sampling significant contamination events for optimal sensor placement in water distribution systems. *Journal of Water Resource Planning and Management* **143** (9), 1–10.
Troutman, S. C., Schambach, N., Love, N. G. & Kerkez, B. 2017 An automated toolchain for the data-driven and dynamical modeling of combined sewer systems. *Water Research* **126**, 88–100.
Vonach, T., Tscheikner-Gratl, F., Rauch, W. & Kleidorfer, M. 2018 A heuristic method for measurement site selection in sewer systems. *Water* **10** (2), 1–16.
Weickgenammt, M., Kapelan, Z., Blokker, M. & Savic, D. A. 2010 Risk-based sensor placement for contaminant detection in water distribution systems. *Journal of Water Resources Planning Management* **136** (6), 629–636.
Yazdi, J. 2018 Water quality monitoring network design for urban drainage systems, an entropy method. *Urban Water Journal* **15** (3), 227–233.

First received 9 February 2019; accepted in revised form 14 July 2019. Available online 13 September 2019

Assessing causes and associated water levels for an urban flood using hydroinformatic tools

J. L. S. Pinho IWA, L. Vieira, J. M. P. Vieira IWA, S. Venâncio, N. E. Simões, J. A. Sá Marques and F. S. Santos

ABSTRACT

Flood events are dependent on meteorological conditions but also depend on several other factors that are case specific, with relevance for reservoir operation. Hydrological and hydrodynamic models are valuable tools for understanding complex river hydrodynamics during flood events. These tools have been applied to improve understanding of the causes for an urban flood event that occurred between 9 and 11 January 2016 in the Mondego river basin, at Coimbra city (Portugal). Seven different factors that can, independently, influence the river flow at the study site were identified: three of them can be associated with the operational discharge schemes of the three upstream dams; two factors with the runoff flows from uncontrolled contributing sub-basins; another one related to discharge measurement uncertainty at a downstream dam; and finally, the seventh studied factor was sedimentation occurring in the main channel of the flooded river stretch. Hydroinformatic tools were applied in different scenarios allowing the characterization and identification of each one of the identified key factors responsible for the flood event. A proposal for a flood early warning system is presented based on the knowledge resulting from the studied flood event.

Key words | floods modelling, hydroinformatic tools, reservoirs operation

J. L. S. Pinho IWA (corresponding author)
L. Vieira
J. M. P. Vieira IWA
Centre for Territory, Environment and Construction, Department of Civil Engineering,
University of Minho,
4710-057 Braga,
Portugal
E-mail: *jpinho@civil.uminho.pt*

S. Venâncio
Department of Civil Engineering,
Federal University of Triângulo Mineiro,
Uberaba,
Brazil

N. E. Simões
J. A. Sá Marques
INESCC, Institute for Systems Engineering and Computers at Coimbra, Department of Civil Engineering,
University of Coimbra,
Coimbra 3030-788,
Portugal

F. S. Santos
Department of Civil Engineering,
University of Coimbra,
Coimbra 3030-788,
Portugal

INTRODUCTION

Regulations have been established in order to control or mitigate flood consequences since they usually are associated with severe social, economic, and environmental impacts. An EU Flood Directive (EU 2007) requires the development of flood hazard maps, which may include information on hydrological and hydrodynamic characteristics of vulnerable regions, i.e., inundated areas, river discharges and water levels. However, the occurrence of floods depends on complex hydrological processes and anthropic actions whose individual contribution to the water level rise at particular locations is hard to distinguish. High rainfall events are not the unique factor that explains the occurrence of floods. Lack or inadequate urban planning, resulting in buildings and other infrastructures being built on floodplains, is responsible for major losses (Adeloye & Rustum 2011).

Deficient design or operation of the urban drainage systems can also be a cause for urban floods. Several additional threats, like increasing of impervious areas and climate change can aggravate this problem (Hammond *et al.* 2015; Palla *et al.* 2018). Drainage systems frequently discharge into rivers that are embedded in the urban environment as a consequence of cities growing (Phong 2015). Floods are one of the main instigators of the interaction between rivers and cities, leading to artificialization of river banks, installation of defence structures and, in many other cases,

doi: 10.2166/hydro.2019.019

to the construction of dams to control flood flows during extreme precipitation events (FitzHugh & Vogel 2011). Proper operational dam rules during flood events should be guaranteed to defend urban areas located near by river banks.

Methodological approaches to deal with flooding problems recommend the incorporation of decision support tools in the process to better investigate which planning options and remedial actions need to be taken at river basin scale and which operational management actions are more effectively protecting urban populations (Werner *et al.* 2004; Demir & Krajewski 2013; Henonin *et al.* 2013; Filatova 2014; Sharma *et al.* 2014; Goharian & Burian 2018). These decision support systems should be based on adequate hydroinformatic tools capable of processing, analysing and simulating the relevant hydrological data required for potential flood assessment in at-risk locations.

Uncertainty related to hydrodynamic models is usually associated with model inputs, model structure and calibration (Deletic *et al.* 2012; Efstratiadis *et al.* 2014; Dimitriadis *et al.* 2016; Papaioannou *et al.* 2016; Bellos *et al.* 2017; Teng *et al.* 2017). Nevertheless, software packages for two-dimensional simulation of water levels and flow characteristics during flood events have been found to be efficient (e.g., Stelling & Verwey 2005; Liang 2010; Tsakiris & Bellos 2014). Néelz & Pender (2010) presented a sound performance analysis consisting of the comparison of software packages in the simulation of different inundation tests. SOBEK (Deltares 2011a) was one of the evaluated software packages. This software showed the capabilities to capture the detail of the flow field for all the tests involving complex geometries of the floodplain. When applied in operational management, the computational time costs, depending on the spatial resolution, could be a restriction for some simulations. Nevertheless, capabilities to simulate operations adopted at hydraulic structures are fundamental in order to properly consider discharges at dams during flood events. The real-time control module is a quite distinctive feature of this software solution that allows simulation of complex river flow scenarios. Moreover, it includes a hydrological module, that together with the hydrodynamics (1D and 2DH) modules, constitutes an adequate solution for river flood modelling.

Pinho *et al.* (2018) have identified the operational discharge scheme at two of the upstream dams as the main cause for a flood event that occurred near the city of Coimbra (Portugal), in January 2016. Seven different factors that can, independently, influence the river flow at the study site were identified: three of them can be associated with the discharges at the three upstream dams; two factors with the runoff flows from uncontrolled contributing sub-basins; another one related to the operation of a downstream dam; and finally, sedimentation occurring in the main channel that could cause water levels to rise during flood events.

This relatively complex case study was properly monitored, thus constitutes a rare opportunity to assess the performance of a set of actual software packages for river flood and inundation modelling (Teng *et al.* 2017). Modelling of monitored historical intense inundation events at river basin scale is quite rare since floods are controlled at monitored basins and normally for inundations at uncontrolled basins the hydrometric data are not sufficient.

Application of modelling tools to anticipate inundations associated with floods or assess flood mitigation measures (e.g., Aparicio *et al.* 2009; Popescu *et al.* 2010; Schwanenberg *et al.* 2013; Gibertoni *et al.* 2014) is quite common. Moreover, development of hydroinformatic platforms for flood or inundation warnings must be based on robust models, including realistic capacity to simulate the hydrology and the river's hydrodynamics within the basin, and particularly for the flood-prone areas. These platforms are responsible for the synchronization and management of all the required data in the case of flood forecasting and automation of model running (Werner *et al.* 2012).

This paper presents a detailed description of a set of hydrological, hydrodynamic and morphodynamics models used to study that flood event. The set of available data, not only from the river system (topography, bathymetry, dam operation, etc.), but also from the flood event, such as flow, water depth and rainfall, provide a unique opportunity to validate and improve the knowledge of river floods in rivers controlled by dams. Seven different factors that influence river water levels at Coimbra city are identified and a flood forecast early warning system for the Mondego river basin is proposed, based on the knowledge acquired from the flood event study. Dam operations' auditing and improvement, and responsibility attribution for security purposes are among the potential applications of this suite of tools to flood events.

METHODS

Study area

Mondego river basin is a Portuguese basin with an approximate area of 6,660 km^2 characterized by a Mediterranean climate with a strong seasonal variation of flow associated with flood and dry events, with an average annual influx of about 1,213 hm^3 and 4,032 hm^3 in a dry and in a wet year, respectively. The river extends from its spring in the Serra da Estrela mountains with an altitude of 1,500 m, to the river mouth at Figueira da Foz, in the Atlantic Ocean, for 258 km. Much of its course develops in a very embedded valley, entering the area of Coimbra in an alluvium basin. The main tributaries are the rivers Alva, Dão and Ceira. In this basin, there are several dams whose operation influences the river flow characteristics near the city of Coimbra: Aguieira (the greatest storage capacity reservoir), Raiva, Fronhas and Coimbra. The water stored at Fronhas dam can be transferred to the Aguieira reservoir. Flows discharged or turbined to Raiva dam may be pumped to the Agueira reservoir (Figure 1).

Coimbra dam has a small reservoir volume (1.6 hm^3 for water level at 18.0 m above sea level (asl)) but is responsible for controlling upstream levels for river discharges up to 930 m^3/s. For higher discharges, according to this infrastructure's operational rules, its floodgates are completely opened. In this scenario, river water levels are only conditioned by the cross-section contractions caused by the bridge and mobile gate support pillars and the downstream river section.

The study focuses on the intermediate stretch of the Mondego river located between the main upstream dams and Coimbra city. The upstream dams (Aguieira, Raiva and Fronhas) are also included in the analysis developed in this work.

The river reach comprises three distinct segments. An initial segment, located downstream of Raiva dam, formed by relatively narrow river cross sections with about 80 m of surface width and a wider floodplain used for agricultural activities, both located in an embedded valley. In this segment, the longitudinal profile presents a slope of the order of 0.08%, and the river includes several small weirs along its course. The following segment, that presents a cross

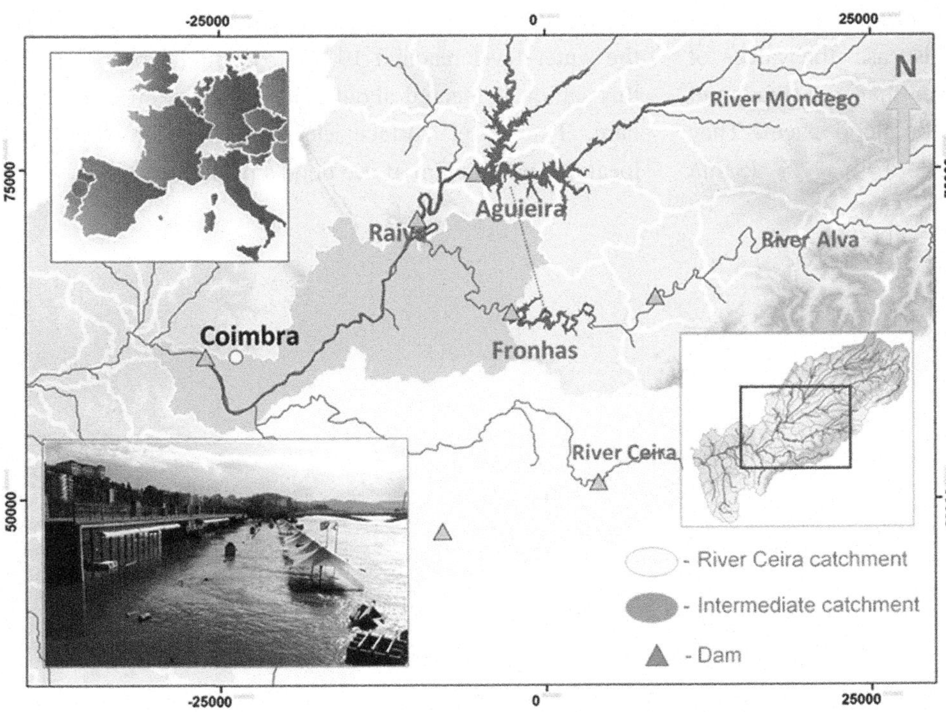

Figure 1 | Study site location, Mondego river basin with the location of the dams and photo of Coimbra inundation that occurred in January 2016 (Pinho et al. 2018).

section with a surface width of about 200 m under frequent river discharges, makes the transition between the initial one and the third where the river banks are artificially fixed with river walls. The top of these walls can be inundated at certain locations during flood events.

January 2016 flood event

On 11 January 2016, a severe inundation took place at Coimbra city affecting several local infrastructures. Rainfall was recorded at seven meteorological stations in the Mondego river basin during the flood event (for the period between 1 and 12 January). Hourly records were obtained from the National Water Resources Information System at the following meteorological stations: Alagoa, Maçainhas, Mangualde, Pombal, Santa Comba Dão, Sátão and Soure. Recorded daily maximum values for the flood event were compared with the historical daily maximums at each station. The recorded values on 11 January 2016 vary between 29% (Maçainhas station) and 77% (Santa Comba Dão station) of the historical maximums. A comparison between accumulated rainfall between 1 and 12 January 2016 (Figure 2) and the maximum value of the accumulated values in consecutive periods of 11 days of the historical records was also carried out. In this case, the values of accumulated precipitation were lower than the maximum historical ones for the period of the flood event. They varied from 30% (Maçainhas station) to 77% (Santa Comba Dão station) of the historical accumulated maximum values (411 mm and 424 mm, respectively).

Hourly hydrometric data were made available by the Portuguese Environment Agency (Pinho *et al.* 2018); water levels immediately upstream and downstream of Coimbra dam (Figure 1), at the upstream reservoirs (Aguieira, Raiva, Fronhas) and at Santa Clara Bridge hydrometric station located in the vicinity of the inundated area. Inflows and outflows at all the dams were also made available. Moreover, discharges near the river Ceira mouth and at Coimbra dam were estimated based on rating curves available for these locations.

The water level at Aguieira reservoir started rising on 1 January, 2016, from 117.89 m elevation until reaching a value of 120.12 m on 10 January, at midnight, around the beginning of the occurrence of the flood event. During this event, a maximum value of 124.66 m was reached (Figure 3). At Fronhas reservoir there was an increase of water level from 118.59 m (1 January 2016, 00:00) and a maximum water level of 131.94 m (12 January 2016, 06:00). This variation resulted from the storage of almost all of the reservoir inflows during the initial phase of the period under analysis. During the flood event, part of the volume was transferred to the Aguieira reservoir. At Santa Clara Bridge hydrometric station (located immediately downstream of the bridge), the water level reached 19.56 m. It should be noted that this station is located about 1,300 m upstream of Coimbra dam. The water surface elevation between these two locations was 1.55 m at the flood's peak time.

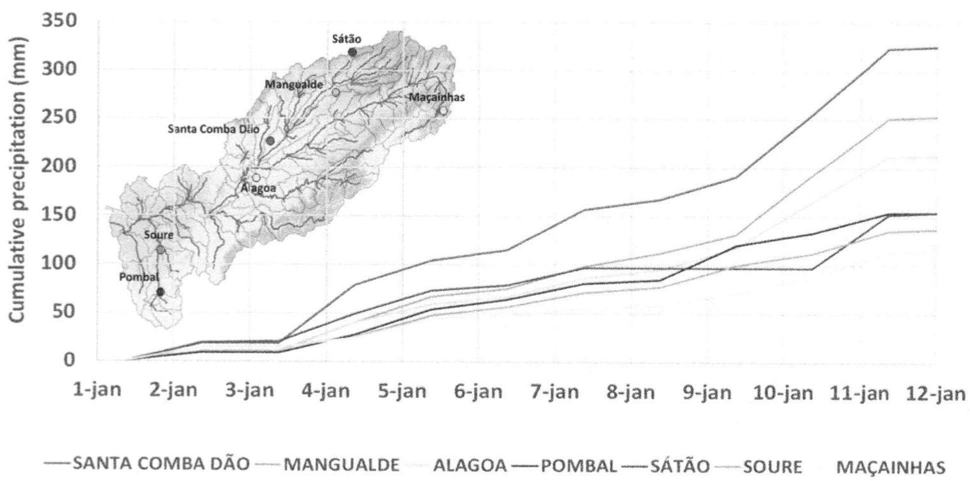

Figure 2 | Cumulative precipitation registered between 1 and 12 January 2016.

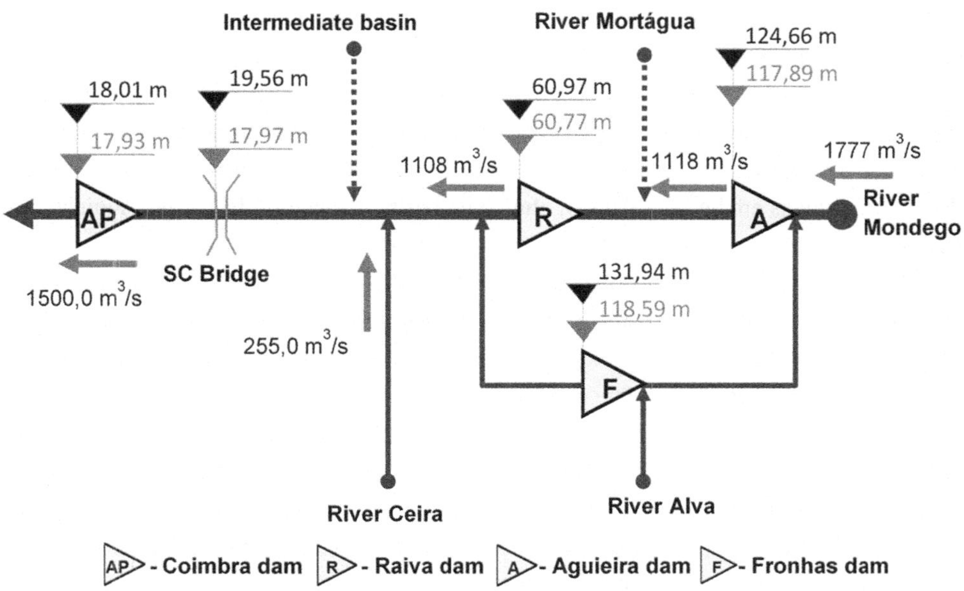

Figure 3 | Hydrometric measurements during the flood event. Water levels on 1 January (grey numbers) and maximum water levels and discharges (black numbers).

The inflow hydrograph at Aguieira reservoir presented a peak of 1,777 m^3/s (11 January 2016, 02:00). Other potential flows to this reservoir (transferred from Fronhas reservoir and/or pumped from the Raiva reservoir) are not considered in this inflow maximum value. The maximum discharge at Aguieira dam reached 1,118 m^3/s (11 January 2016, 12:00). During the initial phase of the period under analysis, the operation of the dam includes turbine and pump periods (records not included in this paper). The maximum discharge at Raiva dam reached a value of 1,108 m^3/s on 11 January 2016 at 12:00. River Ceira flows (at Ponte Conraria hydrometric station) were estimated based on the rating curve, presenting a maximum value of 255 m^3/s on 11 January 2016 at 14:00. At Coimbra dam the estimated peak flow was 1,500 m^3/s. This flow, according to information provided by the Portuguese Environment Agency, was estimated based on the original rating curves that were defined before the construction of Coimbra dam.

Hydrological, hydrodynamic and two-dimensional models

The flood event was analysed using measured hydrological data and numerical models specifically implemented for this work. The river basin hydrology was simulated by a model based on the Sacramento concept implemented with GIS support. The river and reservoirs hydrodynamics were simulated using a model implemented with SOBEK software. Two complementary local models were implemented, with the aid of RMA2 (ERDC 2013) and Delft3D (Deltares 2011b) software packages. These were used in the analysis and verification of the downstream river flow characteristics influenced by the dam pillars' constriction and in the assessment of sedimentation conditions in the inundated fluvial stretch.

Sobek river model

The Sacramento hydrological model, implemented in SOBEK software (Deltares 2011a), allows the quantification of river discharges using precipitation and evapotranspiration data as input (Singh & Woolhiser 2002). The Sacramento model conceptualizes the watershed as a soil column divided vertically into two storage zones which are filled and emptied to simulate infiltration, percolation, baseflow and interflow through the watershed. The upper and lower zones represent the infiltration capacity of shallow soils and the underlying aquifer, respectively. Runoff is computed as the net excess volume remaining from precipitation after interception and infiltration have been satisfied. Rates of infiltration and soil water volume capacities are represented with conceptual parameters which correspond closely to physical values associated with soil properties such as void space ratio and saturated hydraulic conductivities (Burnash & Ferral 2002). This model was implemented for two sub-basins (Figure 1) based on the available hydrometric and meteorological stations' data. An optimization tool (CRC 2004) was applied for the estimation of the 17 required parameter values for each sub-basin.

River and dam hydraulic structures were modelled with SOBEK software package (Deltares 2011a). It is based on a robust numerical method that allows solutions to be obtained for complex model set-ups. The governing equations are based on one- and two-dimensional version formulations of the Saint Venant equations and solved numerically by a finite difference method known as the Delft scheme (Verwey 2001). Main inputs to the model are the river stretches and associated cross sections and roughness coefficients. Different combinations of boundary conditions are allowed and a complete set of hydraulic structures can be inserted in the river schematization. The spatial discretization of the model was performed using GIS map data of the river basin. It extends from the Aguieira and Fronhas dams (upstream boundaries) to a section located downstream of Coimbra dam (downstream boundary). All the tributaries present in this intermediate reach of the Mondego river were considered in this model (Figure 4). The model was discretized by 479 calculation points, 12 open boundaries and 27 controlled discharges. Simulation of hydraulic structures was implemented applying the real-time control module. The river channel geometry was defined considering 147 cross sections, based on bathymetric surveys and Lidar data.

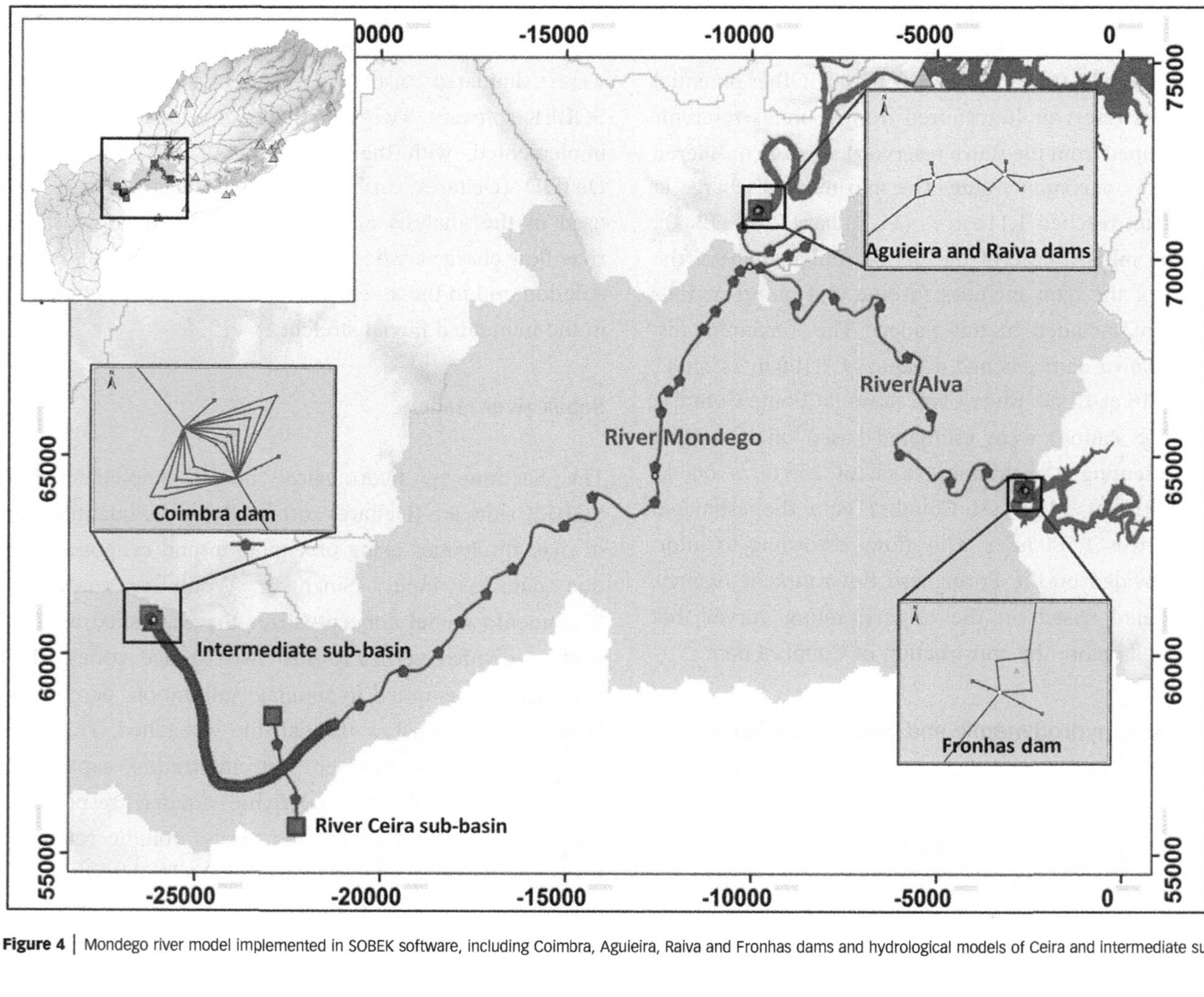

Figure 4 | Mondego river model implemented in SOBEK software, including Coimbra, Aguieira, Raiva and Fronhas dams and hydrological models of Ceira and intermediate sub-basins.

Hydraulic structures at dams were modelled in order to consider all discharges that occurred during the flood event at the upstream reservoirs. They include flood gates, bottom outlets and turbines that were discretized through orifice or pump nodes. Water levels within the river stretch under analysis depend on the operation of the upstream dams, hydraulic structures and on the operation of the Coimbra dam floodgates. Thus, the real-time control module available in SOBEK software, was crucial to define reliable hydrodynamic scenarios to analyse the observed inundation conditions.

RMA2 local model

Water discharges at Coimbra dam are estimated based on rating curve and measured water levels. This rating curve (Figure 5) was defined before the construction of the dam so that, in the case of extreme events, all of its gates should open. However, the river cross-section contraction resulting from the dam pillars could modify the flow characteristics at this stretch. A local hydrodynamic model for the verification of the flow characteristics during flood events at Coimbra dam was implemented using RMA2 software. This software, based on the finite element numerical method, solves the shallow water equations in unstructured meshes, that are suitable for discretization of the river geometry at the dam cross section. The domain was discretized with an unstructured mesh with 6,839 quadratic triangular elements and 14,076 nodes. Bathymetry was defined based on the geometric characteristics of the dam and the river bathymetry downstream (Figure 5).

Delft3D local morphodynamic model

Considering the uncertainty of flood water levels arising from the lack of an updated bathymetry of the river stretch near the city of Coimbra (the most recent bathymetric survey was completed in 2008), a two-dimensional morphodynamic model was implemented with Delft3D software (Deltares 2011b). For this work, the 2DH version of the

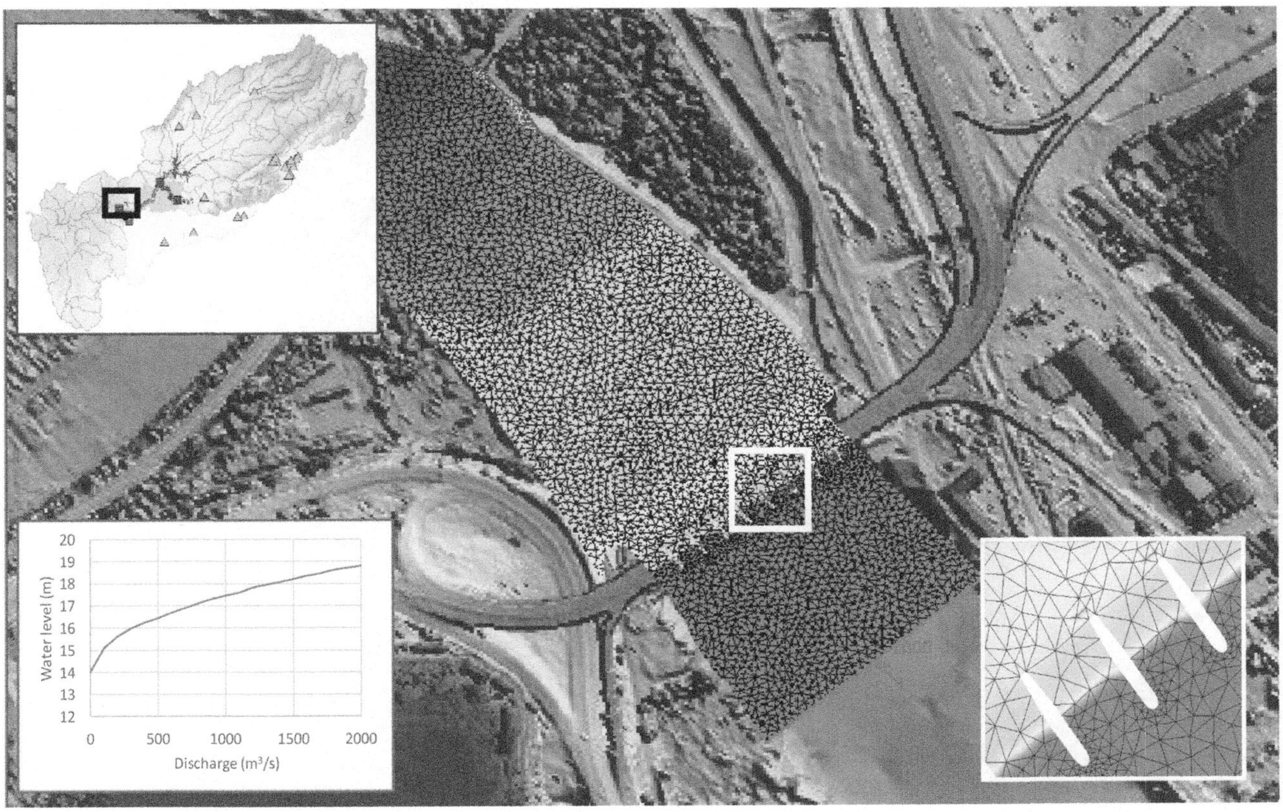

Figure 5 | Local RMA2 two-dimensional hydrodynamic model.

Delft3D modelling suite was selected, which is able to accurately simulate the hydrodynamics and morphodynamics during flooding events. This numerical model solves depth-averaged free surface flow equations, taking into consideration several processes such as bed friction, turbulence, supercritical and subcritical flows, density effects related to horizontal temperature and salinity gradients, and drag forces created by vertical structures.

This model was discretized with a high spatial resolution grid (~5 m × 5 m). The model was set up to analyse the river bed morphology between 2001 and 2008 (Figure 6). During this period the river showed a sedimentation tendency, registering the formation of sediment bars that reached 1 m above the bed level of 2011 at several places.

The simulation period was artificially defined based on extreme events that took place in the period of analysis. From March 2001 to January 2008, the following occurrences were registered according to the average daily flow data: flow of 500.0 m^3/s in 4 days, flow rate of 700.0 m^3/s in 2 days and river flow of 900.0 m^3/s in 2 days. The sediments of the bed are composed of coarse sand. In the adopted simulation a characteristic dimension of 2 mm was considered.

Model calibration

Hydrological model

The hydrological model was calibrated for the river Ceira sub-basin based on hydrometric station data available at the downstream river reach before the river mouth and in the recorded rainfall data registered during the flood event. The Sacramento model parameter values were obtained using an optimization tool. A Nash–Sutcliffe efficiency coefficient value of 0.71 was obtained for this event (Figure 7(a)). The same Sacramento parameter values were adopted for the intermediate sub-basin.

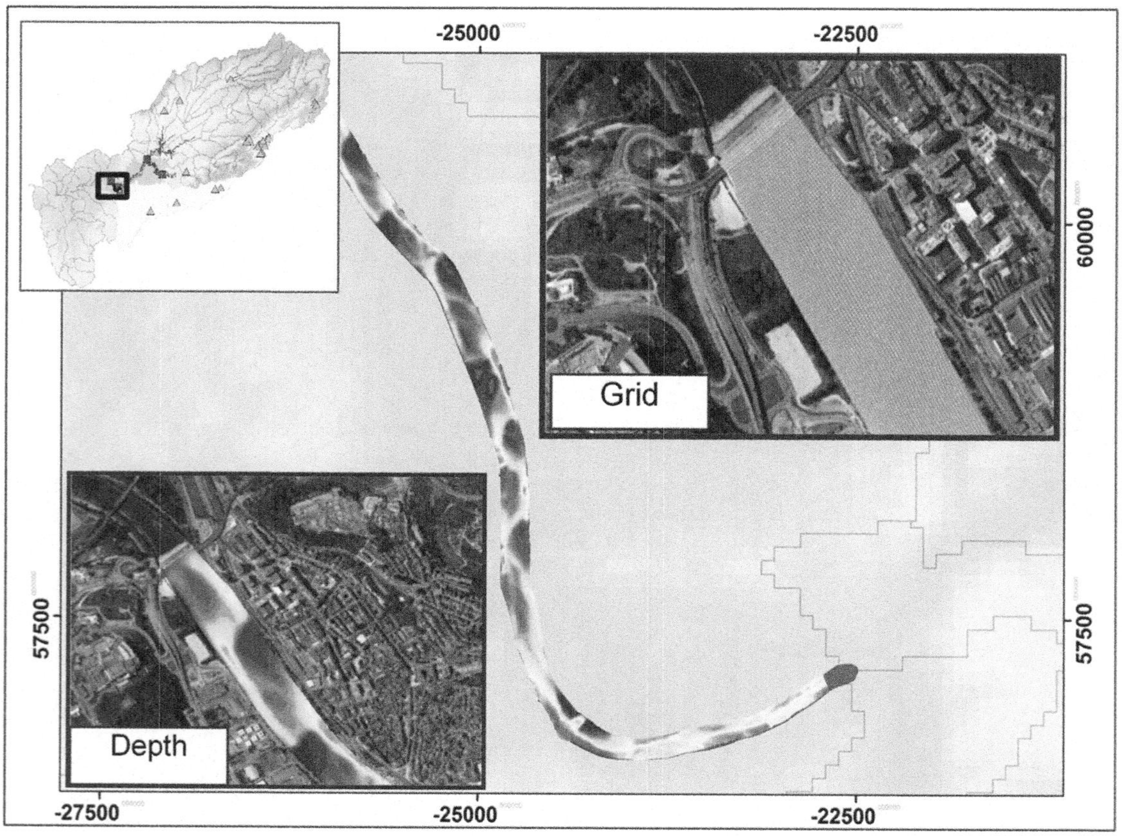

Figure 6 | Local Delft3D two-dimensional morphodynamic model.

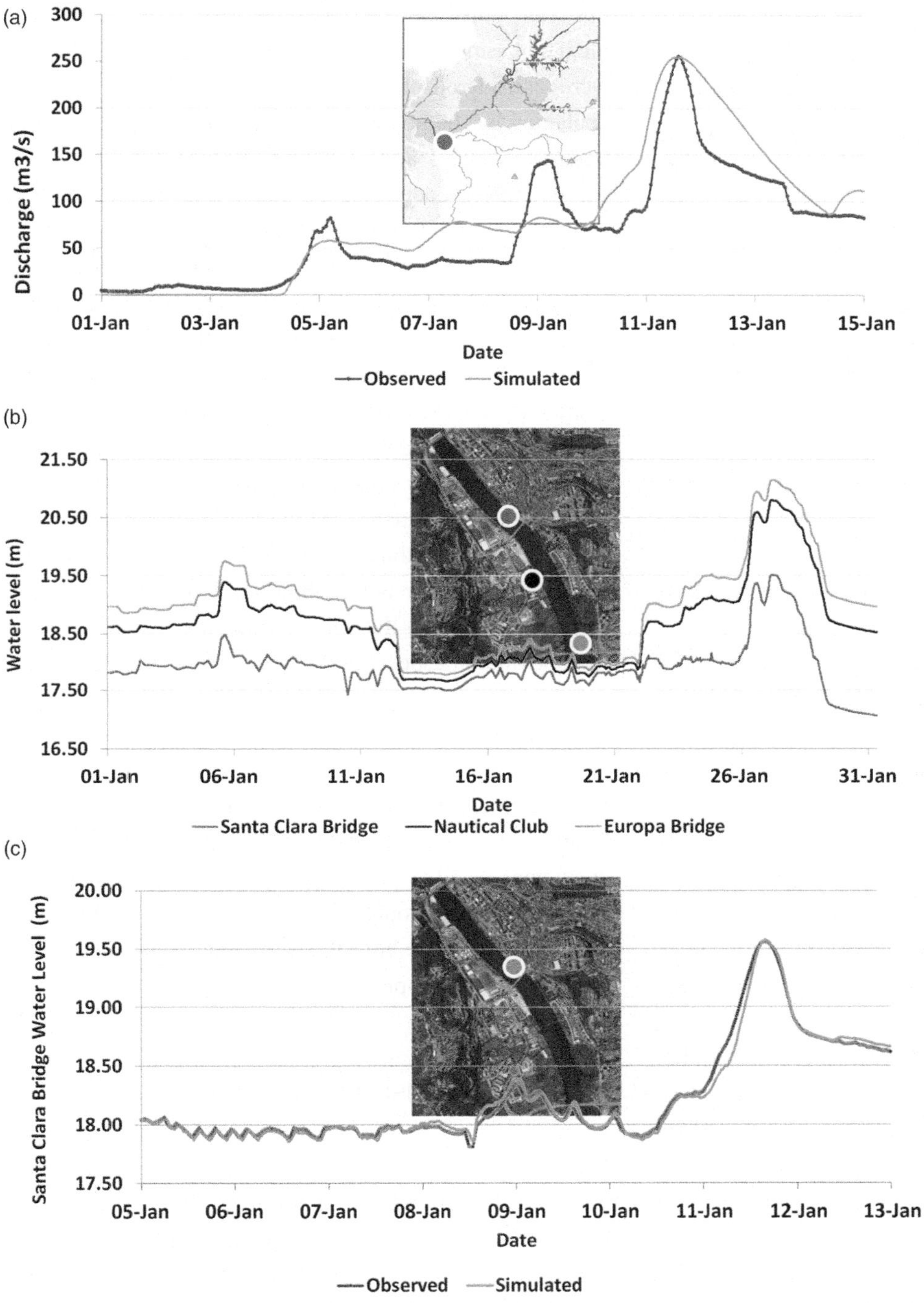

Figure 7 | Model calibration results: (a) comparison between the observed flow rates registered at Ponte de Conraria hydrometric station and flow rates simulated by the hydrological model; (b) 2001 flood water levels results at three different locations; (c) measured and simulated water levels at Coimbra dam for the January 2016 flood event.

River model

The one-dimensional hydrodynamic model was calibrated and validated considering the observed water level information registered at Santa Clara Bridge hydrometric station. This station has a location within the domain under study. Two different flood scenarios were defined in the calibration procedure: (i) in the first one the Manning roughness values were defined in order to approximate the simulated values to the ones registered at three locations

Table 1 | Comparison between water levels obtained by Sobek model simulation of 2001 flood event and water levels obtained by Santos *et al.* (2002)

Water level, 1DH model (m)			Water level, Santos *et al.* (2002)		
Santa Clara Bridge	**Nautic Club**	**Europa Bridge**	**Santa Clara Bridge**	**Nautic Club**	**Europa Bridge**
19,51	20,80	21,10	19,50	20,86	21,05

for the 2001 flood event and using available bathymetric data for 2001; (ii) in the second scenario the hydrological conditions of the January 2016 flood were adopted together with the bathymetry of 2008.

In Figure 7(b), results regarding water levels are presented. They were obtained for three different study locations: Santa Clara Bridge, Nautic Club and Europa Bridge. Maximum water levels are depicted in Table 1 and compared with the ones obtained in a study of the 2001 flood at the Mondego river (Santos *et al.* 2002).

The differences found result from the uncertainties (Dimitriadis *et al.* 2016) associated with different geometric discretization, techniques and modelling parameters and a possible approximation on the exact site location used for the comparison. The order of magnitude of the differences found is compatible with the accuracy that is intended for the present work. It should also be noticed that the morphological variations that occur associated with the flood events with the magnitude of the simulated flood event (flood peak river flow around 2,000 m^3/s) may present a higher order of magnitude than the differences found, which can be reflected in water levels. Only hydrodynamic and morphodynamic coupling modelling may allow the evolution of the flood levels to be simulated more rigorously. Thus, only a bathymetric survey prior to the floods of 2001 could allow an integrated analysis of the evolution of the morphology and water levels during this event.

Results of water levels at Santa Clara bridge for the 2016 flood event simulation, considering all the relevant variables of the fluvial system, are presented in Figure 7(c). For this event the following metrics (Willmott 1982; Krause *et al.* 2005) were obtained: (i) Nash–Sutcliffe efficiency coefficient value of 0.98; (ii) statistical bias of 0.026 m; (iii) root mean square error of 0.057 m; and (iv) scatter index of 0.003. In this simulation, the opening sequence of the Coimbra dam floodgates, the discharged and turbinated flows in the upstream dams, the contribution of Ceira river and the intermediate basin flows estimated with the hydrological model, were considered. The quality of the obtained results is explained by the application of the appropriate calibration parameter values and by considering, in an explicit way, the opening sequence of the floodgates whose operation was defined with time series controllers.

A comparison between observed and simulated water levels registered at the upstream reservoirs included in the 1DH model for the January 2016 flood event is shown in Figure 8. The reservoirs were modelled considering the respective characteristic curves (volume – surface area) defined, considering the historical hydrometric information registered in their respective hydrometric stations. In Figure 8, also considered are the flow rates associated with the controllers of the simulated hydraulic structures. The results obtained for the calibration of the hydraulic structures indicate that the model simulates almost perfectly the operational aspects of the upstream dams.

Local models

For the local two-dimensional hydrodynamic model, a Manning coefficient of 0.01 $m^{-1/3}$ s and an eddy diffusion coefficient of 20 m^2/s were adopted. For the downstream morphodynamic model a value of 0.05 $m^{-1/3}$ s for roughness and 1 m^2/s for both the horizontal eddy viscosity and diffusivity were adopted.

RESULTS AND DISCUSSION

Results in terms of water level change at Coimbra city are presented in Table 2 for each one of the seven identified factors: (i) discharges of upstream Aguieira and Raiva dams; (ii) discharges at upstream Fronhas dam; (iii) river Ceira sub-basin flows; (iv) intermediate sub-basin flows; (v) operation of floodgates at Coimbra dam; (vi) uncertainty associated with river discharge measurements at Coimbra; and (vii) sedimentation at the downstream stretch.

The influence of each one of these factors was analysed based on different simulation scenarios applying the SOBEK model. The Mondego river flow discharge at Coimbra city is the result of the contribution of the river

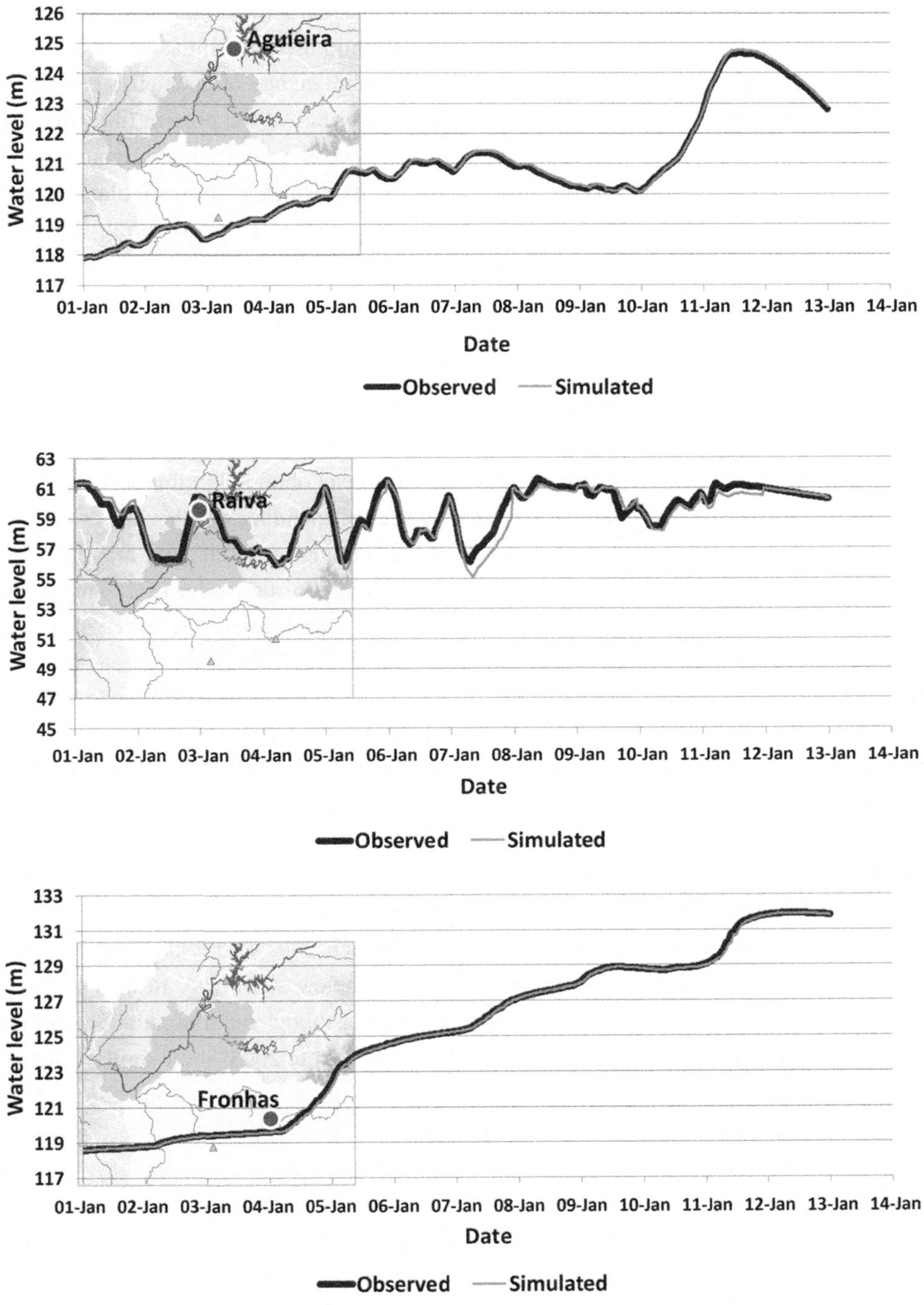

Figure 8 | Calibrated water level model results at upstream reservoirs.

Ceira, the intermediate sub-basin and from the upstream dam's discharges. The water level variation at Coimbra city was quantified to analyse the effect of factors (i) to (iv). In each one of these scenarios all the conditions that took place during the flood event were considered with the exception of the one that is being analysed in the corresponding scenario (considered with a null contribution). The variation in terms of water level at Coimbra city was registered and is presented in the first four rows of Table 2. It is clearly demonstrated that the discharges that

Table 2 | Simulated river water level variations according to different factors affecting inundations in Coimbra city

Inundation factors	Water level variations at Santa Clara bridge (m)
(i) Absence of upper Aguieira and Raiva dam discharges	−1.36
(ii) Absence of upper Fronhas dam discharges	−0.10
(iii) Absence of River Ceira sub-basin flow discharge	−0.36
(iv) Absence of intermediate sub-basin flow discharge	−0.12
(v) Operation of gates at Coimbra dam: one gate closed	+0.20
(vi) Uncertainty associated with river discharge measurements at Coimbra dam	±0.10 to 0.20
(vii) Estimated sedimentation occurred in a period of 8 years	+0.50

result from the operation of the Aguieira and Raiva dams were the main cause for the water level rise and associated flood at Coimbra. This way, if a different discharge schedule was adopted during the flood event the inundation would be avoided or its impact mitigated.

During the flood event, all the Coimbra dam floodgates were opened as the river flow discharge increased. This operation is automatically implemented with a PID controller at the dam. If any of these gates were kept closed the effect of this operation (factor (v)) would imply an increase of +0.20 m in the water level at Coimbra city according to simulated results.

Regarding factor (vi), uncertainty associated with river discharge measurements at Coimbra dam, free-flow (flow with all gates opened) was simulated using the two-dimensional local model of the dam river stretch (Figure 5). It was verified that the measured upstream and downstream water levels during the flood event are compatible with the values obtained by the model when considering river discharges identical to those reached during the event, and it was possible to verify that the pillars' constriction is not responsible for a significant increase in the water level at this location. Indeed, for high river discharges (above 900 m^3/s) water levels at the dam seems to be conditioned by a downstream constricted river section. This way, variations in the relationship between levels and flows are expected due to the method used to estimate the discharges. This depends on the maintenance of geometric conditions and roughness of the river bed at the present time identical to those for which the water velocity measurements, that served as the basis for its definition. Two simulations with the SOBEK model were performed, considering in one of them all measured flows discharged during the flood event (dams) and estimated with the hydrological model (sub-basins), and in the other, omitting the contribution of the intermediate sub-basin that is the sole sub-basin which is dependent on the estimated discharges based on the rating curve. Obtained differences are in the order of 100 m^3/s. It turns out that, for each level obtained in this section, different flow rates may occur depending whether the estimation is made during the ascending or descending branch of the flood event. Flow variations of the order of 100 m^3/s estimated in the dam imply variations of the order of 0.10 m to 0.20 m at Santa Clara Bridge. Thus, we can say that this is the order of magnitude in the water level measurements at Santa Clara Bridge resulting from uncertainty associated with discharge estimated at the downstream Coimbra dam.

For the last analysed factor, (vii), the available data from two bathymetric surveys in a time interval of seven years (2001 and 2008) were used. A simulation with the local morphodynamic model was implemented. It should be noted that in the period between 2001 and 2008 there were no floods (river discharges below 900 m^3/s). Results obtained with the morphodynamic model for a simulation defined for the analyzed period (2001 to 2008) show similar patterns of bed morphology to those derived from the surveys analysed (Figure 9).

It should be noted, however, that similar values of sedimentation may occur during intense flood events. In a simulation involving a river discharge of 1,500 m^3/s and the same sediments, deposition values of the order of 0.70 m are reached in the locations identified in Figure 9, in a short time interval (few hours). Once the order of magnitude of the morphological variations occurring in the period between 2001 and 2008 was confirmed, and in order to estimate the influence of eventual bed level variations on water levels measured at Santa Clara Bridge, a simulation with the SOBEK model was defined, maintaining the river bed roughness parameters and inducing a

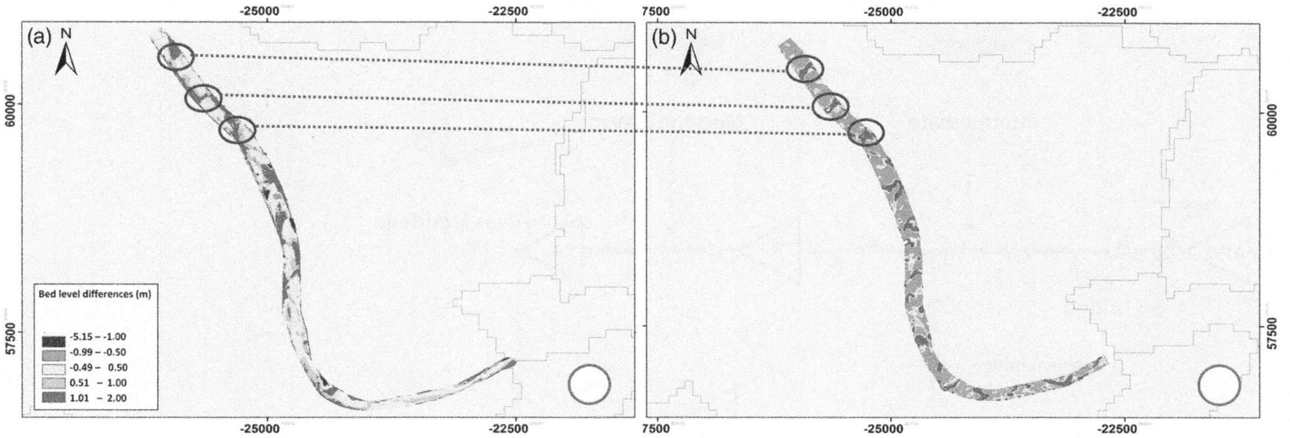

Figure 9 | Morphodynamic model results: (a) bed level differences between 2008 and 2001; (b) Delft3D model results for cumulative bed level change.

bathymetric variation similar to the one that occurred between 2001 and 2008. From the simulated results an increase of water level at Santa Clara bridge of 0.50 m was obtained, for this extreme and unlikely scenario since it is expected that the river bathymetry tends to an equilibrium profile in the absence of intense floods and under an almost constant fluvial discharges regime.

Obtained results for the factors influencing water levels at Coimbra city show that for the January 2016 flood event, the factor that had the greatest impact on the Mondego river water level at Coimbra (Santa Clara bridge) was the discharges at Raiva and Aguieira dams.

Several operational scenarios of the upstream dams were simulated with the calibrated SOBEK model. In one of those scenarios it was assumed that the discharges at Aguiera dam were limited to 670 m^3/s when the reservoir water level reached 120.12 m. In this case, the maximum water level will be 125.25 m (instead of 124.66) and instead of a water level of 19.56 m at Santa Clara bridge, it would have been 18.94 m. In this scenario the river discharges at Coimbra dam would be 1,090 m^3/s.

A more secure situation would be achieved if the initial water level at Aguieira reservoir had been 117.00 m (as established by the dam operating rules during flood events) and the dam discharges were limited to 670 m^3/s. In this case, the maximum water level would be 122.78 m and the water level at Santa Clara bridge would remain the same as the simulated in the previous situation.

Another scenario considered that the water level at Aguieira dam at the beginning of the flood event should be 117.0 m and discharges were limited to the maximum capacities of its turbines (450 m^3/s). In this case, the water level at Coimbra city would be 18.5 m (much lower in comparison with the maximum registered level of 19.36 m during the flood event).

Operation and early warning system

Results showed that flood control and dam operations are too complex, but the use of adequate tools can improve the understanding of the phenomena and help the decision-maker to manage these extreme situations.

Organizations responsible for flood management at Mondego river basin base their decisions on measured precipitation data, discharges observed at dams and hydrometric stations along with precipitations forecasts. A hydrological model is also available. However, this model was implemented about two decades ago and needs to be updated. The emergency plan for floods of Coimbra municipality specifies an alert for water levels based on criteria that result from recorded precipitation values, measured river flows at Coimbra dam and river discharges of Ceira river.

These approaches are not based on forecasts of hydrological and hydrodynamic models and thus, it is not possible to predict the water levels in the fluvial system that result from the operation of the dams and precipitation events with the appropriate accuracy and detail. They do not allow flood events to be foreseen with the desired anticipation and accuracy. Indeed, according to the current

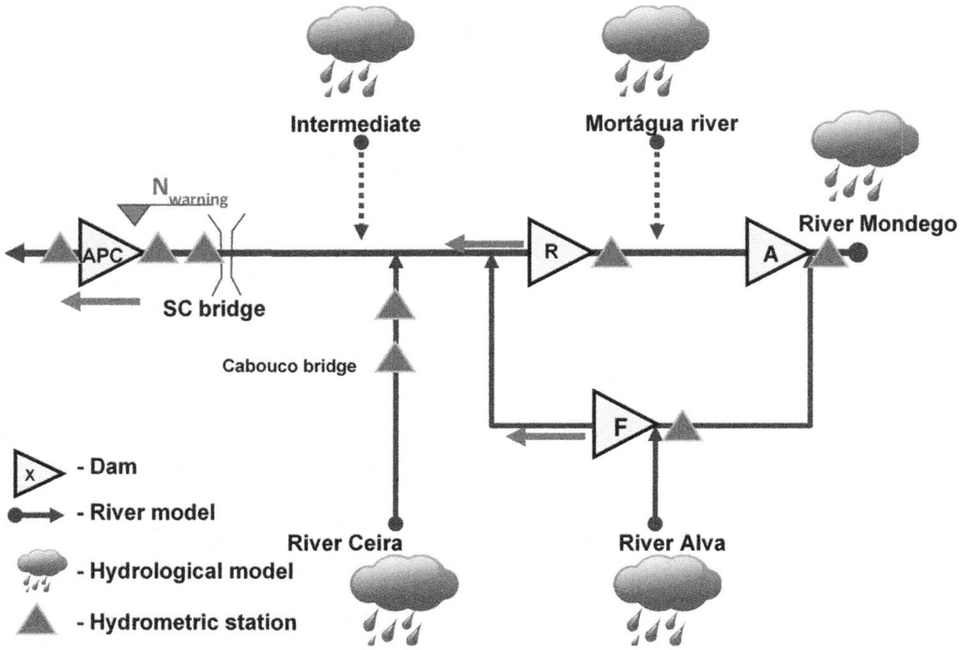

Figure 10 | Conceptual scheme of Mondego river basin flood early warning system.

predictive capacity of the most advanced technologies, limited to 1 or 2 days, the use of this type of technology considerably increases the capacity to plan operations to minimize the impacts of floods.

In an artificial system such as the Mondego river, operational rules should be clear and preferably monitored through technological platforms that facilitate and integrate relevant information and are shared by all stakeholders in the decision-making process. They should include communication tools with automatic early warnings. An early warning flood forecasting system for the Mondego river basin should include a hydrological and hydrodynamic modelling of the main sub-basins and rivers, as identified in the conceptual scheme proposed in Figure 10.

The hydrological models should consider diversified sources of precipitation forecasts and should be implemented with techniques that allow efficient assimilation of monitored data. Hydrodynamic models should be integrated with hydrological models to allow water level forecast at locations vunerable to flood. The models should also consider the simulation of the operating rules of the hydropower plants through methodologies based on real-time control techniques.

CONCLUSIONS

The results obtained showed that the applied hydroinformatic solutions constitute an excellent instrument to analyse a complex system under the influence of different hydraulic structures and natural hydrological processes during a flood event in a river basin.

The key factors responsible for inundations that occurred at Coimbra city were analysed through a comprehensive and state-of-the-art set of modelling tools. It was concluded that for the January 2016 flood event, the upstream dam discharges were a determinant for the water levels verified at the city. Before the beginning of the flood event the reservoir water levels were not limited to a maximum of 117.0 m, as recommended in the established operational rules, and this fact was crucial for the difficulties verified in accommodating the inflows to the reservoir during the flood event.

Even considering the water levels observed in the reservoir of the Aguieira dam during the flood event, based on simulated operational alternatives, the flood event that occurred in January 2016 could have been minimized if another operating scheme had been adopted.

In an artificial river system such as the Mondego river basin, operational rules should be clear and preferably monitored through technological platforms that facilitate and integrate relevant information and are shared by all stakeholders in the decision-making process. They should include communication tools with automatic warnings and preferably based on forecast models.

The current hydroinformatic technologies allow substantial improvement of flood forecasting capabilities with the implementation of a decision support system for the Mondego river basin that allows the use of precipitation forecasts from a variety of sources, uses models with the potential to simulate hydraulic structures and their respective controllers and that also can allow the efficient assimilation of monitoring data.

ACKNOWLEDGEMENTS

The authors acknowledge the financial support given by the Portuguese Environmental Agency and the Portuguese National Association of Engineers.

REFERENCES

Adeloye, A. & Rustum, R. 2011 Lagos (Nigeria) flooding and influence of urban planning. *Journal of Urban Design and Planning* **164** (3), 175–187.

Aparicio, J., Martinez-Austria, P. F., Guitron, A. & Ramirez, A. I. 2009 Floods in Tabasco, Mexico: a diagnosis and proposal for courses of action. *Journal of Flood Risk Management* **2**, 132–138.

Bellos, V., Kourtis, I., Moreno-Rodenas, A. & Tsihrintzis, V. 2017 Quantifying roughness coefficient uncertainty in urban flooding simulations through a simplified methodology. *Water* **9** (12), 944.

Burnash, R. & Ferral, L. 2002 *Conceptualization of the Sacramento Soil Moisture Accounting Model. NWSRFS User Manual Documentation, National Weather Service*. NOAA, Silver Spring, MD, USA.

CRC 2004 *Rainfall Runoff Library - User Guide*. www.tolkit.net.au/rrl (accessed December 2013).

Deletic, A., Dotto, C., McCarthy, D., Kleidorfer, K., Freni, G., Mannina, G., Uhl, M., Henrichs, M., Fletcher, T., Rauch, W., Bertrand-Krajewskf, J. & Tait, S. 2012 Assessing uncertainties in urban drainage models. *Physics and Chemistry of the Earth, Parts A/B/C* **42–44**, 3–10.

Deltares 2011a *Sobek Reference Manual*. Delft Hydraulics, Delft, Netherlands.

Deltares 2011b *Delft3D-Flow User Manual – Simulation of Multi-Dimensional Hydrodynamic Flows and Transport Phenomena, Including Sediments*. Deltares, Delft, Netherlands, 64 p.

Demir, I. & Krajewski, W. F. 2013 Towards an integrated flood information system: centralized data access, analysis, and visualization. *Environmental Modelling and Software* **50**, 77–84.

Dimitriadis, P., Tegos, A., Oikonomou, A., Pagana, V., Koukouvinos, A., Mamassis, N., Koutsoyiannis, D. & Efstratiadis, A. 2016 Comparative evaluation of 1D and quasi-2D hydraulic models based on benchmark and real-world applications for uncertainty assessment in flood mapping. *Journal of Hydrology* **534**, 478–492. doi:10.1016/j.jhydrol.2016.01.020.

Efstratiadis, A., Koussis, A. D., Koutsoyiannis, D. & Mamassis, N. 2014 Flood design recipes vs. reality: can predictions for ungauged basins be trusted? *Natural Hazards and Earth System Sciences* **14**, 1417–1428. doi:10.5194/nhess-14-1417-2014.

ERDC 2013 *ERDC: Engineering Resource Development Center*. http://www.erdc.usace.army.mil/ (accessed May 2017).

EU 2007 *Directive of the European Parliament and of the Council 2007/60/EC, Establishing A Framework for the Assessment and Management of Flood Risks, Official Journal 2007 L 288/27*, European Commission, Brussels, Belgium.

Filatova, T. 2014 Market-based instruments for flood risk management: a review of theory, practice and perspectives for climate adaptation policy. *Environmental Science and Policy* **37**, 227–242.

FitzHugh, T. W. & Vogel, R. M. 2011 The impact of dams on flood flows in the United States. *River Research and Applications* **27** (10), 1192–1215.

Gibertoni, R. F. C., Pinho, J. L., Muhlenhoff, A. P., Kan, A., Schwanenberg, D. & Reis, A. A. 2014 Sacramento model calibration using different methodologies: application to the Itabapoana-MG river basin. In: *6th International Conference on Flood Management-ICFM6*. Brazilian Water Resources Association, San Paulo, Brazil.

Goharian, E. & Burian, S. J. 2018 Developing an integrated framework to build a decision support tool for urban water management. *Journal of Hydroinformatics* **20** (3), 708–727.

Hammond, M. J., Chen, A. S., Djordjević, S., Butler, D. & Mark, O. 2015 Urban flood impact assessment: a state-of-the-art review. *Urban Water Journal* **12** (1), 14–29.

Henonin, J., Russo, B., Mark, O. & Gourbesville, P. 2013 Real-time urban flood forecasting and modelling – a state of the art. *Journal of Hydroinformatics* **15** (3), 717–736.

Krause, P., Boyle, D. P. & Base, F. 2005 Comparison of different efficiency criteria for hydrological model assessment. *Advances in Geosciences* **5**, 89–97.

Liang, Q. 2010 Flood simulation using a well-balanced shallow flow model. *ASCE Journal of Hydraulic Engineering* **136** (9), 669–675.

Néelz, S. & Pender, G. 2010 *Benchmarking of 2D Hydraulic Modelling Packages*. SC080035/R2 Environment Agency, Bristol, UK.

Palla, A., Colli, M., Candela, A., Aronica, G. T. & Lanza, L. G. 2018 Pluvial flooding in urban areas: the role of surface drainage efficiency. *Journal of Flood Risk Management* **11**, S663–S676.

Papaioannou, G., Loukas, A., Vasiliades, L. & Aronica, G. T. 2016 Flood inundation mapping sensitivity to riverine spatial resolution and modelling approach. *Natural Hazards* **83**, 117–132.

Phong, L. H. 2015 The relationship between rivers and cities: influences of urbanization on the riverine zones – a case study of Red River zones in Hanoi, Vietnam. *WIT Transactions on Ecology and the Environment* **193**, 27–43.

Pinho, J. L. S., Vieira, L., Vieira, J. M. P., Venâncio, S., Simões, N., Marques, J. & Seabra Santos, F. 2018 Application of hydroinformatic tools for reservoirs discharges rules assessment during a flood event. HIC 2018. In: *13th International Conference on Hydroinformatics*, Palermo, Italy.

Popescu, I., Jonoski, A., Van Andel, S. J., Onyari, E. & Moya Quiroga, V. G. 2010 Integrated modelling for flood risk mitigation in Romania: case study of the Timis–Bega river basin. *International Journal of River Basin Management* **8** (3–4), 269–280. doi: 10.1080/15715124.2010.512550.

Santos, F. J. S., Marques, J. A. S. & Mendes, P. A. 2002 *Estudo Hidráulico e Hidrológico do Rio Mondego na Zona de Intervenção do Programa Polis em Coimbra Relatório Final.* (Hydraulic and Hydrological Study of the Mondego River in the Polis Program Intervention Zone at Coimbra) Final report, University of Coimbra, Coimbra, Portugal.

Schwanenberg, D., Talsma, J., Pinho, J. L., Assis dos Reis, A., Rodrigues Bessa, M. & Kuwajima, J. 2013 Short-term reservoir optimization for mitigation downstream flood risks. In: *8th International Conference of EWRA Water Resources Management in an Interdisciplinary and Changing Context*, Porto, Portugal.

Sharma, C. S., Mishra, A. & Panda, S. N. 2014 Assessing impact of flood on river dynamics and susceptible regions: geomorphometric analysis. *Water Resources Management* **28**, 2615–2638.

Singh, V. P. & Woolhiser, D. A. 2002 *Mathematical modeling of watershed hydrology. Journal of Hydrologic Engineering* **7** (4), 270–292.

Stelling, G. S. & Verwey, A. 2005 *Numerical Flood Simulation. Encyclopedia of Hydrological Sciences*, Vol. 1. John Wiley & Sons, Chichester, UK, pp. 257–270.

Teng, J., Jakeman, A. J., Vaze, J., Croke, B. F., Dutta, D. & Kim, S. 2017 Flood inundation modelling: a review of methods, recent advances and uncertainty analysis. *Environmental Modelling & Software* **90**, 201–216.

Tsakiris, G. & Bellos, V. 2014 A numerical model for two-dimensional flood routing in complex terrains. *Water Resources Management* **28**, 1277–1291.

Verwey, A. 2001 Latest developments in floodplain modelling-1D/2D integration. In: *Proceedings of the 6th Conference on Hydraulics in Civil Engineering: the State of Hydraulics.* Australia: Institution of Engineers, Hobart, Tasmania, p. 13.

Werner, M., van Dijk, M. & Schellekens, J. 2004 Delft FEWS: An open shell flood forecasting system. In: *6th International Conference on Hydroinformatics*, Singapore, pp. 1205–1212.

Werner, M., Schellekens, J., Gijsbers, P., Van Dijk, M., Van Den Akker, O. & Heynert, K. 2012 The Delft-FEWS flow forecasting system. *Environmental Modelling & Software* **40**, 65–77.

Willmott, C. 1982 Some comments on the evaluation of model performance. *Bulletin of the American Meteorology Society* **63** (11), 1309–1369.

First received 15 January 2019; accepted in revised form 4 May 2019. Available online 3 June 2019

Case study of the cascading effects on critical infrastructure in Torbay coastal/pluvial flooding with climate change and 3D visualisation

M. J. Gibson, A. S. Chen, M. Khoury, L. S. Vamvakeridou-Lyroudia IWA, D. Stewart, M. Wood, D. A. Savić IWA and S. Djordjević IWA

ABSTRACT

Critical infrastructures (CIs) are commonly designed, built and maintained based on rigorous standards in order to withstand the climate and weather-related pressures. However, shifts in climate characteristics may result in increases of the magnitude and frequency of potential risks, or expose specific CI to new or increased risks not previously considered. As vital components of the normal functioning of modern societies, their resilience encompasses the operational elements, their structural integrity and the capacity to maximise business output under climate stressors. In this work, we apply an integrated and participatory methodological approach to assess the risk and enhance the resilience of interconnected CIs to urban flooding under climate change. The proposed methodology has been applied to an extended case study in Torbay to extend previous works, which seeks to protect coastal communities from future events through using the proposed methodology to justify future investment in coastal defences, as a part of the validation of EU-CIRCLE projects developed methodologies.
Key words | cellular-automaton, climate, coastal, critical-infrastructure, hazard, pluvial

M. J. Gibson (corresponding author)
A. S. Chen
M. Khoury
L. S. Vamvakeridou-Lyroudia IWA
D. A. Savić IWA
S. Djordjević IWA
Centre for Water Systems,
University of Exeter,
Harrison Building, North Park Road,
Exeter EX4 4QF,
UK
E-mail: *m.j.gibson@exeter.ac.uk*

D. Stewart
M. Wood
Torbay Council,
Town Hall, Torquay TQ1 3DR,
UK

D. A. Savić
KWR Water Cycle Research Institute,
Nieuwegein,
The Netherlands

INTRODUCTION

Flood hazards are a major threat to urban systems and societies (Gibson *et al.* 2018). A better understanding of the potential impacts caused by flood hazards under various climate and socioeconomic scenarios is essential to developing resilience strategies for critical infrastructures (CIs) planning, design, operation and management (Boin & McConnell 2007). Various studies have developed different approaches to evaluate flood risks (Hall *et al.* 2003; Apel *et al.* 2006; Chen *et al.* 2016) and/or to quantify resilience for decision supporting to develop climate adaptations (Engle *et al.* 2014; Vojinovic 2017; Hammond *et al.* 2018). Such analyses often consider not only direct tangible impacts but also indirect (e.g. economic losses (Andreoni & Miola 2014; Kreibich *et al.* 2017) and/or intangible consequences (e.g. health; (Alderman *et al.* 2012; Mark *et al.* 2015) of natural climate-related hazards.

In addition, society in the modern age, living in cities requires a number of infrastructures which are critical to continued function of cities and modern society, for example: clean water supply and sewer treatment system, public transportation/road/rail networks, energy generation and transmission networks and telecommunication/ICT networks. With each of these services having become critical to modern society, the increasing interconnectivities among these CIs have formed a more complex system than any individual network. A disruption in one infrastructure has a ripple or cascading effects into other infrastructures. As large-scale systems, whose lifetime

doi: 10.2166/hydro.2019.032

exceeds the order of decades, infrastructures are heavily exposed to natural disasters, often with devasting consequences to the society, economy and the environment. The increasingly dependent, interdependent and interconnected nature of CIs may expose societies to previously unseen risks, new vulnerabilities and opportunities due to disruption across multiple CI networks. Meanwhile, the spatiotemporal evolution of hazard impacts is critical for crisis management and strategic planning to prevent cascading impacts (Mazzorana *et al.* 2019). Existing applications might be able to display such a relationship conceptually or at an aggregated high level (Araya-Munoz *et al.* 2017), and details regarding where, when and how such a knock-on effect propagates during disasters are still unavailable.

Acknowledging that infrastructure's vulnerabilities and impacts go far beyond physical damage (Hokstad *et al.* 2012), the work presented here is concerned with an assessment of the impacts on the services provided by CI. We considered not only the impacts associated with the repair and/or replacement of services but also included the externalities of the CIs operation, societal costs, environmental effects and economic consequences due to suspended activities. Interdependencies among infrastructures dramatically increase the overall complexity of the 'systems of systems'. There is therefore a need to consider multiple interconnected infrastructures and their interdependencies in a holistic manner (Galbusera *et al.* 2014).

The EU H2020 funded project EU-CIRCLE (a pan-European framework for strengthening CI resilience) developed a holistic framework for identifying the risks of multi-climate hazards to heterogeneous interconnected and interdependent CI (Sfetsos *et al.* 2017). In the work presented in this paper, the resilience framework was applied for urban flooding. For this purpose, it was combined with fast flood modelling to evaluate possible flood scenarios, with the active participation of local authorities providing the scenarios. Thus, vulnerable CIs were identified, as well as the consequences when their services are disrupted by flooding. High visualisation techniques were used to facilitate the understanding and communication of the results, while the outcomes were used for selecting suitable adaptation measures by local stakeholders. The whole approach and methodology have been applied to the case study of Torbay in South West UK and are presented in this paper.

METHODOLOGY

The EU-CIRCLE methodology is detailed in the literature (Sfetsos *et al.* 2018), but the primary features are also presented here. Shown in Figure 1 is the resilience framework developed in the EU-CIRCLE project and is applied in this work. The generic resilience framework has three main sections which are interconnected and allows for tailoring to the local specific CI, or the area or challenges of interest in the given case study it is applied to. The generic resilience framework consists of three main interacting and interconnected sections (shown in Figure 1):

1. A step-by-step precedure climate risk management framework;
2. An interactive risk and risk modelling framework;
3. Outputs to the end-users.

A key to the implementation of this framework is the interaction of stakeholder and end-users at both the development and tailoring of the framework to the given case and the output required. This framework is very flexible and allows for the inclusion of multiple or different modelling methodologies given the stakeholder and end-users interests in the given target area.

The climate risk management framework begins with establishing relevant climate change resilience policies and decisions which are of interest to stakeholders and end-users in the given area. What policies are in place and what questions are to be answered? These issues may have life spans of many years and may cross many different service providers. Climate data may come from any number of sources, including global climate models at higher and lower resolutions, statistically downscaled climate information and historical data. Given the target area, it is possible to establish which infrastructures are critical and which systems interconnect these, and what the risks to these systems are. Some areas are prone to forest fires, or typhoons/hurricanes, and others as in the case study presented in this paper are susceptible pluvial and coastal

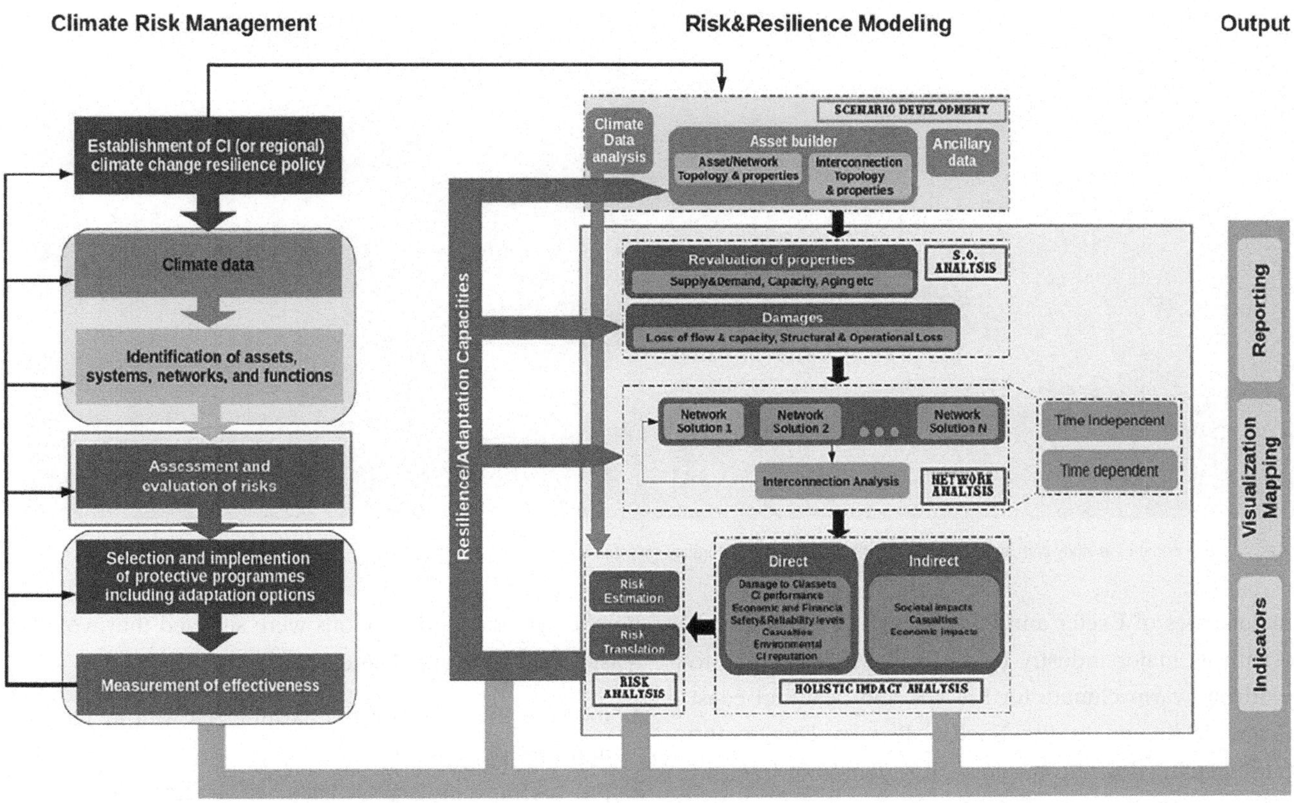

Figure 1 | Generic resilience framework for the CIs for the EU-CIRCLE framework.

flooding. Each area will have different risks stemming from the combination of current risks and how climate change may exacerbate these, for example, increased sea levels or more extreme storms. At this stage, any protective programmes, including adaptation options to be investigated as part of the framework, are established.

The risk and resilience modelling section details the structure of risk modelling and its different components and interactions. A number of scenarios are constructed using the climate data and projections, which are combined with a list of relevant CI and risks established in the climate risk management section. The system overview analysis establishes damage curves for the various different types of CI. With a number of different scenarios, it is important to select an appropriate level of computational modelling such that a large number of scenarios can be modelled in a timely manner. The resulting holistic impact analysis includes direct and indirect impacts to CI, and here the role of stakeholders and end-users is critically important in the validation of the results. Further information on the overall EU-CIRCLE methodology can be found in Sfetsos *et al.* (2017) and Sfetsos *et al.* (2018).

The last section involves the display/visualisation of analysis and result and is connected to all stages, as stakeholder and end-user involvement is so critical. To this end, there is a large focus on using modern computing techniques to fully visualise the results and analysis in order to engage and effectively communicate to stakeholder and end-users. The work in this paper presents a single case study using the EU-CIRCLE methodology.

CASE STUDY SETUP

Torbay case study background

Torbay is a major tourist location on the South Devon coast line (Figure 2), UK, which covers approximately 62 km^2, bordering Teignbridge and the South Hams. Lying between

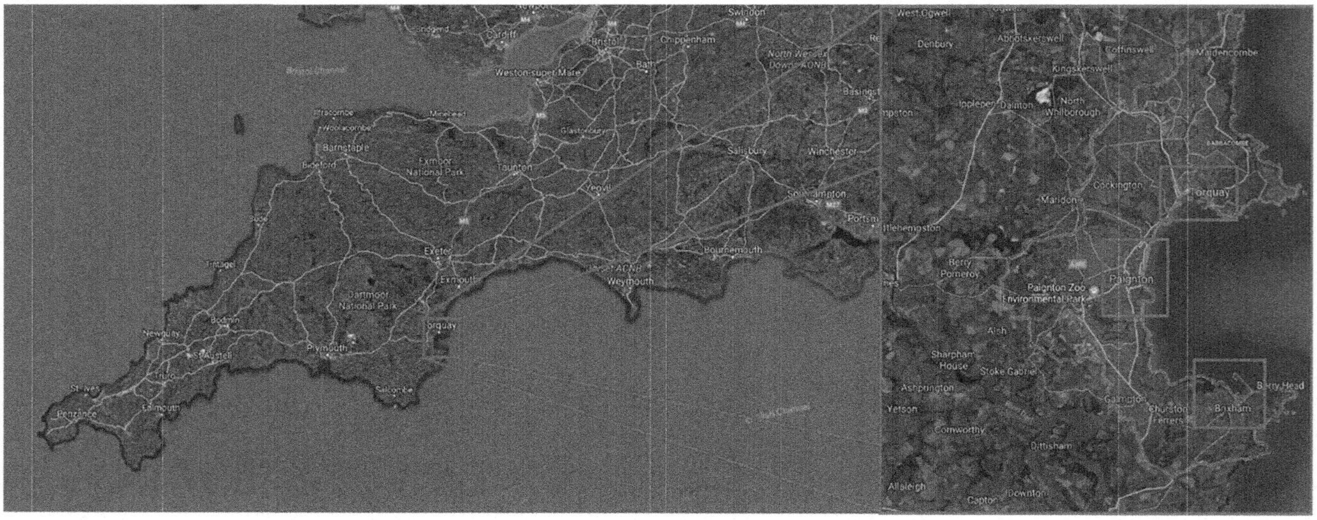

Figure 2 | Location of the case study (left) in the South West of England, Torbay – Torquay, Paignton and Brixham (right).

the major cities of Exeter and Plymouth, Torbay has developed with its major industry being that of tourism, due to its relatively warm climate for the UK and beautiful coast lines, and is known as the English Riviera. Due to the steep hill leading down to the coast, it suffers from frequent pluvial (surface) flooding, but also from extreme storms causing the sea to overtop or breach the coastal defences. There are three large towns of Torquay, Paignton and Brixham which constitute the majority of the catchment area, meaning that the vast majority of the catchment is urban in nature.

There are three target areas in Torbay centred around the three main towns: Torquay, Paignton and Brixham (shown in Figure 2). In order to process comprehensive climate change/adaptation scenarios for each of three target areas in a tractable amount of time for analysis, a fast modelling approach is needed. The CADDIES framework (Guidolin *et al.* 2012, 2016) and associated caFlood(Pro) application provide a fast means for 2D flood modelling and visualisation, which are utilised in the study. The caFlood application is based on the cellular automata system and uses the weighted model (Guidolin *et al.* 2016) to distribute water flows between each cell of the computational grid. This model was validated against a number of UK Environment Agency test cases, a hypothetical test cases and also compared to the industry standard infoworks model on a test case in Torquay (Guidolin *et al.* 2016). At Paignton, a new secondary sea wall has been proposed. A number of adaptation options were selected that would be feasible and relevant. These options included:

- increase the height of the existing sea wall at Paignton and Preston, Greens
- provide a secondary set back sea wall at Paignton and Preston, Greens at the seaward side
- provide a secondary set back sea wall at Paignton and Preston, Greens at the landward side and
- provide a wave return wall to the existing sea wall at Paignton and Preston, Greens.

Providing a secondary set back wall at the seaward side of Paignton and Preston, Greens was selected as the most effective intervention. Additional simulations that model this secondary defence have been undertaken, for a range of different wall heights and returns periods, given the climate change projects for the given planning period for the construction of the wall. This allows for a cost-benefit analysis to be performed.

The Torbay area is particularly vulnerable to a number of flooding sources, including highway flooding, main river and ordinary watercourse, surface water run-off and coastal overtopping. This coastal overtopping occurs when large storms push waves up and over the existing sea wall defences, which can then threaten the low-lying coastal town areas. This occurs when high tides coincide with storm surges and easterly winds. During these times, main river, sewer, surface water, highway and watercourse flooding are exacerbated due to the lack of the capacity of the surface water outfalls, which discharge to the sea. These flood events threaten a large number of commercial and residential properties on a regular basis and also affect numerous roads to some extent. A number of roads are closed due to flooding from such storm events, and this can have critical knock-on effects on the road network when large capillary roads are closed, resulting in long traffic diversion and delays.

Terrain data/parameters

For each of the three geographical locations considered, a high-resolution 1-m grid was applied to simulate flooding scenarios that include coastal, pluvial and combined conditions for the current and future climate change situations of 20, 50 and 100 years ahead. The UK Environment Agency's (EA) Light Detection and Ranging (LiDAR) digital terrain model (DTM) data were used as the ground elevations for modelling. The LiDAR DTM was filtered from the digital surface model (DSM) (Priestnall *et al.* 2000) using algorithms that smooth the surface and thus remove such superfluous features as vehicles, people, animals and trees, but also remove more permanent features such as buildings. Buildings however are critical to flow paths, along with the kerb on roads which acts as a primary water transport system during flood events. Therefore, the use of the DTM has the problem of temporary features such as traffic, etc., while the DSM has its own problems of lacking critical permanent features such as buildings. The solution to this problem is to further preprocess the DSM, in combination with the buildings and road layouts from the Ordnance Survey's Mastermap, in accordance with the EA approach to surface water mapping (E.A. UK 2013). Road cells, including those which touch a road polygon, are lowered by 12.5 cm from their existing terrain levels, which leaves their slopes intact. However for buildings, it would be expected that there would be a uniform level across each building, i.e. assuming all buildings have flat roofs. Whereas apart from the terrain smoothing, buildings built on the steep slopes tend to take on these characteristics. Therefore, the cells within each building polygon are set to the same uniform level, based on that of the highest elevation plus a 15 cm threshold, which is designed to simulate the doorstep level of the building. After the doorstep level, water is allowed to enter the building polygon, but in order to account for the structural walls and content effects on flow within buildings, further parametrisation is required. To account for the reduced flow into, out of and inside of buildings, the caFloodPro application allows for Manning's roughness factor to set for individual or groups of cells. In this case, Manning's roughness is increased from 0.015 to 0.1, which slows down the flow within building areas.

CaFloodPro also allows for an infiltration rate to be set on any particular cell or group thereof, which is utilised to simulate the effects of the sewer system to drain water from the urban surfaces, while also being used to account for the capacity of the green area to absorb water. The infiltration rates used for these areas are shown in Table 1. Torbay's sewer systems are designed to cope with up to a 1 in 30-year return period pluvial event; however, it is thought due to blockages of vegetable matter, gullies and inlets along roads do not provide the equivalent capacity. Therefore, road drainage is reduced to the equivalent of a 1 in 5-year return period event, as well as a rainfall reduction of 12 mm/h to green areas, which is in line with the EA procedure.

A smaller domain was utilised when modelling the coastal events, as these events only affect the low lying area next to the coastal line, by delineating the areas which are lower than 30 m above sea level. Whereas pluvial events are largely affected by the area upon which they fall, especially when designed storms are utilised where rain falls the same amount everywhere in the given area. Therefore, a terrain analysis is used to determine the area which is capable of contributing to towns flooding.

Table 1 | Infiltration/water loss rates per surface type

Area type	Infiltration (mm/h)
Green (any other)	12 + (rainfall reduction of 12 mm/h for pluvial cases)
Roads tracks and paths	19
Buildings	28

Overtopping conditions

The AMAZON model (Hu 2000; H.U. Ltd 2017) was used to produce the overtopping rates at the defensive sea wall, which was then fed into the CADDIES model as a 2D flood input. The AMAZON model simulates random waves travelling as bores, given the tide levels and storm inputs. Climate change scenarios of the present, 20, 50 and 100 years at a 1 in 200-year return period storm were modelled in AMAZON and then used to assess the 2D flooding in CADDIES. Given that there is a 12-h tide cycle, the highest tides affect the rate at which the waves overtop the sea defences, and this can be seen in the inflows to caddies, as shown in Figure 3.

Pluvial conditions

A designed rainfall was utilised for pluvial events, which uses a spatially uniform distribution across the terrain. A 60-min duration of rainfall is simulated, and in order to allow time after the last rain has fallen for the surface flood to flow, a further 3 h of simulation time is used. The rainfall values for events with different return periods were obtained from the Flood Estimation Handbook (C. f. E. & Hydrology 2013) for the three towns. These rates were scaled up based on the EA's guidance (UK Environment Agency 2016) to account for future climate change scenarios, 10% for 20 years, 20% for 50 years and 40% for 100 years of climate change.

Combined pluvial and overtopping conditions

There is a reasonably low chance of either the extreme pluvial or overtopping conditions to occur independently, and while there is obviously some casual connections in that a storm may create the conditions for both cases, it is still considered to be less likely that the conditions for both exist at exactly the same time. Therefore, a more likely pluvial and coastal set of storms are modelled together, using a 1 in 50-year design rainfall for 1 h which was aligned with the peak of a 1 in 50-year coastal event, at the 36th hour of the simulation.

Critical infrastructure

Shown in Figures 4 and 5 are the footprints and locations, respectively, of CI in the Paignton area used in this case study. For CI which has the coverage of other properties (electrical substation, ambulance stations and pumping

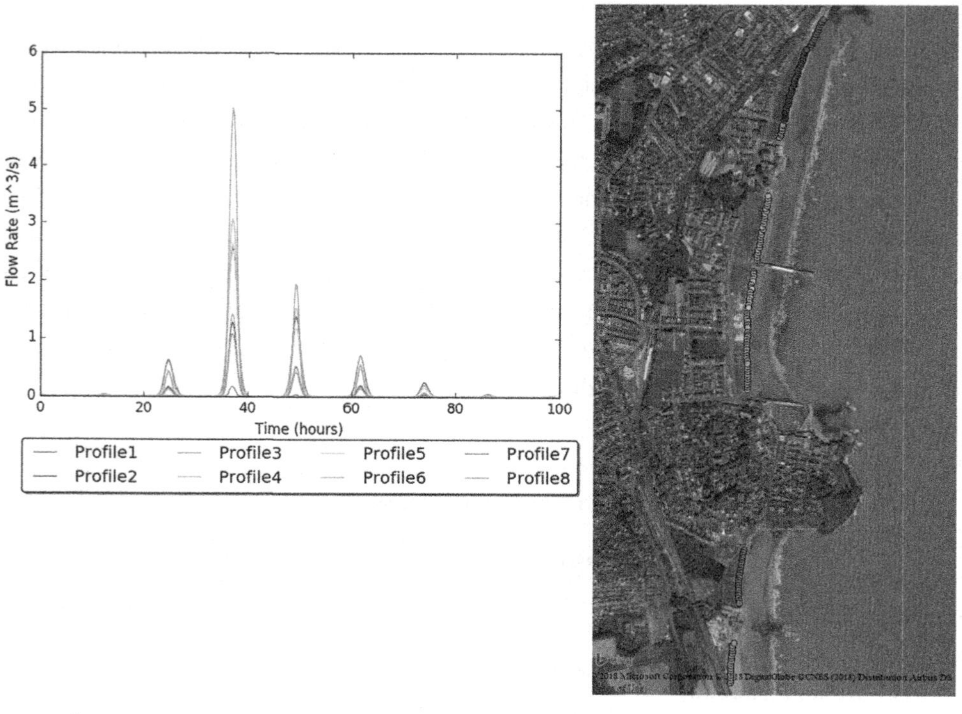

Figure 3 | Discharge profiles for a 1 in 200-year overtopping event (left) for various coastal sections for Paignton (right).

Figure 4 | Shows the footprints of CI in the Paignton area, where the road and rail networks are clearly visible.

stations), Thiessen polygons are used to establish which areas are covered. Shown in Figure 6 are the depth–damage curves utilised to calculate the direct damage, and a rate from the Multi-Coloured-Manual (Penning-Rowsell *et al.* 2005) is used to establish additional damages from the cascading effects of CIs.

RESULTS

Paignton

Figure 7 shows a 1 in 200-year coastal overtopping event (right) and a 1 in 100-year pluvial event (left) for current climate conditions. Figure 7 shows the difference between the coastal risks and pluvial risks to Paignton, both of which affect large areas of the residential area, and the coastal floods particularly affecting Paignton's tourism industry like hotels, cafes, bars and shops which are located near the coast. While obviously coastal flooding is limited to a local area of the coast, depths and damages are more severe. Whereas the pluvial flooding is mainly restricted to the road network and steep valleys of the terrain, it is clearly more widespread than the coastal event. Figure 8 shows a 1 in 100-year pluvial (left) and a 1 in 200-year coastal (right), with 50 years of climate change, where risk is only set to increase. Whereas in Figure 7, the coastal events are largely bound by the rail line, given the 50 years climate change projection, both the line and rail station are completely overwhelmed.

Paignton's coastal flooding represents the largest risk discovered in this case study and can be clearly seen when contrasted to Brixham (Figure 10) and Torquay (Figure 11). An adaptation plan has been proposed to build additional sea defences to mitigate this risk.

Shown in Figure 9 are the resulting flood depths for a 1 in 200-year coastal overtopping event, given 50 years of climate change; firstly on the left, without any additional sea defences, then right showing the largest defensive sea wall. The planned improvements to the sea wall drastically reduce the amount of flow and clearly protect the large

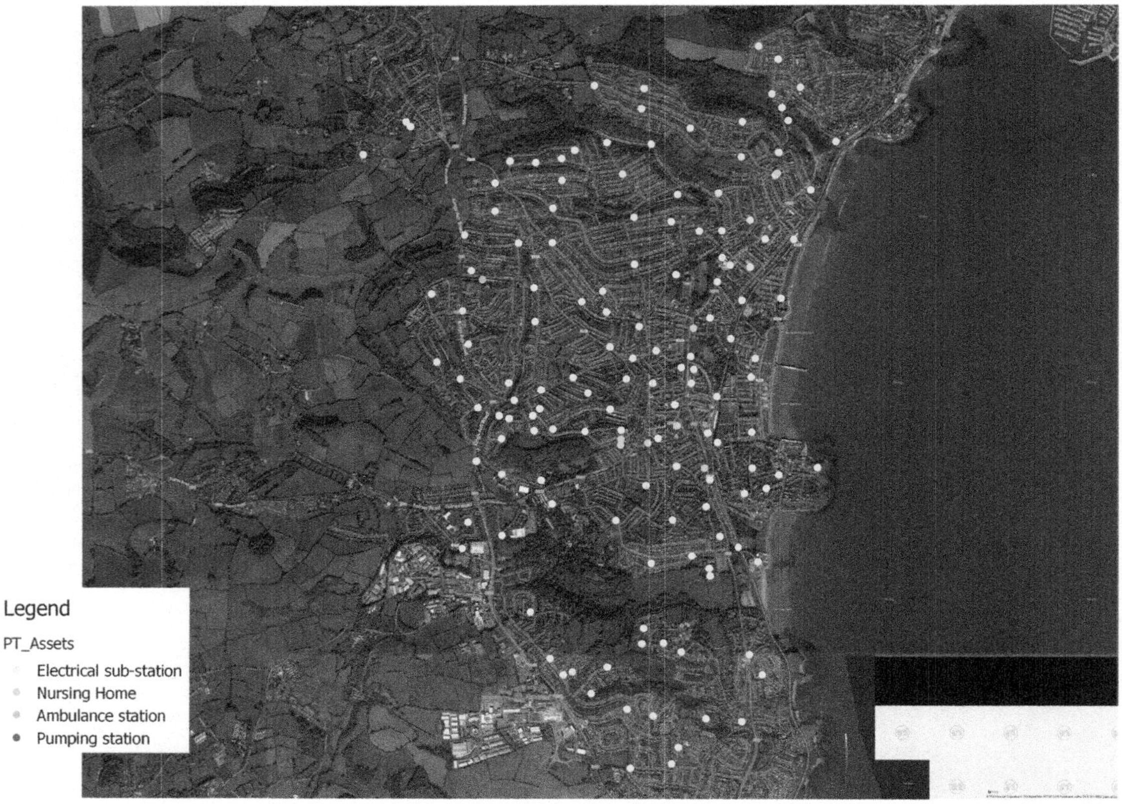

Figure 5 | Shows the locations of small footprint CI in the Paignton area, including electrical substations, nursing homes, ambulance stations and pumping stations.

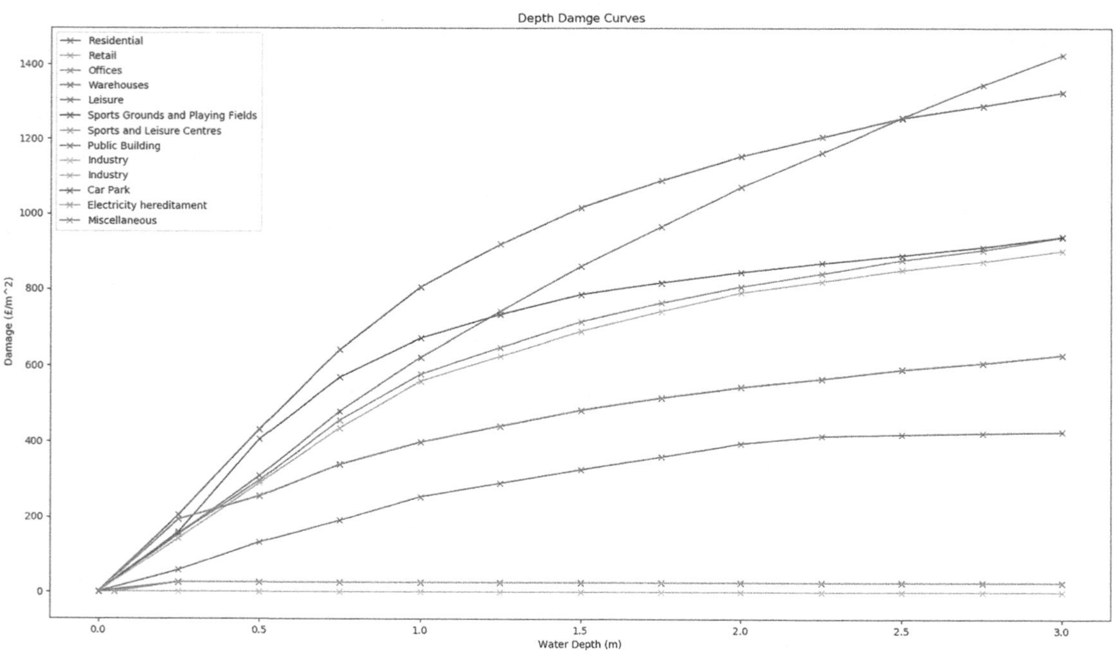

Figure 6 | Shows the depth–damage curves used in this case study for direct damages.

Figure 7 | Flood extents of a 1 in 100-year return period pluvial event (left) and a 1 in 200-year return period overtopping event (right) for the present scenario.

Figure 8 | Flood extents of a 1 in 100-year return period pluvial event (left) and a 1 in 200-year return period overtopping event (centre), and combined 1 in 50-year pluvial and coastal overtopping event (right), each for the 50 years of climate change scenario.

northern area of Paignton from the majority of flooding. However, this does leave a sizable risk of flooding and damage to the critical rail wall line and station.

Brixham

Figure 10 shows Brixham's pluvial 1 in 100-year case on the left and a 1 in 200-year coastal case on the right. Figure 11 shows Torquay's pluvial 1 in 100-year case on the left and the 1 in 200-year coastal case on the right. Unlike the Paignton case (shown in Figures 7–9), Brixham (Figure 10) and Torquay (Figure 11) are at a greater risk of pluvial flooding compared to the coastal, and this is due to the steep terrain in these areas and relatively the low lying area in Paignton which faces directly out into the bay.

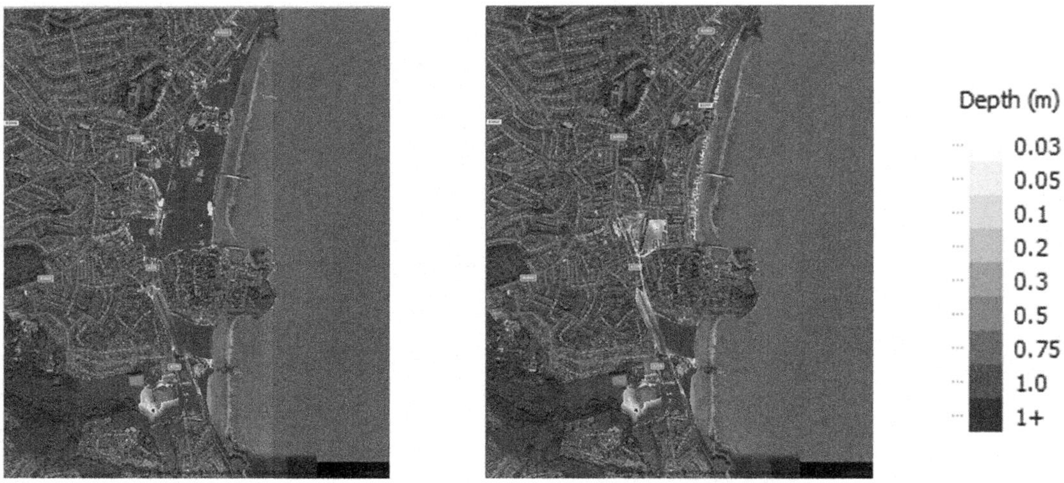

Figure 9 | Flood extents of a 1 in 200-year coastal overtopping event with 50 years climate change: (left) no sea wall, (right) sea wall option.

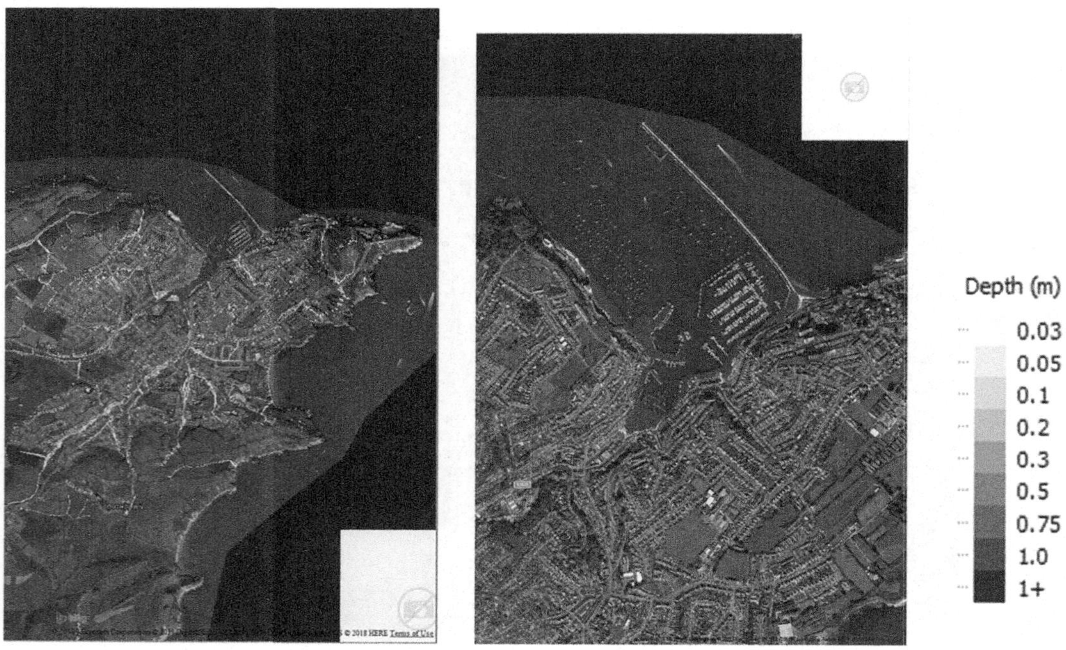

Figure 10 | Flood extents of a 1 in 100-year return period pluvial event (left) and a 1 in 200-year return period overtopping event (right) for the present scenario for the Brixham area.

ANALYSIS

Cascading effects of flooding upon CIs have been assessed using the EU-CIRCLE framework (Sfetsos *et al.* 2017; Chen *et al.* 2018) in the study. The EU-CIRCLE project considers not only the direct flood damage costs based on flood hazards (e.g. depth) but also the cascaded damage to other types of CI and to properties in the area. For example, if flood damages CI assets such as electrical substations or telecommunication, other properties that are not directly affected by the flooding may still lose power due to the failure of substations. Therefore, CIs, such as sewer pumping stations, electricity substation and telecom exchanges, will affect a much larger area beyond their locations when they are flooded beyond a certain threshold depth. A single property may be affected by not only the flood but the failure of other CIs which are affected by the flood.

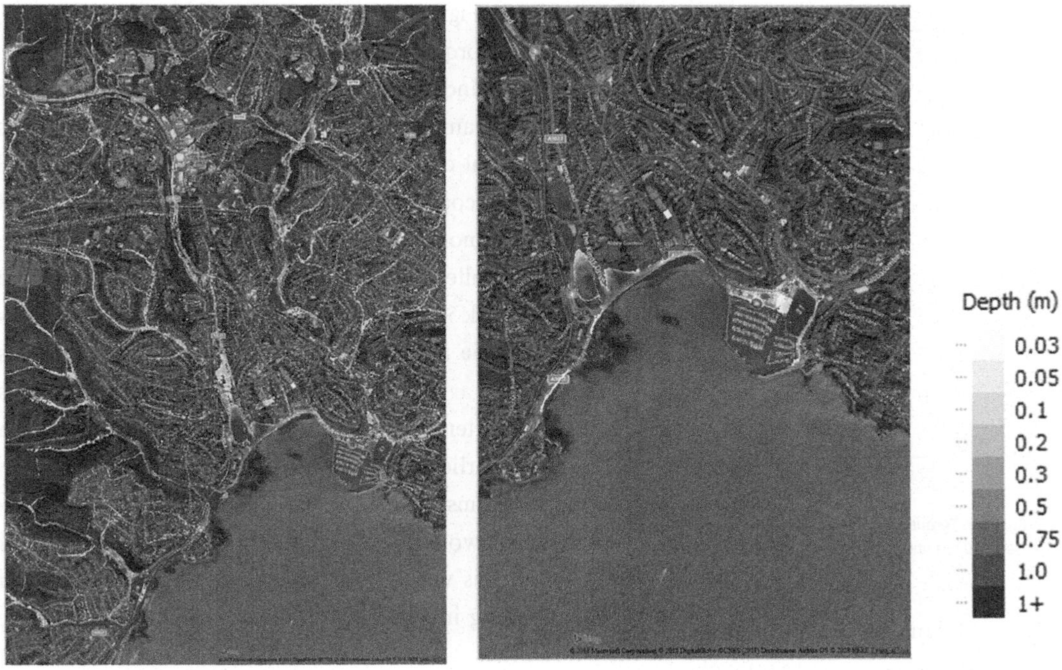

Figure 11 | Flood extents of a 1 in 100-year return period pluvial event (left) and a 1 in 200-year return period overtopping event (right) for the present scenario for the Torquay area.

The resulting flood depths obtained from CADDIES modelling were overlapped with the building layouts, and together with the building use information and the depth–damage relationships from the Multi-Coloured-Manual (Penning-Rowsell *et al.* 2005), these are used to evaluate the direct flood damage of each property.

For CIs, the first level of the cascading effect was evaluated using the algorithm shown in Figure 12. The interdependencies among CIs and other properties were further analysed such that the cascading effects can be assessed using the EU-CIRCLE framework through a looped analysis (Chen *et al.* 2018).

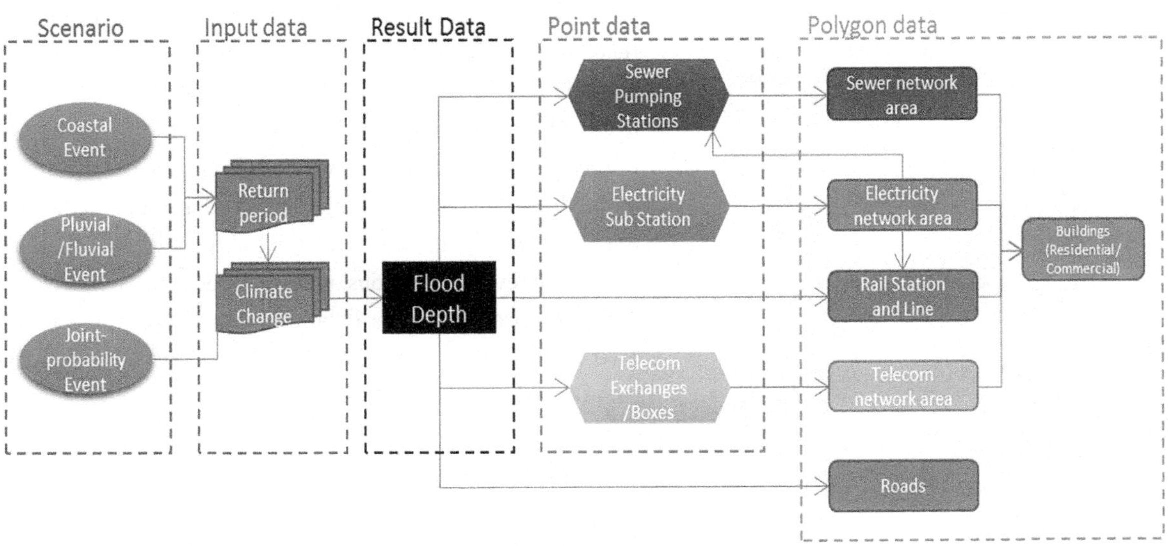

Figure 12 | EU-CIRCLE methodology for assessing flood impact on CIs.

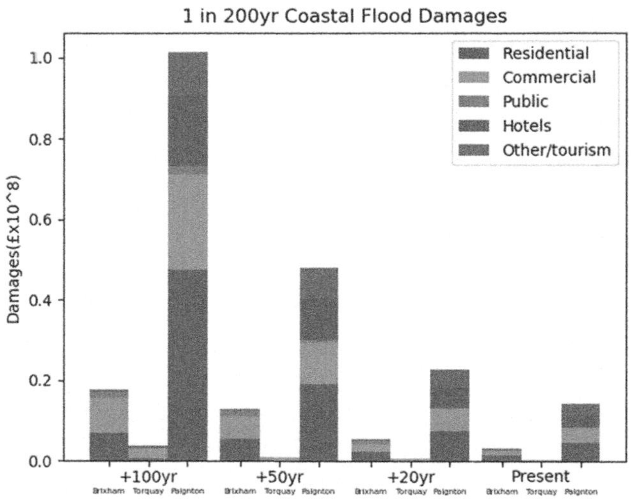

Figure 13 | Calculated flood damages at Brixham, Torquay and Paignton, for a 1 in 200-year coastal flood, for the present, 20, 50 and 100 years climate change conditions.

Figure 13 shows the calculated damages for each sector for the Brixham, Torquay and Paignton areas comparatively when using a 1 in 200-year coastal flood, for the present, 20, 50 and 100 years of climate change. Clearly, Paignton is most at risk from coastal flooding, and this presents an especially high risk to the tourism and hotel industries in the area compared to Brixham and Torquay.

Figure 14 shows the calculated damage for a 1 in 100-year pluvial flood scenario, using the present condition with 2013 data, and 20, 50, and 100 years of climate change, for the three areas Brixham, Torquay and Paignton.

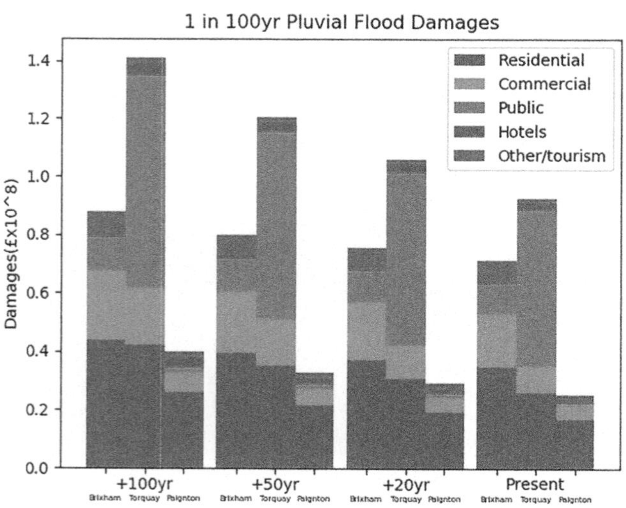

Figure 14 | Calculated flood damage for Brixham, Torquay and Paignton, for a 1 in 100 year pluvial flood, for the present, 20, 50, and 100 years of climate change conditions.

The results in Figure 14 are in stark contrast to Figure 13, showing that Torquay is at greatest risk of damage from pluvial flooding, and that Paignton is the least at risk from pluvial flood damage. The highest overall damage comes from the pluvial cases, due to the steep terrain of the area which limits coastal flooding and exacerbates pluvial cases. Furthermore, Torquay and Brixham have storage tanks to help alleviate this, which was not modelled. The next biggest risk is that of Paignton, which further justifies the work on the additional sea defences. However, this is assuming that a design rainfall occurs, that is rain falls equally for the temporal duration and spatial extent, which as discussed earlier is the maximum size of the catchment. Real-world storms have more variable spatiotemporal distributions which would lead to different spatial distribution of risks. While this work highlights all the areas at risk from pluvial flooding in this fashion, it is likely to overestimate the overall risk.

VISUALISATION

Visualisation of the resulting water depth and damages with realistic, descriptive and accurate mapping is a key to stakeholder engagement with such projects as the EU-CIRCLE framework and requires the collaboration of computer science and hydrological disciplines that form the core of the hydroinformatics discipline. Stakeholders need to be able to see the scale/breadth and be able to easily locate/relate these to real-world locations that can only be provided by a 3D realistic mapping solution. Stakeholders then need to be able to focus on the accurate details of locations of interest of heavy damage, as well as visualising the dynamics of the flood and damages both direct and indirect, in quicker than real time. This is not a trivial challenge, and the EU-circle framework includes methodologies to achieve this high visualisation and is detailed in the following section. Two main software, Google Earth Pro and the Unity3D game engine, are utilised.

Google Earth Pro

To demonstrate the modelling results in a more user-friendly way that improves risk communication, we exported the

Figure 15 | Google Earth Pro visualisation coastal flood (left) and pluvial flood (right) flooding.

flood depth data to Google Earth Pro, allowing stakeholders to explore the study area using the interactive 3D mapping tool. The advantage of such a system (other than being freely available) is that the cloud-based map archive of Google is extensive and includes extruded structures and trees, although this brings the equal disadvantage of containing no permanent features such as cars. Furthermore, the difference between the specific terrain data used in the simulation and online archive can cause erroneous water depths. The application clearly presents the relationship between flood hazards and buildings, as shown in Figure 15, such that stakeholders can intuitively visualise and better understand the threats of flood hazards.

However, the current application can only display static flood information (i.e. maximum flood depth) that cannot demonstrate how flood propagates within the city during an event. It also has limited ability to highlight the level of damage and the impact of cascading effects due to CI failures. Therefore, an advanced visualisation tool has been further developed using the Unity3D Game engine to create a 3D animated visualisation of flood events (Khoury *et al.* 2018).

Using the Unity3D game engine

Unity3D has by default no optimised way to visualise the change in the 3D shape of the flood over a terrain. As such, the Unity3D default terrain object has several resulting shortcomings:

1. The degree of precision used to render the terrain heightmap is insufficient as it cannot render a mesh with subtle differences in height. Presently, the terrain engine only takes 16 bits raw images as input, which limit the precision to 2^{16} or 65,536 unique values. Assuming the terrain height range was set to 2,000 m, the setting would provide an approximate resolution of 3.05 cm (2,000 m/65,536) which is too coarse as a 3 cm difference in the water level is not negligible.
2. The default terrain is unable to display consistently overlapping meshes at different altitude levels as shown in Figure 16. As soon as the user 'zooms out', high detail meshes are substituted with simplified one automatically via a forced optimisation process that is suitable for ordinary games, but not for visualising flood.
3. Finally, the default process by which detailed terrain and flood meshes are animated is too slow to allow real-time changes, especially on machines with a less powerful GPU. In the next section, we describe how our system addresses these shortcomings and then show added functionalities regarding the visualisation of a narrative framework as well as the impact of flood on CIs failure.

Visualisation of 3D animated floods

The 3D rendering engine has been modified to allow the terrain/flood height to be rendered with a higher degree of precision. The height data are provided by standard images in the .png format where each pixel's red, green, blue and alpha values encode in 32 bits the height of the terrain. From a practical point of view, these floating-point data of height are rescaled into large integer numbers, between 0 and 2 (Svensson & Jones 2005) (i.e. allowing the system to

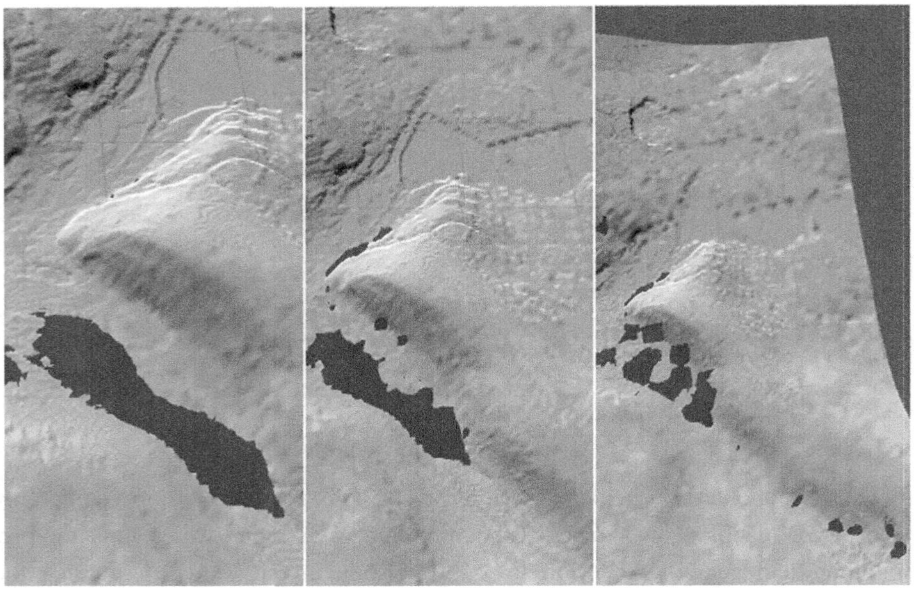

Figure 16 | By default, the further away the observer is from the terrain, the more the Unity3D terrain object does deform meshes by substituting highly detailed meshes with simplified ones.

display the tiniest difference in the height to the sub mm level). Furthermore, by using .png rather than a Unity3D default .raw file format, more compact images are produced that can easily be sent through the internet. Although so far the data are stored locally with the game, whole sequences of images could potentially then be sent from a 'game' server hosting many test cases to provide multiple animation steps through a flood sequence.

Terrain and flood surfaces can now be deformed in real time using advanced Shaders (Lammers 2013) effects exploiting graphic cards computational capabilities. Shaders – programmes implemented in the OpenGL Shading Language to display vertices and visual fragment using the accelerated computational power of the graphic card – are implemented to output in quicker than real time a 3D geometric linear interpolation of the flood 'heightmap' between the starting and the end state by looking at the 'counter' input. Flood animation is implemented by providing as input two images and a 'counter' float number between 0 and 1, expressing how far the animation is from the end stage.

The Shaders are not only engineered to deform 3D meshes efficiently but also change the flood mesh colour gradient from clear blue to dark blue with increasing flood depth (Figure 17). This works seamlessly even on fairly modest machines or small laptops, animating meshes with more than 1,440,000 triangles in real time with ease, as long as they have a DirectX 11 compatible graphic card.

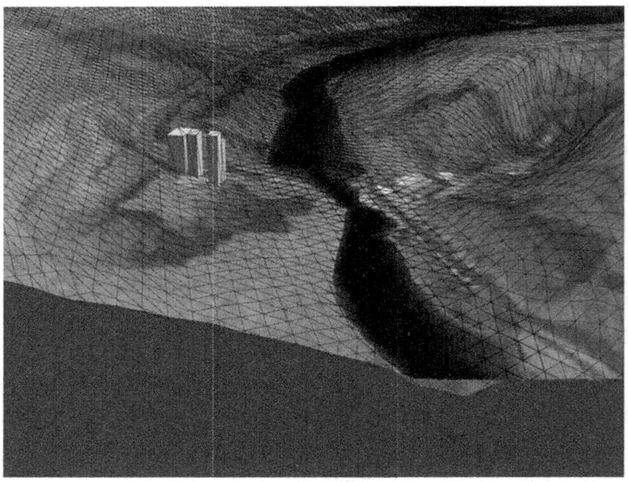

Figure 17 | Zoomed in view of the Shaders built to render in real time an animated 3D flood mesh on top of the terrain and change its colour depending on the water depth. Please refer to the online version of this paper to see this figure in colour: http://dx.doi.org/10.2166/hydro.2019.032.

Visualising consequences of flooding on CIs

All objects such as, for example, 3D buildings can be easily changed in an interactive manner (shown in Figure 18). The

Figure 18 | Substation failures and dependent power outage areas. In yellow are the areas with small risk of the power failure and small water depth. In red are the areas with high risk of the power failure and high water depth. Please refer to the online version of this paper to see this figure in colour: http://dx.doi.org/10.2166/hydro.2019.032.

green/yellow/red buildings indicate the locations of substations. The green ones represent the substations having less than 0.05 m water depth in the nearby area such that they are not affected by flooding. The yellow ones have the flood depth between 0.05 and 0.3 m that there is a small chance of power supply service disruption, and the areas receiving the service from those stations are shaded in yellow as well. The red buildings are the ones with 0.3–1.0 m flood depth that the substations are likely to be damaged and fail to provide service to their servicing areas shades in red. Details of the methodology are described in the related publication (Chen *et al.* 2018). The system allows for direct damage to be displayed on each building and also the area affected by cascade damage to be highlighted (shown in Figure 18).

CONCLUSIONS

In the paper, we presented an extended case study utilising the EU-CIRCLE metholodogy, so as to understand better how future climate regimes might affect the normal operation of interconnected CI in urban areas and how to assess the effectiveness of adaptation measures. The methodology was applied to analyse the flood impacts to Torbay due to pluvial, coastal overtopping events and their combinations under the present and future climate scenarios. Both the direct flood damage and the cascading effects due to CI failures were evaluated using flood hazards information, building characteristics, depth–damage relationship and the interdependencies among CIs and properties. By combining computational modelling with advanced 3D visualisation techniques, we also developed an effective tool for communicating with stakeholders during the decision-making process. The application can help local stakeholders and operators in the co-design of the approach, the assessment and the evaluation of adaptation measures.

ACKNOWLEDGEMENTS

The work presented in this paper was partially funded by the ongoing EC H2020 EU-CIRCLE (GA 653824), the UK NERC FFIR/SINATRA (grant NE/K008765/1), and the UK Royal Academy of Engineering ESPRIT (grant UUFRIP/100024) projects. The development of CADDIES model was funded by the UK Engineering and Physical Sciences Research Council, grant EP/H015736/1 (Simplified Dual-Drainage Modelling for Flood Risk Assessment in Urban Areas). The impact assessment tools were built upon the outcomes from the EU FP7 PEARL (GA 603663) and CORFU (GA 244047) projects. The authors would like to thank the Environment Agency, Ordnance Survey (GB), and Torbay Council for the provision of data. The support of Infoworks ICM license from Innovyze is also appreciated. The authors would also like to thank the Environment Agency, Ordnance Survey (GB), and Torbay Council for the provision of data.

REFERENCES

Alderman, K., Turner, L. R. & Tong, S. 2012 Floods and human health: a systematic review. *Environmental International* **47**, 37–47.

Andreoni, V. & Miola, A. 2014 *Climate Vulnerability of the Supply-Chain: Literature and Methodological Review*. EU Commission JRC, Luxembourg.

Apel, H., Theiken, A. H., Merz, B. & Blöschi, G. 2006 A probabilistic modelling system for assessing flood risks. *Natural Hazards* **38** (1–2), 79–100.

Araya-Munoz, D., Metzger, M. J., Stuart, N., Wilson, A. M. W. & Carvajal, D. 2017 A spatial fuzzy logic approach to urban multi-hazard impact assessment in Concepción. *Science of the Total Environment* **576**, 508–519.

Boin, A. & McConnell, A. 2007 Preparing for critical infrastructure breakdowns: the limits of crisis management and the need for resilience. *Journal of Contingencies Crisis Management* **15**, 50–59.

C. f. E. & Hydrology 2013 *Flood Estimation Handbook*. C. f. E. & Hydrology, Wallingford, UK.

Chen, A. S., Hammond, J. M., Djordjević, S., Khan, M. D. & Veerbeek, W. 2016 From hazard to impact: flood damage assessment tools for mega cities. *Natural Hazards* **82** (2), 857–890.

Chen, A. S., Khoury, M., Vamvakeridou-Lyroudia, L., Stewart, D., Wood, M., Savic, D. & Djordjevic, S. 2018 3D visualisation tool for improving the resilience to urban and coastal flooding in Torbay, UK. *Procedia Engineering* **212**, 809–815.

E.A. UK 2013 *What is the Updated Flood Map for Surface Water?* Environment Agency, Bristol, UK.

Engle, N. L., Bermond, A. d., Malone, E. L. & Moss, R. H. 2014 Towards a resilience indicator framework for making climate-change adaptation decisions. *Mitigation and Adaption Strategies for Global Change* **19** (8), 1295–1312.

Galbusera, L., Giannopoulos, G. & Ward, D. 2014 Developing stress tests to improve the resilience of critical infrastructures: a feasibility analysis. European commission JRC91129 EUR 26971 EN.

Gibson, M. J., Chen, A. S., Khoury, M., Vamvakeridou-Lyroudia, L., Stewart, D., Wood, M. & Savić, D. 2018 Analysing the cascading effects on critical infrastructure in Torbay coastal/pluvial flooding with climate change. In: *HIC 2018. 13th International Conference on Hydroinformatics*, Palermo, Italy.

Guidolin, M., Duncan, A., Ghimire, B., Gibson, M. J., Keedwell, E. C., Chen, A. S., Djordjević, S. & Savić, D. A. 2012 *CADDIES: A new Framework for Rapid Development of Parallel Cellular Automata Algorithms for Flood Simulations*. IWA, Hamburg, Germany.

Guidolin, M., Chen, A. S., Ghimire, B., Keedwell, E. C., Djordjević, S. & Savić, D. 2016 A weighted cellular automata 2D inundation model for rapid flood analysis. *Environmental Modelling and Software* **84**, 378–394.

Hall, J. W., Dawson, R. J., Sayers, P. B., Rosu, C., Chatterton, J. B. & Deakin, R. 2003 A methodology for national-scale flood risk assessment. *Proceedings of the Institution of Civil Engineers – Water and Maritime Engineering* **156** (3), 235–248.

Hammond, M. J., Chen, A. S., Batica, J., Butler, D., Djordjević, S., Gourbesville, P., Manojlović, N., Mark, O. & Veerbeek, W. 2018 A new flood risk assessment framework for evaluating the effectiveness of policies to improve urban flood resilience. *Urban Water Journal* **15** (5), 427–436.

Hokstad, P., Utne, I. B. & Vatn, J. 2012 *Risk and Interdependencies in Critical Infrastructures*. Springer, London.

Hu, K. 2000 *High-Resolution Finite Volume Methods for Hydraulic Flow Modelling*. Centre for Mathematical Modelling and Flow Analysis, Manchester Metropolitan, Manchester, UK.

H.U. Ltd. 2017 'Torbay coastal Defences – Climate Change Adaptation – Quick Winds Study'.

Khoury, M., Chen, A. S., Gibson, M. J., Vamvakeridou-Lyroudia, L., Steward, D., Wood, M., Savić, D. & Djordjević, S. 2018 A serious game to explore different flooding scenarios and their respective effects on infrastructuers. In: *EPiC Series in Engineering, HIC 2018 13th Internation Conference on Hydroinformatics*, Palermo, Italy.

Kreibich, H., Botto, A., Merz, B. & Schröter, K. 2017 Probabilistic, multivariable flood loss modeling on the mesoscale with BT-FLEMO. *Risk Analysis* **37** (4), 774–787.

Lammers, K. 2013 *Unity Shaders and Effects Cookbook*. Packt Publishing, Birmingham.

Mark, O., Jorgensen, C., Hammond, M. J., Khan, D., Tjener, R., Erichsen, A. & Belwigh, B. 2015 A new methodology for modelling of health risk from urban flooding exemplified by cholera – case Dhaka, Bangladesh. *Flood Risk Management* **11** (s1), 28–42.

Mazzorana, B., Picco, L., Rainato, R., Iroume, A., Ruiz-Villanueva, V., Rojas, C., Valdebenito, G., Iribarren-Anacona, P. & Melnick, D. 2019 Cascading processes in a changing environment: disturbances on fluvial ecosystems in Chile and implications. *Science of the Total Environment* **655**, 1089–1103.

Penning-Rowsell, E., Johnson, C., Tunstall, S., Tapsell, S., Morris, J., Chatteron, J. & Green, C. 2005 *The Benefits of Flood and Coastal Risk Management: A Handbook of Assessment Techniques*. Flood Hazard Research Centre, Middlesex University, London, UK.

Priestnall, G., Jaafar, J. & Duncan, A. 2000 Extracting urban features from LiDAR digital surface models. *Computer, Environment and Urban Systems* **24** (2), 65–78.

Sfetsos, A., Vamvakeridou-Lyroudia, L., Chen, A. S., Khoury, M., Savic, D., Djordjevic, S., Eftychidis, G., Leventakis, G., Gkotsis, I. & Karavokrros, G. 2017 Enhancing the resilience of interconnected critical infrastructures to climate hazards. In: *15th International Conference on Environmental Science and Technology*, Rhodes, Greece.

Sfetsos, T., Katopodis, T., Varela, V. & Karozis, S. 2018 EU-CIRCLE methodological approach for assessing the resilience of the interconnected critical infrastructures of the virtual city scenario to climate change. *Energetika* **64** (1), 23–31.

Svensson, C. & Jones, D. A. 2005 *Joint Probability: Dependence Between Extreme Sea Surge, River Flow and Precipitation: A Study in South and West Britain*. Defra Flood Management Division/Environment Agency, London, UK.

UK Environment Agency 2016 Flood risk assessments: climate change allowances.

Vojinovic, Z. 2017 A toolkit for holistic/multiple risk and impact/damage assessment at strategic and operational level (No. D3.3) PEARL project.

First received 31 January 2019; accepted in revised form 14 October 2019. Available online 29 November 2019

Optimising wastewater treatment solutions for the removal of contaminants of emerging concern (CECs): a case study for application in India

Zara Visanji IWA, Seyed M. K. Sadr, Matthew B. Johns, Dragan Savic IWA and Fayyaz A. Memon

ABSTRACT

The aim of this study was to produce optimal wastewater treatment solutions to calculate the removal of different contaminants of emerging concern (CECs) found in developing countries. A new methodology was developed for a decision support tool (WaStewater Decision support OptiMiser, WiSDOM), which focuses on producing treatment solutions suited to treating water for reuse to Indian Water Quality Standards. WiSDOM-CEC analyses the removal of CECs through different treatment solutions and was also used to evaluate the performance of each treatment train solution in terms of the removal of conventional pollutants using multi-objective optimisation and multi-criteria decision analysis. Information was collected on different CECs across different regions of India, and the removal of 18 different CECs through 42 wastewater treatment unit processes for five different regions of India was analysed. Comparisons between similar categories of CECs, such as non-steroidal anti-inflammatory, showed that emerging contaminants all react differently with individual treatment options. For example, the removal of ibuprofen (IBP) and naproxen (NPX) varied from >80% and 0%, respectively, for a solution in Karnataka involving sedimentation, submerged aerated filter, ultrafiltration and nanofiltration. In Tamil Nadu, results ranged from 36.8% to 72% for diclofenac, 10.7% to 66.5% for IBP, and 0% for NPX.

Key words | contaminants of emerging concern, decision support tool, India, water quality, WiSDoM

Zara Visanji IWA (corresponding author)
Seyed M. K. Sadr
Matthew B. Johns
Dragan Savic IWA
Fayyaz A. Memon
Centre for Water Systems, College of Engineering, Mathematics and Physical Sciences,
University of Exeter,
Harrison Building, North Park Road, Exeter, Devon EX4 4QF,
UK
E-mail: *zv208@exeter.ac.uk*

Dragan Savic
KWR Water Cycle Research Institute,
Nieuwegein,
Netherlands

INTRODUCTION

Contaminants of emerging concern (CECs) have been detected since 1965 (Stumm-Zollinger & Fair 1965), when the first steroid hormones were revealed in the aquatic environment. Since this time, there have been advances in technology leading to the improved detection and analysis of increased quantities of CECs at levels ranging from microgram per litre (μg/L) to nanogram per litre (ng/L) using a range of techniques (Calvo-Flores *et al.* 2018). CECs are frequently defined as naturally occurring, synthetic or anthropogenic chemicals/substances, which are not regularly monitored (Visanji *et al.* 2018). These substances also have a negative impact on the environment and human health; however, further toxicological studies are required to determine the full toxicity implications of these substances. The main categories of CECs commonly reported in the literature are: (i) pharmaceuticals, (ii) personal care products, and (iii) endocrine disruptors (Fischer *et al.* 2017; Montes-Grajales *et al.* 2017; Tran *et al.* 2018). More recently, an increasing number of publications have also included other categories such as steroid hormones, surfactants (i.e., nonylphenol and octylphenol), perfluorinated compounds (i.e., perfluorooctane sulfonate and perfluorooctanesulfonic acid), flame retardants (i.e., polybrominated biphenyl ethers), industrial additives and

doi: 10.2166/hydro.2019.031

agents, illicit drugs, UV filters, and artificial sweeteners (Baalbaki *et al.* 2017).

The dominant pathway that allows CECs to enter the aquatic environment is via the effluent of wastewater treatment plants, with other pathways including veterinary locations and surface runoff from agricultural areas (Lapworth *et al.* 2012). Human bodies are unable to fully metabolise pharmaceutical compounds causing CECs to be excreted via urine and faeces (Anumol *et al.* 2016). Unfortunately, wastewater treatment plants were not originally designed to remove these compounds (Tran *et al.* 2018). Therefore, CECs can pass through as parent compounds or metabolites (a transformed product) and continuously enter the environment. In some locations, water is abstracted downstream of wastewater treatment plants and processed through a water treatment plant for drinking water. These water treatment plants are also unable to efficiently remove CECs allowing them to contaminate drinking water. Therefore, it is important that the removal of CECs through existing and new technologies is thoroughly explored. The design of effective treatment solutions requires such analysis to further reduce the concentration circulating through the environment and humans.

Environmental regulations for CECs are limited to developed countries, with only a few monitoring processes put in place for some developing countries. Unlike developing countries, the European Union (EU) has implemented a watch list (EU 2015/495) on priority substances classed as CECs (European Union 2015). The EU has also introduced and applied legislations such the Regulation of Registration, Authorisation and Restriction of Chemicals (REACH), which focuses on monitoring chemical substances when manufactured or imported (REACH 2016). Research programmes in the UK have also been carried out to better understand and monitor the problem at hand, including the Chemicals Investigation Programme (ALS Environmental 2015) and the National Implementation Plan (DEFRA 2017). The USA have produced the Contaminants Candidate List which focuses on unregulated drinking water contaminants (US EPA 2016).

Developing countries such as South Africa, Brazil, China, and India currently have no official legislation surrounding CECs; however, an increase in publications and research has allowed for a clearer indication of the problem in these countries. In Brazil, around 50% of wastewater does not pass through a treatment facility (Machado *et al.* 2016) further reducing the chances of CECs being removed from the effluent due to less treatment occurring in the first place. In the past 19 years in Brazil, levels of CECs have reached the following concentrations: 6,806 ng/L for hormones, 20,960 ng/L for acetyl salicylic acid concentrations in rivers, 14,955 ng/L for caffeine, up to 785 ng/L for diclofenac (DCF), 5,896 ng/L for cocaine and for paracetamol concentrations have exceeded 30,000 ng/L (Starling *et al.* 2018).

India is currently one of the top pharmaceutical emerging markets in the world, and one of the largest global providers of medicines (drugs) accounting for 20% of global exports. Proper wastewater management techniques do not exist in India, and conventional treatment plants are inefficient at the complete removal of CECs, with sewage treatment plants discharging their effluent to rivers. The Bureau of Indian Standards is not currently addressing CECs and not all regions of India have carried out investigations on CECs (Philip *et al.* 2018); therefore, it has become essential for the creation of baseline data to act as a framework for any future research or regulatory initiatives. The latest reports show that sewage generated from towns and cities in India is not being fully treated, with only 11,787 million litres per day (MLD) receiving treatment from 38,255 MLD (Philip *et al.* 2018). The concentration of certain CECs such as antibiotics are seen to be 40 times higher in Indian wastewater treatment plants than other countries (Balakrishna *et al.* 2017), suggesting that the effluent from wastewater treatment plants is contaminated with CECs even though treatment has occurred.

Balakrishna *et al.* (2017) explored the occurrence of pharmaceuticals and personal care products in the influent and effluent of wastewater treatment plants in India and detected 73 different CECs. High concentrations of artificial sweeteners were seen in both the influent and the effluent ranging from 143,000 (ng/L) to 389,000 ng/L and 6,020 to 379,000 ng/L, respectively. Recorded caffeine concentrations were also high ranging from 19 ng/L to 102,840 ng/L (Balakrishna *et al.* 2017), whereas levels in the European Union were seen to reach 3,000 ng/L (Loos *et al.* 2012). A more recent review carried out by Philip *et al.* (2018) documents the extensive occurrence of CECs throughout the different regions of India with the detection of 166 different CECs belonging to 36 main categories. The highest recorded concentration in surface waters of India is noted as ciprofloxacin, reaching 14,000,000 ng/L. In Northern India, lower levels of antibiotics were found in rivers during the monsoon season; however, a study by Mohapatra *et al.* (2016) found high levels of antibiotics during the monsoon season in comparison to winter and summer. High levels of antibiotics were also recorded during the monsoon season in the South of India, along with pharmaceuticals due to an increased use during this time (Philip *et al.* 2018). Caffeine values reaching <150,000 ng/L were seen during summer months, with concentrations in winter reaching <40,000 ng/L (Mohapatra *et al.* 2016).

With the above understanding of the problem of CECs and the need for proper wastewater management in less developed countries, this paper aims to produce optimal wastewater treatment solutions for developing countries. The selection of wastewater treatment solutions involves intricate decision-making, incorporating all elements of treatment (from preliminary options through tertiary treatment and disinfection), to ensure a suitable relationship exists between the biological, physical, and chemical processes needed to treat wastewater to the required levels (Poch *et al.* 2014). Sadr *et al.* (2018) proposes that for developing countries, the selection process becomes even more complex as additional considerations are needed due to a variety of socio-economic and environmental factors that exist in these countries. WiSDOM (WaStewater Decision support OptiMiser), which is a decision support tool, was created to include these additional parameters (capital, operational and maintenance costs, energy consumption, chemical requirement, land requirement, and reliability). A range of decision objectives and criteria have been incorporated into the tool as a process of formulating wastewater treatment solutions (Sadr *et al.* 2018). This will allow stakeholders and decision makers to implement sufficient wastewater treatment solutions for proper wastewater management to treat wastewater to desirable levels. This paper describes the development and application of a new methodology WiSDOM-CEC, which was encapsulated as a software programme (as an add-in for a stand-alone user-friendly decision support tool – WiSDOM) used to calculate the removal rates of CECs during different wastewater treatment processes. India is used as a case study, with scenarios developed to demonstrate WiSDOM-CEC's application when combined with an existing decision support tool. The scenarios focused on five different regions of India (North, South Western, South, North Eastern, and Central) to analyse the removal efficiencies for antibiotics, nonsteroidal anti-inflammatory drugs, and other categories of CECs (hormones, stimulants, personal care products, and insect repellents).

METHODS

The main aim of this study was to analyse the performance of different treatment unit trains to determine their efficiency at removing CECs. This section describes the methodology for WiSDOM-CEC employed to analyse the removal of CECs from different treatment processes generated in India, using an existing decision support tool, WiSDOM. The tool currently does not incorporate the removal of CECs; therefore, a new software programme (Excel Spreadsheet Programme (ESP)) was designed to work alongside WiSDOM. The ESP calculates the removal of CECs once treatment solutions were populated through WiSDOM depending on specific user requirements and objectives. The methodology for the ESP was originally developed to calculate the removal of 39 CECs for 42 WiSDOM-generated wastewater treatment trains (Visanji *et al.* 2018). However, this paper focuses on the removal of CECs commonly recorded within India. Due to the limited published data on the removal of CECs in India, removal rates for different treatment unit processes were taken from publicly available sources to ensure a complete dataset.

WiSDOM: a decision support tool

WiSDOM is a user-friendly tool designed to aid in the formulation of wastewater treatment trains for the removal of conventional pollutants in different contexts. WiSDOM evaluates the performance of each solution with respect to different objectives. The tool also calculates the removal of the following conventional pollutants by each generated treatment train: (i) chemical oxygen demand (COD), (ii) biochemical oxygen demand (BOD), (iii) total suspended solids, (iv) total nitrogen (TN), (v) total phosphorus, (vi) faecal coliform (FC), (vii) turbidity, (viii) intestinal nematode eggs (INEs), and (ix) *Escherichia coli*.

WiSDOM was chosen to generate treatment solutions as it determines the optimal treatment options considering sustainability indicators and ensuring that the removal of conventional pollutants meets the different water reuse standards in India (Table 1). At the core of the software is a technology library that contains detailed information on a wide range of wastewater treatment processes applicable within the context of India (Sadr *et al.* 2018). The tool uses the technology library and a multi-objective optimisation (MOO) algorithm to generate optimal wastewater treatment trains which are then processed by a multi-criteria decision analysis (MCDA) technique to narrow down the resultant non-dominated solution set. The user is given two choices of MOO algorithm; the Non-Dominated Sorting Genetic Algorithm II (NSGA-II) (Deb *et al.* 2002) and the Omni-optimizer (Omni) (Deb & Tiwari 2008). The two algorithms have shown to handle the vagaries of practical optimisation problems well and prove suited to the formulation of wastewater treatment trains. The user has full control over the formulation of the problem; from defining which objectives are being considered for optimisation to the hydraulic, water quality, and design constraints.

The available optimisation objectives are as follows: (1) Capital Expenditure (CAPEX), (2) Operational and Maintenance Expenditure (OPEX), (3) energy consumption, (4) sludge production, (5) land requirement, and (6) labour requirement, all of which are minimised by the optimisation process. Following the MOO, compromise programming (CP), a MCDA technique is used to assess the solutions according to user-defined weighting of various criteria spanning a range of design aspects including technical, environmental, and social and economic considerations (Sadr *et al.* 2018). CP calculates the distance each solution is from the theoretical ideal. The solutions are then ranked according to this 'comprise distance' and displayed to the user. The ESP (described further below) was created as an add-in to WiSDOM, using the results from the tool depending on the scenario or the context defined by the user. For the purpose of this study, the MOO objectives, parameters, and MCDA criteria weight settings were set to their relevant default settings (Sadr *et al.* 2018). For each set of data, NSGA-II is applied with a termination criterion of 12,000 fitness evaluations, population size of 75, a cross-over rate of 0.85 and a mutation rate of 0.15. These parameter values were obtained through experimentation to ensure near optimal algorithm performance on the problems in question (Sadr *et al.* 2018).

It should be noted that although separate regions of India were chosen for the tool's application, assumptions were made regarding the input factors (found in Table S2 in the Supplementary Information) to ensure that the results focused on the removal of CECs:

(1) For the scenario, the genetic algorithm objectives, parameters, and MCDA criteria weight were set to either urban or rural settings. This was dependent on where most of the population lived. For example, in Delhi, 97.5% of the population live in urban regions, whereas in Karnataka 61.3% live in rural areas (Census 2011). Therefore, Delhi would be set to the urban default settings and Karnataka as rural.

Table 1 | Indian water quality standards for reuse application (Sadr *et al.* 2017)

Contaminants	Toilet flushing	Vehicle exterior washing	Horticulture	Non-edible crops	Edible crops — raw	Edible crops — cooked
COD (mg/l)	16	16	16	30	16	30
BOD (mg/l)	10	10	10	20	10	20
Suspended solids (mg/l)	40	35	40	30	25	30
TN (mg/l)	10	10	10	10	10	10
Phosphorus (mg/l)	1	1	2	5	2	5
FC (no./100 ml)	0	0	0	230	0	0
Turbidity (NTU)	<2	<2	<2	<2	<2	<2
INEs (no./100 ml)	<2	<2	<2	<1	<1	<1

(2) The treatment train solution was set to include the following number of each unit process: 1 × preliminary treatment, 1 × primary treatment, 2 × secondary treatment, 2 × tertiary treatment and 1 × disinfection.
(3) Due to limited data available, the influent concentrations (found in Table S1 – 1, S1 – 2 and S1 – 3, Supplementary Information) for the wastewater influent conventional contaminants were all set to WiSDOM's default values.
(4) Certain unit processes (listed below under the ESP) were removed from WiSDOM as a potential solution due to the lack of data on CECs, therefore, eliminating them as a treatment train solution.

It is important to note that WiSDOM is used as a wastewater reuse tool, whereby treatment solutions that are provided are aimed at treating the water to a higher water quality standard than regular wastewater treatment plants. For example, urban areas were set to be reused for toilet flushing, whereas rural areas had an intended reuse application for non-edible crops, therefore resulting in the level of treatment being higher than if the end use was aimed at discharge into a river. Incorporating a water reuse function allows the treatment trains to provide a higher water quality as the effluent. However, in some cases, the unit processes selected by the tool, such as reverse osmosis, are not always the most efficient process to treat CECs, especially in the conditions of rural India, where some of the suitable treatment options of advanced oxidation are not a practical option for the surrounding conditions.

More details on WiSDOM can be found in Sadr *et al.* (2018). (All work with WiSDOM has been built upon original research carried out by Joksimovic *et al.* (2008, 2006).)

New methodology: WiSDOM-CEC

WiSDOM-CEC was formed from a new methodology encapsulated as a software programme, the ESP. The ESP used to determine the removal of CECs from different treatment unit processes, previously generated in WiSDOM, and was created from three different separate worksheets. Removal efficiencies were researched for each treatment option taken from the WiSDOM tool. Three worksheets were combined using functions and formulas to allow for a user-friendly software programme; this is explained further in Visanji *et al.* (2018). A database was created containing over 500 recorded CECs with data present from many countries. The database was used to gather information on 39 CECs; however, only 17 were chosen for the final study. The database included abbreviations of CECs, their chemical abstract service number, and recorded minimum and maximum concentrations from surface water, groundwater, untreated wastewater, drinking water, and treated wastewater.

To produce WiSDOM-CEC, different components and datasets were pulled from other tables/spreadsheets and populated into a singular user-friendly ESP. The names of the CECs from Spreadsheet B along with minimum and maximum initial concentration values (ng/L) were inputted into the first four columns. The top three rows consisted of the stage of treatment, including drop-down options allowing the user to select the ID number for a unit process (from Spreadsheet A). The ESP matches the treatment unit processes used within WiSDOM; therefore, the ESP contained the same stages and unit processes' ID numbers. The drop-down option for the unit processes was taken from Spreadsheet A, as the ID number is changed the treatment changes to the name corresponding process from Spreadsheet C. This allows the ESP to take the relevant removal percentage from Spreadsheet C of each individual CEC, depending on the treatment process chosen. The main section of the ESP involved an equation (Equation (1)), which calculated the removal of CECs throughout different unit processes which have been selected. Equation (1) was used across the ESP which takes the removal rates associated with a unit process from Spreadsheet C and calculates the new concentration (ng/L) after that treatment unit. If the concentration after a treatment stage reaches the desired level (defined by the tool user), then the words 'No Further Treatment' will appear, showing the end user where the CEC was fully removed. The inbuilt 'IF' function in Excel is used to change the information regarding removal rates (from Spreadsheet C), depending on the unit process ID number selected by the user within the drop-down options on the ESP.

$$Y_m = I_m \times \prod_{k=1}^{S} (1 - R_{m,u,k})$$

where m is the contaminant ID; k the stage of treatment; S the maximum number of stages considered in the proposed treatment train; I_m the influent quality with respect to concentration (ng/L) of m; Y_m the effluent quality with respect to concentration (ng/L) of m and $R_{m,u,k}$ the contaminant removal rate (%) of the unit process u in treatment stage k.

An example demonstrating the components of the ESP and Equation (1) are demonstrated below for a treatment train with four stages (see Equations (2) and (3)) for removal of DCF. As shown in Figure 1, in the first stage, a grit chamber process is selected (used as a preliminary/primary treatment). This process, however, does not remove any of the DCF in the wastewater. A membrane bioreactor (MBR) was applied in the second stage; this unit process can remove around 40% of DCF (Luo *et al.* 2014). The MBR is followed by a nanofiltration (NF) and an ultraviolet (UV) process with DCF removal rates of 60% and 40%, respectively. The initial concentration of DCF was 9,520 ng/L and after going through all the four stages is 1,370 ng/L, which means that the total DCF removal efficiency of this treatment train is 85%.

$$Y_{\mathrm{DCF}} = I_m \times \prod_{k=1}^{S=4} 1 - R_{m,u,k}$$

$$Y_{\mathrm{DCF}} = 9,520\,\mathrm{ng/L} \times [(1-0)\times(1-0.4)\times(1-0.6)\times(1-0.4)]$$
$$\rightarrow Y_{\mathrm{m}} = 1,370\,\mathrm{ng/L}$$

WiSDOM-CEC considered a range of assumptions to allow for a more complete dataset of removal rates for different treatment processes. Previous research (Luo *et al.* 2014; Petrie *et al.* 2015; Tran *et al.* 2018) focuses on the overall removal rate of CECs through different treatment trains and does not focus on individual treatment unit processes. To date and to the best of the authors' knowledge, information surrounding the breakdown of the removal efficiencies within the effluent at different stages of treatment is unavailable. In addition, insufficient data exist for each emerging contaminant and each unit process which has been chosen. Consequently, where no data were found for an individual treatment process, a removal rate of 0% was inputted into the cells to produce a complete dataset, allowing for the calculations within the ESP to effectively run. The treatment options set to 0% removal were: bar screen, grit chamber, coarse screen, fine screen, Actiflo, enhanced biological phosphorus removal, P-precipitation and soil aquifer treatment, which are all within the categories of preliminary or primary treatment.

The results from WiSDOM-CEC are displayed in tabular and graphical format. A bar chart of the CECs final concentration in the effluent for both minimum and maximum removal rates is displayed. Results are shown for the final concentration after each treatment train and at the end of each individual unit process. For this study, maximum influent concentrations are used, and graphs are displayed for both minimum removal and maximum removal. This is due to the ESP having two removal settings that allow for

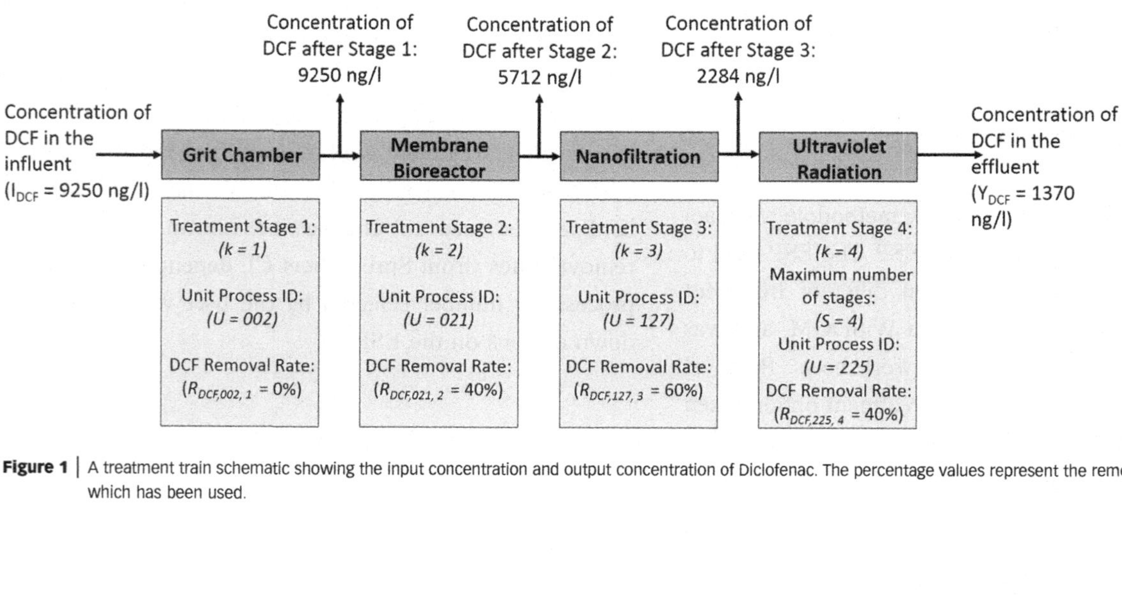

Figure 1 | A treatment train schematic showing the input concentration and output concentration of Diclofenac. The percentage values represent the removal rates for the unit process which has been used.

different removals to be calculated as both data on minimum and maximum removal rates are present. For example, Diazinon is able to be removed from 75% to 100% when advanced oxidation is used (Samadi *et al.* 2007); therefore, the results would show the removal for both these rates.

Scenario development: India

For this study, an initial review was carried out on existing literature published within India, to find the most commonly recorded CECs (see Supplementary Information, Table S1). Following these, 17 CECs that were present within the ESP were chosen to further explore their removal efficiencies within wastewater treatment plants. The CECs chosen in the study can be seen in Table 2 and consisted of pharmaceuticals, a hormone, a stimulant, personal care products, and an insect repellent.

The scenario used to test the application of the tool and ESP looked at investigating the different regions of India and wastewater treatment technologies that would be suitable for removing the concentrations of CECs found in the North, North East, South, South West, and Central India. This was done by focusing on the removal efficiencies for CECs with similar properties categorised by different drug classes.

a. To compare and find optimal treatment solutions for the removal of different antibiotic classes (penicillin, fluoroquinolone, macrolide, quinolone, sulphonamides and trimethoprim).

Table 2 | Concentrations of CECs recorded in different regions of India (the data in this table are adapted from Table S2 and can be found in the Supplementary Information, including all references)

Region	North		South Western	South	North Eastern	Central
Location	Utter Pradesh	Delhi	Karnataka	Tamil Nadu	Bihar	Nagpur
Emerging contaminant	Concentration (ng/L)					
Pharmaceuticals						
Ampicillin	–	104.2	–	–	–	–
Atenolol	–	–	192–2,900	110–2,440	1.3–1,010	–
Carbamazepine	4.49–650	–	1–726	75–840	82–270	–
Ciprofloxacin	4.8–16	20.06	–	–	7–28.8	12,900
Diclofenac	1.68–360	–	15–412	170–540	1.41–2.55	–
Erythromycin	–	–	–	12	–	–
Ibuprofen	1,430–2,380	–	17–4,460	2,140	686–1,130	–
Naproxen	1.67–120	–	14–235	23–510	2.4–2.62	–
Ofloxacin	–	–	–	212	–	–
Sulfamethoxazole	10.7–27.5	–	5–2,260	3–480	11–17.5	–
Triclosan	5.4	–	892–2,440	2,500	145–450	4,890
Trimethoprim	210–4,010	–	21–180	3–240	33–90	–
Hormones						
Estrone	–	–	10–124	–	–	–
Stimulants						
Caffeine	17.5–40	–	14–61,000	41,000–220,000	16–743	102,840
Personal care products						
Benzophenone	–	–	–	–	–	3,960
Bisphenol A	–	–	59–299	–	–	–
Insect repellent						
N,N-diethyl-meta-toluamide	6.4–22.3	–	15–388	9.1–50	4–4.5	–

b. To compare and find optimal treatment solutions for the removal of nonsteroidal anti-inflammatory drugs (NSAIDs) (DCF, ibuprofen (IBP), and naproxen (NPX)).
c. To compare and find optimal treatment solutions for the removal of the other categories of CECs (hormones, stimulants, personal care products and insect repellents).

RESULTS AND DISCUSSION

Results from the optimisation (in WiSDOM) of the objectives such as CAPEX, OPEX, and land are displayed as a radar plot where the higher the point on the axis the greater the performance of the solution. However, results for the removal of conventional pollutants work in a reverse effect where the higher the point on the axis the worse the solution has performed, as this graph compares the performance of each solution in regard to contaminant removal. However, it should be noted that all treatment solutions have met the required Indian Water Quality Standards and the performance noted is a comparison to the other solutions. The treatment trains outputted in the WiSDOM tool were then processed through the ESP to determine their capability of removing CECs. Data from the ESP showing the removal of CECs are presented as bar charts. The below results display the solutions across different regions in India, with sections also focusing on the removal across the categories of CECs (antibiotics, NSAIDs, hormones, stimulants, personal care products, and insect repellents). Results are displayed for Uttar Pradesh, Delhi, Karnataka, and Tamil Nadu in the main text, with the results for Bihar and Nagpur presented in S3 in the Supplementary Information.

Uttar Pradesh

Results for Uttar Pradesh from the WiSDOM tool populated six optimised treatment solutions which fit the original constraints entered into the tool (found in Table S1, Supplementary Information). Table 3 displays the treatment train selected for each solution generated.

Solution (S) 6 outperformed the other solutions at removing conventional pollutants, and in regard to the sustainability indicators performed poorly for land, and performed highly for CAPEX and OPEX (Figure 2). S5 performed well for the conventional pollutants and like S6 performed well for CAPEX and OPEX. S4 performed in the middle for conventional pollutants, and still had low rates in comparison to S1, S2, and S3. S3 performed the worst in regard to CAPEX and OPEX, and ranked second best in regard to land requirements. S2 had a similar pattern to S3, whereas S4 performed highly in all the categories. Regarding the MOO objectives, the solutions that met most of the requirements were S4 if the focus was on sustainability indicators, or S6 if the focus was on removal of conventional pollutants.

Table 3 | Wastewater treatment solutions taken from the WiSDOM tool for Uttar Pradesh

Solution number	Treatment train
1	Constructed wetlands → advanced oxidation – UV/O_3
2	Constructed wetlands → advanced oxidation – UV/H_2O_2
3	Constructed wetlands polishing → chlorine dioxide
4	Constructed wetlands polishing → ultrafiltration
5	Coarse screen → fine screen → sedimentation → trickling filter + secondary sedimentation → ultrafiltration
6	Grit chamber → dissolved air flotation (DAF) → conventional activated sludge process (CASP) + secondary sedimentation → ultrafiltration

The use of constructed wetlands (CWs) with advanced oxidation techniques was able to remove most of the CECs apart from Ciprofloxacin (CIP) which was left with a 0% removal. S1 was able to remove IBP and Sulfamethoxazole (SMZ) to 100%, and DCF, Triclosan (TCS), caffeine (CAF), and N,N-Diethyl-m-toluamide (DEET) to greater than 90% removal. The use of advanced oxidation H_2O_2 in S2 allowed the removal of DCF to increase from 95% in S1 to 100% removal in S2. However, no other CECs were affected by the change in advanced oxidation. When chlorine dioxide was incorporated in S3, the removal of TCS dropped to 0% removal, showing that there is no impact from CWs. The removal rates of CBZ, CAF, and DEET also dropped to 40%, 59% and 55%, suggesting that a higher quantity of removal was processed at advanced oxidation. S4 showed the lowest removal rates, with four of the CECs having a 0% removal

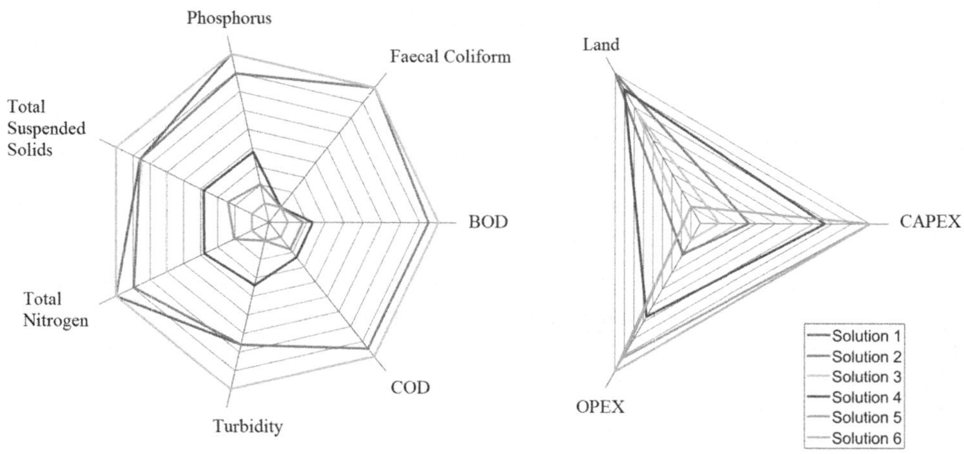

Figure 2 | Performance results from WiSDOM. (Left) Radar plot from the MOO optimisation for the results for the performance of the removal of conventional pollutants for all solutions. The higher the point on the axis the worse the solution has performed. (Right) Radar plot from WiSDOM from the MOO optimisation for the results for the performance of different sustainability indicator objectives (Land, OPEX, and CAPEX). The higher the point on the axis the better the solution has performed.

(Figure 3). S5 was unable to remove NPX and minimal removal rates dropped to 0%; however, TCS and CIP were removed to 97% and 83%, respectively. Lastly, S6 was the only solution able to remove all the CECs; however, the maximum removal rates which reached 100% removal for CIP, DCF, IBP, and SMZ in S3 dropped to 92.3%, 82.5%, 66.5%, and 60%, respectively. Overall, the best solution to remove the majority of the CECs would be S2 or S6; however, S6 was also able to outperform S2 when removing conventional pollutants.

Delhi

Results for Delhi from the WiSDOM tool populated six optimised treatment solutions which fit the original constraints entered into the tool (found in Table S1, Supplementary Information). Table 4 displays the treatment train selected for each solution generated.

Solution (S) 4 performed poorly compared to the other solutions in regard to the sustainability indicators and objectives inputted into WiSDOM (Figure 4), and also performed

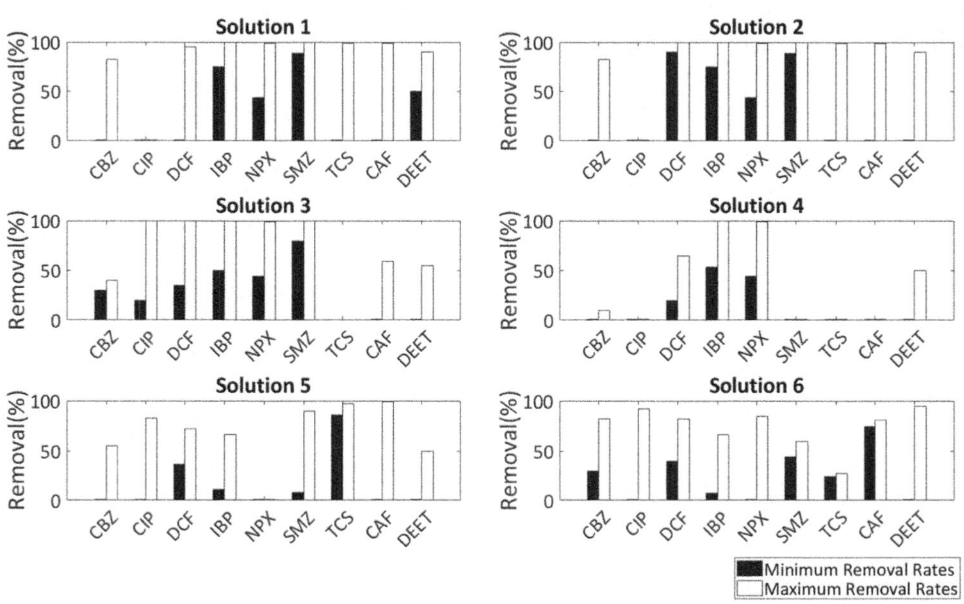

Figure 3 | Percentage of CECs removed from the different treatment solutions produced from WiSDOM-CEC for Uttar Pradesh.

Table 4 | Wastewater treatment solutions taken from the WiSDOM tool for Delhi

Solution number	Treatment train
1	Bar screen → fine screen → CASP → powdered activated carbon (PAC) → reverse osmosis
2	Bar screen → fine screen → CASP → microfiltration → reverse osmosis
3	Sedimentation → trickling filter + secondary sedimentation → PAC → reverse osmosis
4	Bar screen → fine screen → CASP → ultrafiltration
5	Grit chamber → dissolved air flotation (DAF) →trickling filter + secondary sedimentation → PAC → nanofiltration
6	Grit chamber → sedimentation → CASP → microfiltration → nanofiltration

the worst in regard to the removal of conventional pollutants compared to the other solutions. S3 outperformed all solutions when looking at the removal of conventional pollutants (Figure 4) and performed highly in regard to CAPEX and OPEX but low in regard to land requirements. S6 was the best solution at meeting land requirements but had similar results to S5 in regard to conventional pollutant removal. S5 was the second worst solution at both MOO objectives (Figure 4). The best solution to remove conventional pollutants was S3, and this solution also was suited to meeting the requirements for CAPEX and OPEX, although it performed poorly in regard to land requirements. Therefore, this would be the best solution if land requirements were not a preference.

When using the above solutions to determine the best solution for the removal of CECs, it is interesting to see similar results due to the unit processes used. S5 was unable to remove either Ampicillin (AMP) or CIP through this treatment train, showing that the use of either a grit chamber, Dissolved Air Flotation (DAF), trickling filter, PAC or NF does not contribute to the removal of these contaminants. S6 was able to remove CIP by 98.7% at maximum removal, whereas S1, S2, and S4 were only able to remove 92.3%, and S3 only reached 83% removal (Figure 5). For all solutions when the ESP's minimal removal rates were considered, there was no removal of CIP leading to a 0% removal rate. For AMP, only S1, S2, and S3 were able to remove this contaminant to 25.9 ng/L at minimal removal rates causing a 75.1% removal, and to 1.97 ng/L when maximum removal rates were considered leading to a 98.1% removal of this contaminant. The best treatment option for the removal of these two CECs would be S1, S2, and S3 as these were able to remove both CECs. However, from Figure 2, we know that S3 was the solution that outperformed the others in regard to the constraints set for Delhi.

Karnataka

Results for Karnataka from the WiSDOM tool populated six optimised treatment solutions which fit the original constraints entered into the tool (found in Table S1, Supplementary Information). Table 5 displays the treatment train selected for each solution generated.

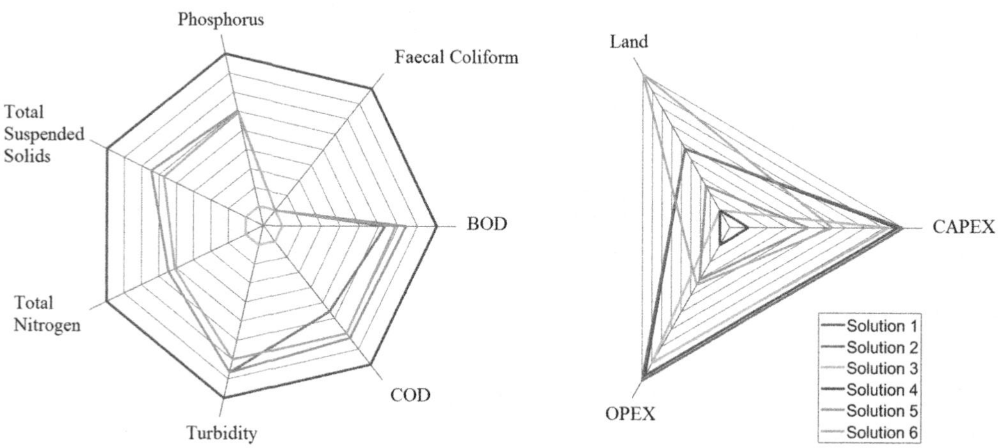

Figure 4 | Performance results from WiSDOM. (Left) Radar plot from the MOO optimisation for the results for the performance of the removal of conventional pollutants for all solutions. The higher the point on the axis the worse the solution has performed. (Right) Radar plot from WiSDOM from the MOO optimisation for the results for the performance of different sustainability indicator objectives (Land, OPEX, and CAPEX). The higher the point on the axis the better the solution has performed.

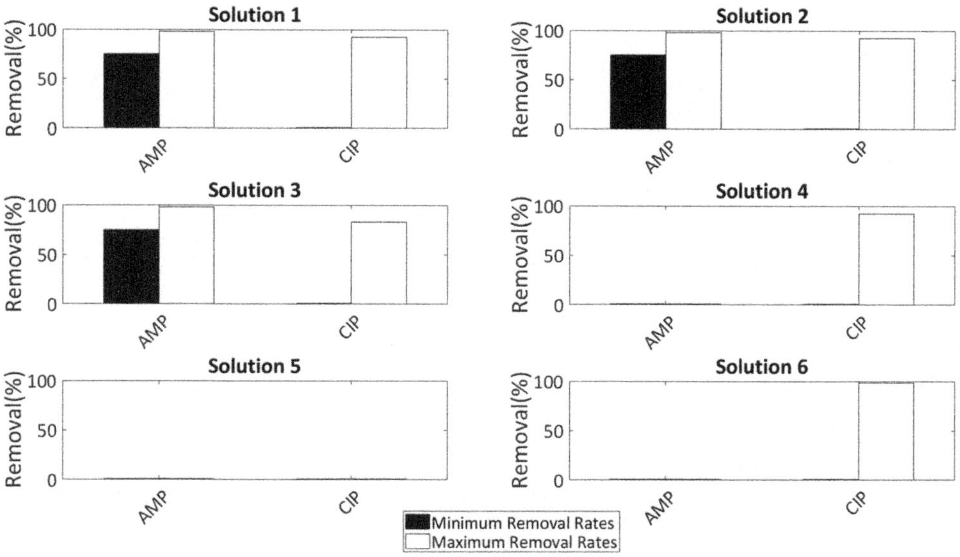

Figure 5 | Percentage of CECs removed from the different treatment solutions produced from WiSDOM-CEC for Delhi.

Table 5 | Wastewater treatment solutions taken from the WiSDOM tool for Karnataka

Solution number	Treatment train
1	Grit chamber → sedimentation → CASP + secondary sediment → ultrafiltration → reverse osmosis
2	Sedimentation → submerged aerated filter → ultrafiltration → nanofiltration
3	Constructed wetlands polishing → chlorine dioxide
4	Constructed wetlands polishing → ultrafiltration
5	Coarse screen → fine screen → sedimentation → trickling filter + secondary sedimentation → microfiltration
6	Grit chamber → DAF → CASP + secondary sedimentation → ultrafiltration

S1 outperformed the other solutions in regard to the removal of conventional pollutants, followed by S2, S6, S4, S5, and finally S3 (Figure 6). However, S3 and S5 were the worst in regard to BOD and COD, whereas all other solutions ranked in the lower scores. On the other hand, regarding the sustainability indicators, S5 performed the worst for all three constraints. S3 and S4 were the best in regard to land requirements; however, their scores for CAPEX and OPEX were also low. S1 and S2 ranked highly for all the MOO objectives. Therefore, the best solution that met all the requirements and was able to remove conventional pollutants was S1.

From Figure 7, it is clear that S1 and S6 were the two solutions which were capable of removing the majority of the CECs found in the influent. The main difference between these two solutions was the addition of reverse osmosis and the use of DAF instead of sedimentation in S6. S1 had higher removal rates for all the CECs and was able to remove CAF, BPA, and DEET to 100%. However, both S1 and S6 were only able to remove TCS from 24% to 26%, unlike S5 was able to remove this contaminant from 86% removal to 97% removal. On the other hand, S5 was unable to remove NPX and had lower removal rates for the hormone Estrone (EST). The solution best suited to the removal of CECs and the removal of conventional pollutants is S1.

Tamil Nadu

Results for Tamil Nadu from the WiSDOM tool populated six optimised treatment solutions which fit the original constraints entered into the tool (found in Table S1, Supplementary Information). Table 6 displays the treatment train selected for each solution generated.

S6 outperformed all the solutions in regard to the removal of conventional pollutants, with S2 having similar results, apart from a slightly higher result for turbidity (Figure 8). However, this was still low in comparison to

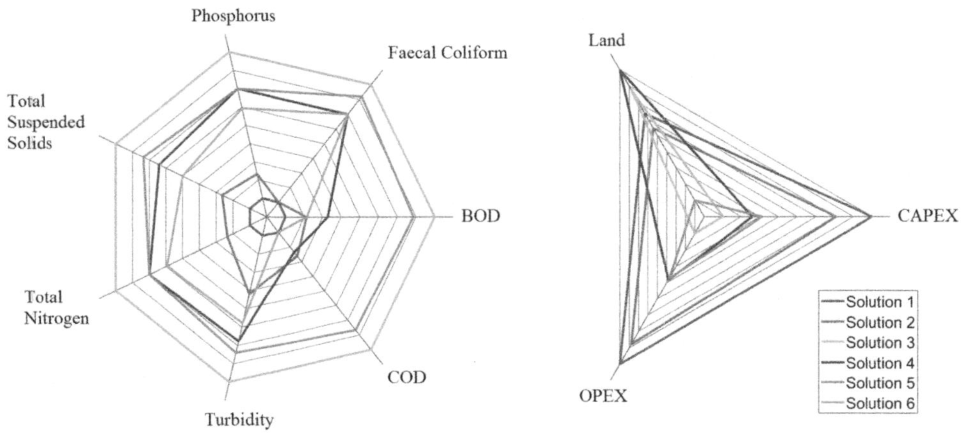

Figure 6 | Performance results from WiSDOM. (Left) Radar plot from the MOO optimisation for the results for the performance of the removal of conventional pollutants for all solutions. The higher the point on the axis the worse the solution has performed. (Right) Radar plot from WiSDOM from the MOO optimisation for the results for the performance of different sustainability indicator objectives (Land, OPEX, and CAPEX). The higher the point on the axis the better the solution has performed.

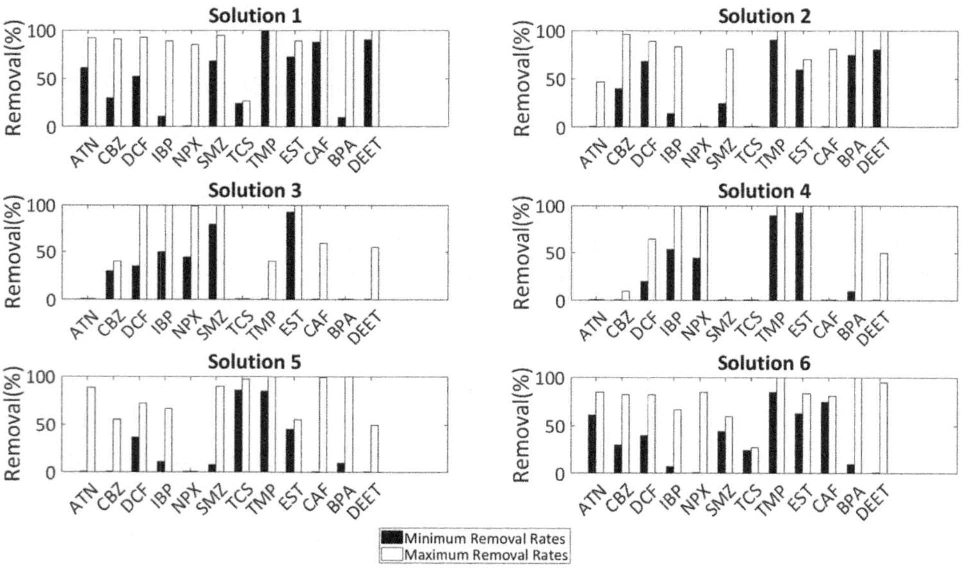

Figure 7 | Percentage of CECs removed from the different treatment solutions produced from WiSDOM-CEC for Karnataka.

S1, S3, and S4. S3 performed the worst having high values for all conventional pollutants, with S1 following close behind. S4 and S5 performed at the higher end of the scale, with S4 being slightly higher for all parameters. Regarding sustainability indicators S2, S5, and S6, all performed highly for CAPEX and OPEX with low-land requirement results. Whereas S3, S1, and S4 had a high-land requirement but low results for CAPEX and OPEX in the order listed. If land requirement is not as important as CAPEX and OPEX, then S4 would be an ideal solution in regard to meeting the removal of conventional pollutants. However, if CAPEX and OPEX are not important, then S6 or S2 would then be the best solution in removing conventional pollutants and meeting the requirements set as constraints in the WiSDOM tool.

Figure 9 displays the removal through each treatment train of each EC which was found in the influent in Tamil Nadu. S1 and S2 were the only solutions able to remove all the CECs; however, S1 had low removal rates for ERY, reaching only 15% removal, whereas S2 was able to

Table 6 | Wastewater treatment solutions taken from the WiSDOM tool for Tamil Nadu

Solution number	Treatment train
1	Constructed wetlands → advanced oxidation – UV/O_3
2	Grit chamber → sedimentation → activated sludge process + secondary sedimentation → microfiltration → nanofiltration
3	Constructed wetlands polishing → chlorine dioxide
4	Constructed wetlands polishing → ultrafiltration
5	Coarse screen → fine screen → sedimentation → trickling filter + secondary sedimentation → ultrafiltration
6	Grit chamber → sedimentation → trickling filter + secondary sedimentation → microfiltration → nanofiltration

remove 72% of this contaminant. S4 was an ideal solution for meeting the requirements set in the WiSDOM tool but was unable to remove ATN, ERY, SMZ, TCS, and CAF. Therefore, S4 is not an ideal solution when it comes to the main objective of removing CECs. S5 and S6 were unable to remove NPX and OPX, and for the other four solutions, removal rates were the same ranging from 44% removal to 99%. S1 would be the ideal solution to remove CECs and still was able to remove conventional pollutants to the required standards (Figure 8). However, the solution that fits all constraints to a better standard and was still able to remove CECs would be S2. This solution was able to remove all CECs and had a lower removal for Ofloxacin (OFC) of 45% and for TCS reaching a maximum removal of 27%.

Removing contaminants of emerging concern

CECs in the same category would be expected to behave in the same way due to having similar properties. However, each individual CEC has distinct physical and chemical properties resulting in a unique response to breaking down in unit processes or reacting to certain treatment options. The following section characterises the two different drug classes of pharmaceuticals (NSAIDs and antibiotics) and other categories explored in this study to determine if CECs in the same category have the same removal rates.

Antibiotics

In this study, both CIP and OFX were not detected in the same location; however, one can look at their removal through different unit processes. Across all sites, CIP was infectively removed through the use of CWs, trickling filters, ultrafiltration, or advanced oxidation, whereas the greatest removal rates were achieved for OFX. The addition of chlorine dioxide as a tertiary treatment removal saw close to 100% removal. Greater than 50% removal was also seen when primary and secondary sedimentation was used, alongside activated sludge processes or powdered activated carbon. OFX on the other hand showed little removal during these techniques. The same results are reflected in

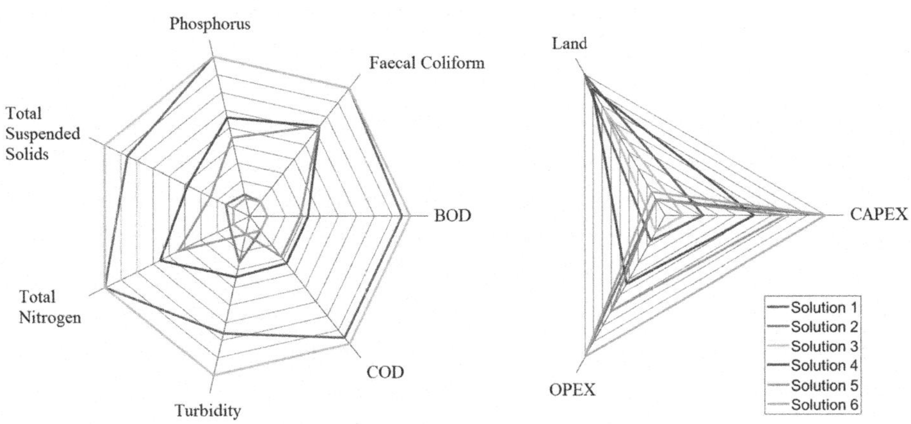

Figure 8 | Performance results from WiSDOM. (Left) Radar plot from the MOO optimisation for the results for the performance of the removal of conventional pollutants for all solutions. The higher the point on the axis the worse the solution has performed. (Right) Radar plot from WiSDOM from the MOO optimisation for the results for the performance of different sustainability indicator objectives (Land, OPEX, and CAPEX). The higher the point on the axis the better the solution has performed.

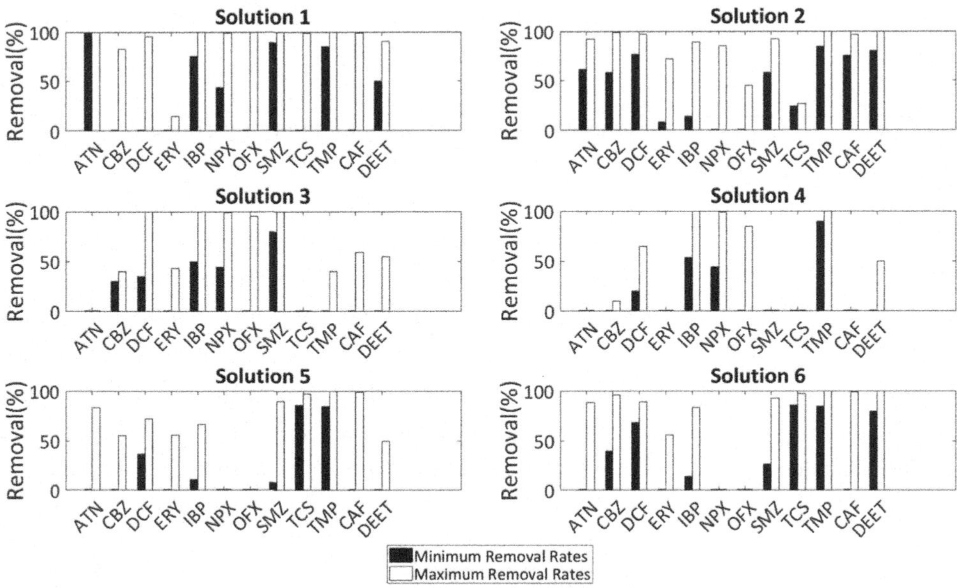

Figure 9 | Percentage of CECs removed from the different treatment solutions produced from WiSDOM-CEC for Tamil Nadu.

a study by Michael *et al.* (2013) who found that the best removal for OFX used advanced oxidation techniques which led to 83% removal and up to 85% removal was achieved through the use of CWs. The study also supported the removal of CIP showing higher removal rates from primary sedimentation (78%), secondary sedimentation (83%), and powdered activated carbon (100%). On the other hand, a study by Jia *et al.* (2012) found similar removal rates between the two fluoroquinolones through a sewage treatment plant with an overall removal of 64% for CIP and 66% for OFL (Jia *et al.* 2012), although a lower initial concentration of CIP was detected in the raw sewage.

Nonsteroidal anti-inflammatory drugs

DCF, IBP, and NPX are all classed as anti-inflammatory substances. In the case of Uttar Pradesh, the three substances exhibited similar removal rates exceeding 90% for solutions involving CWs as a unit process, combined with advanced oxidation or chlorine dioxide. The levels of DCF dropped to below 70% removal when ultrafiltration was combined with CWs. DCF and IBP had similar removal rates when sedimentation and trickling filters were used; however, NPX was not removed during this process. All three contaminants were removed between 50% and 80% when activated sludge was combined with sedimentation and ultrafiltration. These low removal rates for activated sludge are not reflected in recent studies in the literature where NPX had a 94–99% removal efficiency, IBP was removed to 99%, and DCF had the lowest removal rate from 92% to 98%. Similarly, DCF had low removal rates of 77–98% for tertiary treatment, and 9–21% for primary treatment, NPX followed with 91–98% and 17–55%, respectively. IBP had a higher minimal removal rate recorded for the primary treatment of 25%; however, the maximum removal only reached 53% (Larsson *et al.* 2014). 100% removal was achieved for tertiary treatment as displayed in the solutions for Uttar Pradesh. In Karnataka, NPX was not removed when sedimentation, trickling filters, and other filtration techniques were used; otherwise, the three contaminants displayed similar removal efficiencies within 10% of each other. This was also the case for the other solutions when alike unit processes were used as previously mentioned. In the literature, varying removal rates exist for each NSAID, suggesting that other controlling factors might be present as the studies were carried out in different locations. For NPX, varying removal rates have been found for different experiments where Snyder *et al.* (2007) found a removal via PAC of 50%; however, Nam *et al.* (2014) found a removal efficiency of 95% (Nam *et al.* 2014; Snyder *et al.* 2007). CWs seemed a prominent treatment option for removing NSAIDs, suggested by a greater than 90% removal in this study. However, Luo *et al.* (2014) found a removal from 44% to 99% for NPX, 40% for IBP, and 50% for DCF, although when used as a final treatment option such as a polishing pond, the removal increased to greater than 90%.

Removal of hormones, stimulants, personal care products, and insect repellents

EST, found in Karnataka, was able to be removed via the unit processes outputted in the WiSDOM tool. CWs and advanced oxidation are a successful treatment solution for this hormone, shown in S1, S3, and S4 in Figure 7. These solutions have removal rates of 100% and 98%, respectively, in comparison to 26% to 75% removal available with the use of sedimentation, CASP, and filtration methods.

High concentrations of CAF were found across the South, West, and Central regions of India, with concentrations reaching 220,000 ng/L (Table 2). With high concentrations being found in these regions, it is important to produce a treatment solution effective at removing the contaminant. Many studies have been carried out on the removal of CAF through wastewater treatment plants, proving that effective solutions should incorporate the following unit processes. Sedimentation is able to remove around 81% and this proved to be the main removal process along with reverse osmosis which was able to remove 50–80% (Luo *et al.* 2014). Many studies which have been carried out have focused on the removal throughout the whole treatment train and not the removal through individual unit processes (Sui *et al.* 2010; Froehner *et al.* 2011).

Two personal care products, BEN and BPA, were found in the influent wastewater of Nagpur and Karnataka, respectively. The treatment trains that were ideal to the area of Nagpur in regard to the MOO objectives were not suited to removing BEN (Figure S3 – 4, Supplementary Information). BEN is described as one of the most common UV filters with endocrine-disrupting effects; therefore, it is important to determine a unit process able to remove this compound. Most of the existing published literature has found that removal varied from 40% to 100%, with the use of advanced oxidation (Zúñiga-Benítez *et al.* 2016). This substance has high lipophilic properties; therefore, further investigations should be carried out on unit processes where sorption onto solids or biodegradation occurs (Gong *et al.* 2015). On the other hand, BPA was able to be removed during five of the chosen treatment trains in Karnataka with submerged-aerated filters able to remove up to 95% of the contaminant. Zielińska *et al.* (2016) found that BPA is difficult to remove during conventional biological methods, but found microfiltration and NF to have positive removal rates leading to 100% removal (Zielińska *et al.* 2016). However, other research has found lower removal rates for these unit processes reaching up to only 10% (Zhu & Li 2013).

The highest influent concentration recorded for DEET was found in Karnataka with a concentration of 388 ng/L and the lowest recorded concentration was 4 ng/L in Bihar. High removal rates were found for DEET and most of the treatment solutions produced were able to remove up to 90% of DEET, with NF and ultrafiltration able to remove 50–100% (Sui *et al.* 2010).

The treatment train solutions created from the WiSDOM tool for each region in India were effectively produced to suit the original constraints entered into the tool. These solutions showed the ability to remove conventional pollutants in order to allow for water reuse application in India. The chosen treatment trains were also able to remove the CECs found in the influent; however, in some cases, removal via other treatment options would be more suited, such as for BEN, where no treatment train chosen was able to remove the contaminants. However, in some cases, CECs in the same categories exhibited similar removal rates across treatment solutions; in most cases, each CEC exhibited different removal rates. When compared to other studies, no two CECs in a category or drug class (e.g., NSAIDs) display the same properties or exact removal efficiencies. This is because each contaminant, whether hydrophilic or hydrophobic, displays individual physical, chemical, and biological characteristics resulting in an individual breakdown of the pollutant.

In terms of the highest recorded concentration across India, CAF was effectively removed during sedimentation and advanced oxidation treatment options; however, for other CECs, this was not an effective removal technique. Advanced oxidation techniques and the use of reverse osmosis, although practical and efficient, are not suited to rural areas due to the cost associated with this treatment method, both CAPEX and OPEX. CWs were effective at removing the majority of recorded emerging contaminants and are a sustainable, cost-effective solution. However, in some locations of India where land requirement is a restriction, such as urban areas, this is not a feasible solution resulting in the selection of other unit processes.

Future applications

Currently, this research focuses on determining effective solutions for the removal of CECs in developing countries such as India. This is because, for most developed countries, the infrastructure is already in place for wastewater treatment and it would be costly to recreate treatment train solutions for application. However, a decision support tool could be developed that analyses the existing treatment unit processes and provides an additional unit process that could be implemented to achieve the required results. Additionally, the focus was applied to developing countries as the decision support tool WiSDOM was originally created to generate water reuse solutions in India. Therefore, data were collected on emerging contaminants in India. Since the information included in the technology library is not case-specific, future application could involve applying the tool to calculate the removal of CECs in developed countries.

To further this work, the functionality of the add-on worksheet can be integrated directly into WiSDOM by expanding the source code. Additionally, the ESP removal model can be incorporated directly into WiSDOM by imposing a limited number of CECs as constraints (based on the user's preference) in addition to the conventional contaminants. This will allow the tool to decide on best solutions by also incorporating the removal of the chosen CECs selected by the user. Furthermore, as the ESP can be easily altered to suit a different developing country if data are available regarding the removal of CECs. This could then be combined with different decision support tools that currently do not incorporate the removal of CECs.

CONCLUSIONS

The aim of this study was to determine the effectiveness of a decision support tool for producing optimal wastewater management solutions for developing countries. This was done by using an existing decision support tool for India, WiSDOM, and building an ESP that works alongside the tool, taking the treatment train solutions from WiSDOM and using these to calculate the removal of CECs through the unit processes. Different regions of India were analysed to show varying constraints and input data. With CECs becoming an increasing cause of concern, the ESP was designed to calculate the minimal and maximum removal of CECs through the produced treatment solutions. Natural processes, such as CWs, are the most effective at removing CECs along with being a more sustainable solution. However, due to the constraints such as land requirements along with cost, these are not always the best solution as shown by the low incidence of this unit process being chosen in the WiSDOM tool. Although CECs classed under the same category have similar physical, chemical, and biological parameters, these contaminants are not always removed in the same way. For example, the pharmaceutical group NSAIDs which showed different removal rates between DCF, IBP, and NPX when passed through the same treatment solution. This can be seen in Uttar Pradesh, Karnataka, Tamil Nadu, and Bihar (in S3 in the Supplementary Information). In Tamil Nadu, removal rates ranged from 0% to 95% for DCF, 75% to 100% for IBP, and 44% to 99% for NPX for the first solution (CWs, advanced oxidation, and UV/O_3). In S5 (coarse screen, fine screen, sedimentation, trickling filter and secondary sedimentation, and ultrafiltration), the removal rates showed a greater difference with removal rates ranging from 36.8% to 72% for DCF, 10.7% to 66.5% for IBP, and 0% for NPX. This demonstrates that although it is expected that CECs under the same class of pharmaceutical or personal care product, for example, should be removed in the same way, this is not always the case as each substance has its own properties.

ACKNOWLEDGEMENTS

This work was supported by the EPSRC in the UK via grant EP/R512254/1 awarded for the Water Informatics, Science and Engineering (WISE) centre for doctoral training. Part of this study describes work being carried out under EC FP7-funded projects SARASWATI (grant agreement no. 308672) and NERC project on the Fate and management of Emerging Contaminants (grant no. NE/R003548/1).

The project sponsors were not involved in the study design, collection, analysis, and interpretation of data or in the writing of the manuscript.

SUPPLEMENTARY MATERIAL

The Supplementary Material for this paper is available online at https://dx.doi.org/10.2166/hydro.2019.031.

REFERENCES

ALS Environmental 2015 Chemicals Investigation Programme 2 – Overview.

Anumol, T., Vijayanandan, A., Park, M., Philip, L. & Snyder, S. A. 2016 Occurrence and fate of emerging trace organic chemicals in wastewater plants in Chennai, India. *Environ. Int.* **92**, 33–42. https://doi.org/10.1016/j.envint.2016.03.022.

Baalbaki, Z., Sultana, T., Metcalfe, C. & Yargeau, V. 2017 Estimating removals of contaminants of emerging concern from wastewater treatment plants: the critical role of wastewater hydrodynamics. *Chemosphere* **178**, 439–448. https://doi.org/10.1016/j.chemosphere.2017.03.070.

Balakrishna, K., Rath, A., Praveenkumarreddy, Y., Guruge, K. S. & Subedi, B. 2017 A review of the occurrence of pharmaceuticals and personal care products in Indian water bodies. *Ecotoxicol. Environ. Saf.* **137**, 113–120. https://doi.org/10.1016/j.ecoenv.2016.11.014.

Calvo-Flores, F. G., Isac-Garcia, J. & Dobado, J. A. 2018 *Emerging Pollutants: Origin, Structure, and Properties*, 1st edn. Wiley-VCH Verlag GmbH & Co., Germany.

Census 2011 Census Data 2011 [WWW Document]. https://www.census2011.co.in (accessed 12 November 2018).

Deb, K. & Tiwari, S. 2008 Omni-optimizer: a generic evolutionary algorithm for single and multi-objective optimization. *Eur. J. Oper. Res.* **185**, 1062–1087. https://doi.org/10.1016/j.ejor.2006.06.042.

Deb, K., Pratap, A., Agarwal, S. & Meyarivan, T. 2002 A fast and elitist multiobjective genetic algorithm: NSGA-II. *IEEE Trans. Evol. Comput.* **6**, 182–197. https://doi.org/10.1109/4235.996017.

DEFRA 2017 *National Implementation Plan for the Stockholm Convention on Persistant Organic Pollutants: United Kingdom of Great Britain and Northern Ireland.*

European Union 2015 Commission Implementing Decision (EU) 2015/495.

Fischer, A., ter Laak, T., Bronders, J., Desmet, N., Christoffels, E., van Wezel, A. & van der Hoek, J. P. 2017 Decision support for water quality management of contaminants of emerging concern. *J. Environ. Manage.* **193**, 360–372. https://doi.org/10.1016/j.jenvman.2017.02.002.

Froehner, S., Piccioni, W., Machado, K. S. & Aisse, M. M. 2011 Removal capacity of caffeine, hormones, and bisphenol by aerobic and anaerobic sewage treatment. *Water Air Soil Pollut.* **216**, 463–471. https://doi.org/10.1007/s11270-010-0545-3.

Gong, P., Yuan, H., Zhai, P., Xue, Y., Li, H., Dong, W. & Mailhot, G. 2015 Investigation on the degradation of benzophenone-3 by UV/H_2O_2 in aqueous solution. *Chem. Eng. J.* **277**, 97–103. https://doi.org/10.1016/j.cej.2015.04.078.

Jia, A., Wan, Y., Xiao, Y. & Hu, J. 2012 Occurrence and fate of quinolone and fluoroquinolone antibiotics in a municipal sewage treatment plant. *Water Res.* **46**, 387–394. https://doi.org/10.1016/j.watres.2011.10.055.

Joksimovic, D., Kubik, J., Hlavinek, P., Savic, D. & Walters, G. 2006 Development of an integrated simulation model for treatment and distribution of reclaimed water. desalination. *Integr. Concepts Water Recycling* **188**, 9–20. https://doi.org/10.1016/j.desal.2005.04.097.

Joksimovic, D., Savic, D. A., Walters, G. A., Bixio, D., Katsoufidou, K. & Yiantsios, S. G. 2008 Development and validation of system design principles for water reuse systems. desalination. *AQUAREC 2006* **218**, 142–153. https://doi.org/10.1016/j.desal.2006.04.091.

Lapworth, D. J., Baran, N., Stuart, M. E. & Ward, R. S. 2012 Emerging organic contaminants in groundwater: a review of sources, fate and occurrence. *Environ. Pollut.* **163**, 287–303. https://doi.org/10.1016/j.envpol.2011.12.034.

Larsson, E., al-Hamimi, S. & Jönsson, J. Å. 2014 Behaviour of nonsteroidal anti-inflammatory drugs and eight of their metabolites during wastewater treatment studied by hollow fibre liquid phase microextraction and liquid chromatography mass spectrometry. *Sci. Total Environ.* **485–486**, 300–308. https://doi.org/10.1016/j.scitotenv.2014.03.055.

Loos, R., Carvalho, R., Comero, S., Antonio, D., Ghiani, M., Lettieri, T., Locoro, G., Paracchini, B., Tavazzi, S., Gawlik, B. M., Blaha, L., Jarosova, B., Voorspoels, S., Schwesig, D., Haglund, P., Fick, J. & Gans, O. 2012 *EU Wide Monitoring Survey on Wastewater Treatment Plant Effluents.* Joint Research Centre, Italy.

Luo, Y., Guo, W., Ngo, H. H., Nghiem, L. D., Hai, F. I., Zhang, J., Liang, S. & Wang, X. C. 2014 A review on the occurrence of micropollutants in the aquatic environment and their fate and removal during wastewater treatment. *Sci. Total Environ.* **473–474**, 619–641. https://doi.org/10.1016/j.scitotenv.2013.12.065.

Machado, K. C., Grassi, M. T., Vidal, C., Pescara, I. C., Jardim, W. F., Fernandes, A. N., Sodré, F. F., Almeida, F. V., Santana, J. S., Canela, M. C., Nunes, C. R. O., Bichinho, K. M. & Severo, F. J. R. 2016 A preliminary nationwide survey of the presence of emerging contaminants in drinking and source waters in Brazil. *Sci. Total Environ.* **572**, 138–146. https://doi.org/10.1016/j.scitotenv.2016.07.210.

Michael, I., Rizzo, L., McArdell, C. S., Manaia, C. M., Merlin, C., Schwartz, T., Dagot, C. & Fatta-Kassinos, D. 2013 Urban wastewater treatment plants as hotspots for the release of antibiotics in the environment: A review. *Water Res.* **47**, 957–995. https://doi.org/10.1016/j.watres.2012.11.027.

Mohapatra, S., Huang, C.-H., Mukherji, S. & Padhye, L. P. 2016 Occurrence and fate of pharmaceuticals in WWTPs in India and comparison with a similar study in the United States. *Chemosphere* **159**, 526–535. https://doi.org/10.1016/j.chemosphere.2016.06.047.

Montes-Grajales, D., Fennix-Agudelo, M. & Miranda-Castro, W. 2017 Occurrence of personal care products as emerging chemicals of concern in water resources: a review. *Sci. Total Environ.* **595**, 601–614. https://doi.org/10.1016/j.scitotenv.2017.03.286.

Nam, S.-W., Jo, B.-I., Yoon, Y. & Zoh, K.-D. 2014 Occurrence and removal of selected micropollutants in a water treatment plant. *Chemosphere* **95**, 156–165. https://doi.org/10.1016/j.chemosphere.2013.08.055.

Petrie, B., Barden, R. & Kasprzyk-Hordern, B. 2015 A review on emerging contaminants in wastewaters and the environment: current knowledge, understudied areas and recommendations for future monitoring. *Water Res.* **72**, 3–27 (Occurrence, fate, removal and assessment of emerging contaminants in water in the water cycle (from wastewater to drinking water)). https://doi.org/10.1016/j.watres.2014.08.053.

Philip, J. M., Aravind, U. K. & Aravindakumar, C. T. 2018 Emerging contaminants in Indian environmental matrices – a review. *Chemosphere* **190**, 307–326. https://doi.org/10.1016/j.chemosphere.2017.09.120.

Poch, M., Rigola, M., Comas, J. & Rodriguez-Roda, I. 2014 An environmental decision support system for the control and supervison of municipal wastewater treatment plants. In: *Presented at the 8th European Waste Water Management Conference*, AquaEnviro, Manchester.

REACH 2016 C1 Regulation (EC) No. 1907/2006 of the European Parliament and of the Council of 18 December 2006 concering the Registration, Evaluation, Authorisation and Restriction of Chemicals (REACH), establishing a European Chemicals Agency, amending Directive 1999/45/EC and repealing Council Regulation (EEC) No. 793/93 and Commission Regulation (EC) No. 1488/94 as well as Council Directive 76/769/EEC and Commission Directives 91/155/EEC, 93/67/EEC, 93/105/EC and 2000/21/EC.

Sadr, S. M. K., Johns, M. B. & Memon, F. A., 2017 Technology library comprising investigated technologies in earlier work packages and innovations investigated elsewhere. EU FP7 SARASWATI Project Deliverable D5.2.

Sadr, S. M. K., Johns, M. B., Memon, F. A., Morley, M. & Savić, D. 2018 Development and application of a user-friendly decision support tool for optimization of wastewater treatment technologies in India. In: *Presented at the HIC 2018: 13th International Conference on Hydroinformatics*, Palermo, Italy.

Samadi, M., Khodadai, M. & Rahmani, A. 2007 The comparison of advanced oxidation process and chemical coagulation for the removal of residual pesticides from water [WWW Document]. https://doi.org/10.3923/rjes.2011.817.826

Snyder, S. A., Adham, S., Redding, A. M., Cannon, F. S., DeCarolis, J., Oppenheimer, J., Wert, E. C. & Yoon, Y. 2007 Role of membranes and activated carbon in the removal of endocrine disruptors and pharmaceuticals. *Desalination, Wastewater Reclamation and Reuse for Sustainability* **202**, 156–181. https://doi.org/10.1016/j.desal.2005.12.052.

Starling, M. C. V. M., Amorim, C. C. & Leão, M. M. D. 2018 Occurrence, control and fate of contaminants of emerging concern in environmental compartments in Brazil. *J. Hazard. Mater.* https://doi.org/10.1016/j.jhazmat.2018.04.043

Stumm-Zollinger, E. & Fair, G. M. 1965 Biodegradation of steroid hormones. *J. Water Pollut. Control Fed.* **37**, 1506–1510.

Sui, Q., Huang, J., Deng, S., Yu, G. & Fan, Q. 2010 Occurrence and removal of pharmaceuticals, caffeine and DEET in wastewater treatment plants of Beijing, China. *Water Res.* **44**, 417–426 (Emerging Contaminants in water: Occurrence, fate, removal and assessment in the water cycle (from wastewater to drinking water)). https://doi.org/10.1016/j.watres.2009.07.010.

Tran, N. H., Reinhard, M. & Gin, K. Y.-H. 2018 Occurrence and fate of emerging contaminants in municipal wastewater treatment plants from different geographical regions-a review. *Water Res.* **133**, 182–207. https://doi.org/10.1016/j.watres.2017.12.029.

US EPA 2016 *Contaminant Candidate List (CCL) and Regulatory Determination.*

Visanji, Z., Sadr, S. M. K., Johns, M. B., Savic, D. & Memon, F. A. 2018 Emerging pollutants in developing countries: optimising sustainable treatment solutions. In: *EPiC Series in Computing. Presented at the HIC 2018 13th International Conference on Hydroinformatics*, Italy, pp. 2204–2215. https://doi.org/10.29007/txmw

Zhu, H. & Li, W. 2013 Bisphenol A removal from synthetic municipal wastewater by a bioreactor coupled with either a forward osmotic membrane or a microfiltration membrane unit. *Front. Environ. Sci. Eng.* **7**, 294–300. https://doi.org/10.1007/s11783-013-0486-3.

Zielińska, M., Bułkowska, K., Cydzik-Kwiatkowska, A., Bernat, K. & Wojnowska-Baryła, I. 2016 Removal of bisphenol A (BPA) from biologically treated wastewater by microfiltration and nanofiltration. *Int. J. Environ. Sci. Technol.* **13**, 2239–2248. https://doi.org/10.1007/s13762-016-1056-6.

Zúñiga-Benítez, H., Aristizábal-Ciro, C. & Peñuela, G. A. 2016 Photodegradation of the endocrine-disrupting chemicals benzophenone-3 and methylparaben using Fenton reagent: optimization of factors and mineralization/biodegradability studies. *J. Taiwan Inst. Chem. Eng.* **59**, 380–388. https://doi.org/10.1016/j.jtice.2015.09.004.

First received 30 January 2019; accepted in revised form 23 May 2019. Available online 14 June 2019

Uncertainty quantification of water age in water supply systems by use of spectral propagation

Mathias Braun, Olivier Piller, Jochen Deuerlein, Iraj Mortazavi and Angelo Iollo

ABSTRACT

Water distribution networks are critical infrastructures that should ensure the reliable supply of high quality potable water to its users. Numerical models of these networks are generally governed by many parameters for which the exact value is not known. This may be due to a lack of precise knowledge like for consumer demand or due to a lack of accessibility as for the pipe roughness. For network managers, the effect of these uncertainties on the network state is important information that supports them in the decision-making process. This effect is generally evaluated by propagating the uncertainties using the mathematical model. In the past, perturbation, fuzzy and stochastic collocation methods have been used for uncertainty propagation. However, these methods are limited either in the accuracy of the results or the computational effort of the necessary calculations. This paper uses an alternative spectral approach that uses the polynomial chaos expansion and has the potential to give results of comparable accuracy to the Monte Carlo sampling through the definition of a stochastic model. This approach is applied to the hydraulic model of two real networks in order to evaluate the influence of uncertain demands on the water age.

Key words | multimodal distribution, real network, uncertainty quantification, water age, water distribution systems

Mathias Braun (corresponding author)
Olivier Piller
Irstea, Dept. of Water,
Bordeaux Regional Centre, UR ETBX,
50 Ave. de Verdun, Gazinet, F-33612 Cestas, France
E-mail: *mathias.braun@irstea.fr*

Mathias Braun
Angelo Iollo
Inria Bordeaux-Sud-Ouest, Team MEMPHIS,
200 Avenue de la Vieille Tour, 33405 Talence, France

Olivier Piller
Jochen Deuerlein
School of Civil, Environmental and Mining Engineering,
University of Adelaide,
South Australia 5005, Australia

Jochen Deuerlein
3S Consult GmbH,
Albtalstrasse 13, 76137 Karlsruhe, Germany

Iraj Mortazavi
CNAM Paris, EA-7340-M2N-Modelisation Mathématique et Numérique,
75003 Paris, France

Angelo Iollo
Université de Bordeaux, IMB,
33400 Talence, France

INTRODUCTION

Evaluating the quality of drinking water supplied by a distribution system is usually done by using quality indicators like water age or chlorine concentration. These indicators may be determined from mathematical network models. The model of a water distribution network depends on a great number of parameters, like the nodal water demand, pipe roughness and reservoir or tank levels. However, the exact values of these parameters are rarely known and thus introduce parameter uncertainty to the model.

Uncertainties in mathematical models are generally divided into aleatory and epistemic uncertainties. Aleatory uncertainties are the result of inherently random processes and are often described by probability distributions that are modelled on observations. These uncertainties are regarded as irreducible since the underlying process cannot be influenced. In contrast, epistemic uncertainties follow from incomplete insight into the system, which in theory could be eliminated through perfect knowledge. However, epistemic uncertainties are often the necessary effect of simplifications in the modelling process and the numerical evaluation (Der Kiureghian & Ditlevsen 2009; Smith 2013).

Parameter uncertainties are one of the most important aspects of hydraulic modelling as they can result in large prediction errors. Hidden parameters like the pipe roughness that cannot be observed directly are often estimated through calibration (Savic *et al.* 2009; Piller *et al.* 2010).

doi: 10.2166/hydro.2019.017

But, even for a well-calibrated network some uncertainty remains. Other parameters, like for example, the demand, are clearly aleatory in nature.

The objective of uncertainty analysis (UA) is the evaluation and mitigation of such parameter uncertainties on the quantities of interest (QoI). The central part of this process is the uncertainty propagation by means of the mathematical model. Classical approaches contain perturbation, stochastic collocation and fuzzy methods. Stochastic spectral methods are a relatively new addition to the field that has been introduced by Xiu (2010).

Perturbation methods calculate the moments for the distribution of the quantity of interest directly from the system equations by means of a truncated Taylor expansion. Typically, the expansions employed are limited to first- or second-order expansions (Savic *et al.* 2009). This limits their accuracy for highly non-linear models. Stochastic sampling methods like the Monte Carlo simulation are some of the most prominent algorithms for uncertainty propagation. This is due to the easy implementation and the ability to be used for non-linear systems. However, it can be computationally demanding as its rate of convergence is proportional to $1/\sqrt{M}$, where M is the number of simulations (Cacuci *et al.* 2005; Savic *et al.* 2009). The objective of stochastic spectral methods like stochastic Galerkin and non-intrusive spectral projection is the calculation of a spectral representation of the random variables. Utilizing the smoothness requirement of the polynomial basis leads to a more efficient convergence behaviour (Xiu 2010; Smith 2013).

The effect of uncertainties in the context of water distribution networks has been investigated in a number of publications with a wide variety of interests. Special interest lies in the inclusion of parameter uncertainties during the network design. Tsakiris *et al.* (2017) use a fuzzy approach to uncertainty propagation in order to include the uncertainty in parameters like the pipe roughness. The application leads to useful recommendations for the network design, in the case of medium to low flow velocities. Fu & Kapelan (2011) present a fuzzy probabilistic approach for optimal design and rehabilitation of water distribution systems. The approach is tested on two example networks and the results show the solution to be more reliable than the deterministic Pareto optimal solutions. Sumer & Lansey (2009) propose a method for pipe network extension optimization and calibration of pipe roughness at the same time. Uncertainties in the pipe roughness values are specified using FOSM analysis and embedded using a D-optimality criterion. The authors show that the mean values of the unknown parameters converged to their true values and also the cost changes showed a monotonic trend.

A second focus lies in the inclusion of uncertainty information into the management of the network.

Xu & Qin (2014) propose an integrated fuzzy programming and decision analysis approach for water distribution system management under uncertainty. The results indicate that the method helps analyse trade-offs between minimization of operation cost and reliability of running the system and linking optimization model outputs with decision analysis. Raei *et al.* (2018) apply multi-objective optimization for the design of an optimal pressure sensor network for leak detection. Uncertainty in the pressure measurements is considered by application of a range of thresholds. The detection times are calculated for leaks at any node with varying starting times. However, in this approach, pressure dependency is neither considered for the leakage outflows nor for the demands which lead to unrealistic results in real applications.

Studies on the effect of uncertainties on the water quality have been performed with respect to a number of objectives. Di Cristo *et al.* (2015) investigate the effect of parameter uncertainties in the kinetic models for the formation of trihalomethanes (THM). The parameter uncertainties in the chlorine decay parameters and the THM formation parameters are propagated using the first order second moment (FOSM) perturbation approach and Monte Carlo simulation for validation purposes. They conclude that the FOSM is sufficiently accurate for the application and well worth the benefit in computational effort. The study by Hart *et al.* (2019) performs a sensitivity analysis on a water quality model in order to identify important factors for the reduction in uncertainty for the selection of water quality sampling points in the case of a contamination event. They employ a full factorial design of experiments where the influences of parameters like valve status, reaction rate and injection location is investigated as well as the effect of stochastic demand variability. The results show that the injection location is the main source for uncertainty for the size and total area of the contamination plume, whereas the effect of the demand variability is not significant in comparison. The objective of this paper is to analyse the influence of the demand uncertainties on the water age in a network model that has been provided by the partner Veolia in the framework of the French–German project ResiWater (2019). Following this introduction, the example network will be introduced together with the basic equations for the hydraulic state and water age in the next section. This is followed by a short introduction of the PCE. Then the two use-cases are discussed after which the results of the investigation are presented. Finally, the conclusions drawn from the application are presented.

MATERIAL

Hydraulic model

In hydraulic modelling, the simplified topological structure of a water distribution network is described by a directed graph corresponding to Figure 1. In this graph, links represent pipe sections, valves and pumps, and nodes the resource nodes, tanks, demand nodes and connections. The mathematical description of this graph is given by the node-link incidence matrix $A \in M^{nj \times np}$, where nj is the number of junction nodes and np is the number of links (Jungnickel 2005). The coefficients are defined as follows:

$$A_{i,j} = \begin{cases} -1, & \text{if node } i \text{ is the end node of link } j \\ 0, & \text{if link } j \text{ is not connected to node } i \\ +1, & \text{if node } i \text{ is the start node of link } j \end{cases} \quad (1)$$

Water distribution networks, in general, have a looped structure and the system state is described by the potentials at the nodes (heads) and the currents on the links (flow rates). The classical hydraulic model is defined by the first and second Kirchhoff laws (Cross 1936). First, the mass balance at the nodes:

$$\mathbf{Aq} + \mathbf{d} = \mathbf{0}_{nj} \quad (2)$$

where the node-link incidence matrix $\mathbf{A}$ includes only nodes with unknown heads, $\mathbf{q} \in \mathbb{R}^{np}$ is the vector of link flow rates and $\mathbf{d} \in \mathbb{R}^{nj}$ is the vector of demands at consumption nodes. Second, the energy equation:

$$\Delta\mathbf{h}(\mathbf{r},\ \mathbf{q}) - \mathbf{A}^T\mathbf{h} - \mathbf{A}_f^T\mathbf{h}_f = \mathbf{0}_{np} \quad (3)$$

where $\mathbf{A}_f$ describes the incidence matrix of nodes with fixed potential like reservoirs or tanks and $\mathbf{h} \in \mathbb{R}^{nj}$ is the vector containing the unknown piezometric heads. Parameters are given by the head vector $\mathbf{h}_f \in \mathbb{R}^{nf}$ describing fixed heads at special nodes like reservoirs or tanks and the vector $\mathbf{r} \in \mathbb{R}^{np}$ containing the friction coefficients for each link. The function $\Delta\mathbf{h}(\mathbf{r},\mathbf{q})$ describes the loss in head due to the friction between the fluid and the pipe wall.

Water quality indicators

In the literature, the water age or residence time in the network is commonly used as an indicator for water quality. Dandy *et al.* (2013) studied the advantages and the

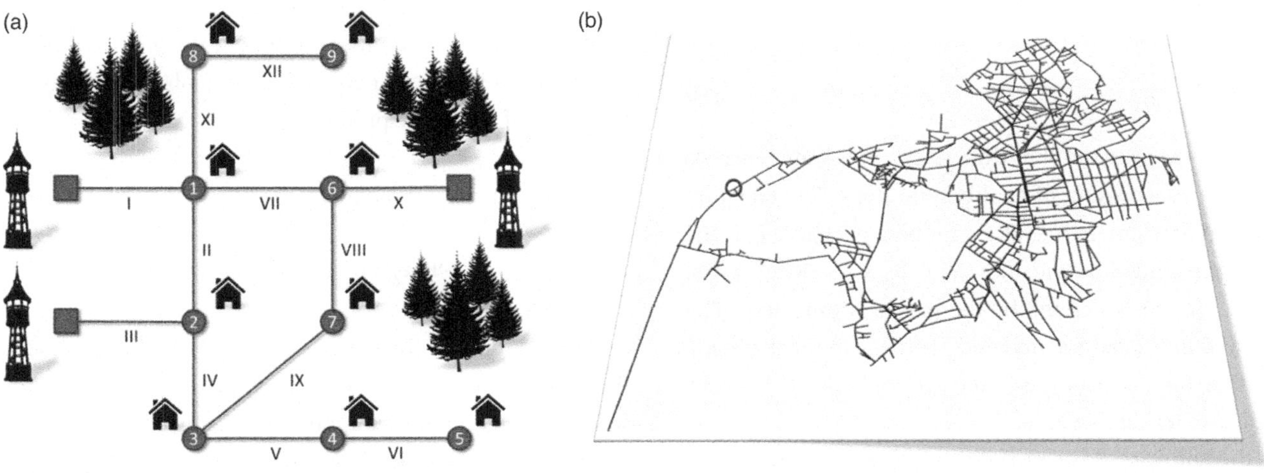

Figure 1 | (a) Strongly aggregated looped network; (b) medium-sized realistic network with the marked evaluation node 15Nf79.

limitations of the use of water age as a surrogate for water quality. They tried to determine the chlorine concentration based on the residence time. They conclude that in certain situations, water age gives an approximate estimate of water quality and disinfection by-product formation. However, this estimate may be inaccurate due to mixing in tanks and at nodes and is inappropriate for systems with several water sources and/or re-chlorination.

The water age c is modelled on the basis of the convection–diffusion equation:

$$\frac{\partial c}{\partial t} = \nabla(D\nabla c) - \nabla(\vec{v}c) + R(t, \vec{x}) = 0 \tag{4}$$

The variable of interest c is usually assumed to be some sort of concentration, D is the diffusion coefficient, $\vec{v}$ the velocity field and R a source term which may be a non-linear function in time and space. For water distribution systems two major simplifications are made. First, an incompressible water flow is assumed, for which extended period simulation (EPS) for the hydraulic is accurate enough (no water hammer or mass oscillation); it follows the hydraulic model is one-dimensional and the velocity vector $\vec{v}$ simplifies to $U(t, x) = U_{i,j}$ which is calculated for every hydraulic time step t_i and constant along x for each pipe p_j. Second, it is assumed that the regime is turbulent as a consequence that the longitudinal dispersion can be neglected (Rossman & Boulos 1996; Fabrie *et al.* 2010). This flow state is often characterized as plug-flow. With these adaptations, the transport equation for water age a may be derived by replacing R with a zero-order source term:

$$\frac{\partial c(t, x)}{\partial t} + U_{i,j}\frac{\partial c(t, x)}{\partial x} - 1 = 0 \tag{5}$$

During each hydraulic time step the water age is resolved by a succession of quality time steps. The equation is solved on the domain $x \in [0, L_j]$ for pipe p_j, where L_j is the pipe length and for the time $t \in [t_i, t_{i+1}]$. The initial condition $c(t_i, x_j)$ is defined by the previous time step. The boundary conditions are updated with the perfect mixing assumption for the average water age, or with the min or max value entering.

In the software Porteau (Piller *et al.* 2011), water age is evaluated in three variables: min, max and flow-averaged ages. Since the software keeps track of these characteristic water ages during the mixing process at the pipe junctions, it is possible to evaluate the extreme values for the residence time as defined by Equation (6):

$$\begin{aligned} c_n^{upper} &= \max_{i \in S_{in}} c_i \\ c_n^{lower} &= \min_{i \in S_{in}} c_i \end{aligned} \tag{6}$$

The third variable, average water age is calculated by the weighted sum of the water age in the water columns mixing at the upstream junction (Piller & Tavard 2014):

$$c_n = \frac{\sum_{i \in S_{in}} |q_i| c_i}{\sum_{i \in S_{in}} |q_i|} \tag{7}$$

UNCERTAINTY PROPAGATION

A recent and promising approach for the propagation of parameter uncertainties is presented in the polynomial chaos expansion (PCE), where a random variable X is expressed as a truncated spectral series expansion of the order N as a function of the basic random variable Z and a set of orthogonal basis polynomials Ψ_k:

$$X_N(Z) = \sum_{k=0}^{N} x_k \Psi_k(Z) \tag{8}$$

where the coefficients of the series are given by x_k. The choice of the basic random variable, also often termed as the germ, and the orthogonal polynomials is joint by the definition of the inner product:

$$\begin{aligned} E[\Psi_i(Z)\Psi_j(Z)] &= \int \Psi_i(Z)\Psi_j(Z) dP_Z(z) \\ &= \int \Psi_i(z)\Psi_j(z) f_Z(z) dz = \delta_i \|\Psi_i\|^2 \end{aligned} \tag{9}$$

Here, E defines the expected value, P_z is the cumulative distribution function and f_z is the probability density function of the germ distribution. The Kronecker delta δ_i is equal to 1 if the indices are equal and 0 for the product of two different polynomials. The coefficients x_k of a random parameter X are calculated by projecting it on the orthogonal basis polynomials:

$$x_k = \frac{\langle X, \Psi_k \rangle}{\|\Psi_k\|^2} \tag{10}$$

As Equation (7) suggests, the choice of the polynomial basis generally depends on the distribution of the basic random variable. In this paper, uncertainties are assumed to follow a Gaussian distribution which allows for the application of the Wiener chaos with a spectral expansion on the basis of the Hermite polynomials (Xiu & Karniadakis 2002). The use of the PCE is not limited to use with Gaussian distributed variables.

First, it is possible to approximate other distributions on the basis of the Hermite polynomials. Second, the extension to the general PCE defines further sets of orthogonal polynomial basis that can be used with different probability distributions (Xiu 2010). One such example would be the spectral expansion of a uniform distribution on the basis of the Legendre polynomials.

Application of the PCE, in general, follows an intrusive or a non-intrusive procedure. The intrusive approach requires a reformulation of the system equations and allows for the direct solution in the parameters of the QoI. Although computationally very efficient, it may pose certain complications since it is not possible to use existing implementations for the solution of the problem. Further, the handling of non-linearities may require additional steps that introduce additional errors. Non-intrusive algorithms use a number of samples similar to the Monte Carlo simulation to evaluate the coefficients. This means that in regard to the computation current software and models may be used. Even though the non-intrusive PCE is more efficient than the MCS for high dimensional parameter spaces the problem may become computationally infeasible for applications with high dimensional parameter spaces.

For the application of the PCE to the hydraulic equations, random variables are substituted by their spectral series expansion. This includes parameters (nodal demands $\mathbf{d}_N$, pipe resistance $\mathbf{r}_N$) for which the coefficients are determined on the basis of their probability density function and the QoI (flow rates $\mathbf{q}_N$, nodal heads $\mathbf{h}_N$) for which the coefficients still have to be determined. This substitution is formulated in the stochastic system equations:

$$\begin{aligned} &\mathbf{A}\mathbf{q}_N + \mathbf{d}_N = \mathbf{0}_{nj} \\ &\Delta\mathbf{h}(\mathbf{r}_N,\ \mathbf{q}_N) - \mathbf{A}^T\mathbf{h}_N - \mathbf{A}_f^T\mathbf{h}_f = \mathbf{0}_{np} \end{aligned} \tag{11}$$

In this system of equations, the random variables have been replaced by their PCE of the order N, following the description in the section 'Material'. There exist two general approaches to evaluate the coefficients of the QoI with their respective benefits and drawbacks.

For the *intrusive* methods, the most prominent approach is the Galerkin projection. It projects the stochastic system equations on the polynomial basis functions Ψ_k to create a new augmented system of equations that has $N+1$ times the number of equations to determine the expansion coefficients of the QoI:

$$\begin{aligned} &\langle \mathbf{A}\mathbf{q}_N + \mathbf{d}_N, \Psi_k \rangle = \mathbf{0}_{nj} \\ &\langle \Delta\mathbf{h}(\mathbf{r}_N,\ \mathbf{q}_N) - \mathbf{A}^T\mathbf{h}_N - \mathbf{A}_f^T\mathbf{h}_f, \Psi_k \rangle = \mathbf{0}_{np} \end{aligned} \tag{12}$$

Similar to the hydraulic equations the partial differential equation for the coefficients of the water age may be determined by the following projection:

$$\left\langle \frac{\partial A^K}{\partial t} + U_{i,j}^N \frac{\partial A^K}{\partial x} - 1, \Psi_m \right\rangle = 0 \tag{13}$$

Here, the uncertain parameter is given by $U_{i,j}^N$ which is derived from the PCE of the flow $q_{i,j}^N$ from the stochastic hydraulic equations and the QoI is water age A^K.

The *non-intrusive* method is a stochastic collocation like Monte Carlo simulation. It uses collocation points for which the deterministic system equations are evaluated. In contrast to the MCS, the direct evaluations in intrusive methods are used to fit the coefficients of the PCE to the data. For a QoI X vector, which could be either velocity or water age, the coefficients are determined by the following equation:

$$\begin{bmatrix} \mathbf{\Psi}_0(\mathbf{z}^{(1)}) & \cdots & \mathbf{\Psi}_N(\mathbf{z}^{(1)}) \\ \vdots & \ddots & \vdots \\ \mathbf{\Psi}_0(\mathbf{z}^{(M)}) & \cdots & \mathbf{\Psi}_N(\mathbf{z}^{(M)}) \end{bmatrix} \begin{bmatrix} \mathbf{x}_1^T \\ \vdots \\ \mathbf{x}_N^T \end{bmatrix} = \begin{bmatrix} \mathbf{x}(\mathbf{y}(\mathbf{z}^{(1)}))^T \\ \vdots \\ \mathbf{x}(\mathbf{y}(\mathbf{z}^{(M)}))^T \end{bmatrix} \tag{14}$$

where $\mathbf{z}^{(i)}$ are realizations of the $\mathbf{Z}$ random vector and $\Psi_k\ (\mathbf{z}^{(i)})$ is the value of the k-th polynomial basis function at realization $\mathbf{z}^{(i)}$. The right-hand side is determined by the simulation result

$\mathbf{x}$ of the deterministic system at the random parameter vector $\mathbf{y}$ that is associated with the realization $\mathbf{z}^{(i)}$ of the germ distribution. $\mathbf{x}_k$ vectors (k = 1, … , N) are the coefficients for the PCE of the state vector, which allows evaluation of the spectral coefficients (Le Maître & Knio 2010).

Like all sampling-based methods, the complexity of the parameter-space influences the amount of evaluations for the collocation points. For the application of Monte Carlo sampling, the Latin hypercube sampling has proven to be far more efficient for multidimensional applications (Helton & Davis 2003) and sparse grid sampling methods are frequently used in the application of the PCE (Xiong *et al.* 2009). Another approach was introduced by Blatman & Sudret (2011) with the least angle regression sampling, which strives to delete insignificant polynomials from the spectral expansion.

NETWORK MODELS

In this article, classical propagation methods are applied to the water quality model in order to evaluate the influence of parameter uncertainties and compare them to the PCE. This is done in two case studies. The first one is the small, simplified graph of a strongly aggregated distribution network, as shown in Figure 1(a). It contains one loop and three elevated tanks. It has a total number of 21 free elements given by 12 pipes and 9 nodes. The model uses one domestic demand pattern for all nodes and the number of consumers per node is given in Table 1. The topology of the second distribution network is illustrated by the graph in Figure 1(b). In order to test the algorithms under practical conditions, this realistic network model has been provided by the project partner Veolia in the context of the ResiWater project. It is made up of 2,175 pipes, 1,822 nodes and one reservoir in the densely looped part of the network. In all, the model contains almost 4,000 graph elements. The model uses two demand patterns that have been defined and calibrated on operational data.

The general setup that is investigated in this paper is applied to both example networks. In each model, a diurnal demand pattern is defined for a selection of free nodes. This pattern contains a series of demand multipliers. For each hydraulic time step, the according demand factor is multiplied by the number of consumers that is defined individually for each node. Through this behavioural aggregation a large portion of the network is influenced by a concentrated number of variables.

Table 1 | Consumer distribution in the strongly aggregated looped network

Node	1	2	3	4	5	6	7	8	9
Consumers	72	108	84	172	6	51	180	83	116

The inherent aleatory uncertainty in the demand is modelled by a normal distribution with a mean of $\mu = 1$ and standard deviation of $\sigma = 0.3$. This way the variation is applied on the deterministic demand estimate and the generation of negative demand values is highly unlikely. From this distribution, a sample of $N = 1e5$ elements is taken to modify the parameters for the Monte Carlo simulation. From the ensemble of these simulations a reduced number of $N = 1e4$ elements are taken to evaluate the PCE. The propagation uses an EPS over the course of 1 week.

RESULTS

Figure 2 shows the mean water age at nodes 3 and 9 in the network, together with the 95% confidence intervals as a function of the simulated time. It can be seen that after roughly 1 day the result stabilizes and is not dependent on the initial conditions any more. In Figure 3, the estimate PDF is visualized for the same nodes after a simulation period of 58 hours with the histogram for the Monte Carlo approach, the dashed line for the PCE and the point dashed line for the FOSM.

Figure 3(a) gives the distribution at node 3 and Figure 3(b) the one for node 9. The values of the mean values and standard deviations are given in Table 2 along with the PCE order. In the evaluation of the results it has become apparent that for the small network no significant difference can be observed between the minimum, maximum and average water age. The behaviour at the two nodes is very different. The distribution at node 3, while clearly being non-linear, has still a single peak and a limited standard deviation. This observation is supported by the coefficients in Table 3 which decrease rapidly after the second coefficient. In contrast, the distribution at node 9 shows two peaks and the expansion coefficients decrease more slowly. This means that the probable water age in the loop node 3 is higher

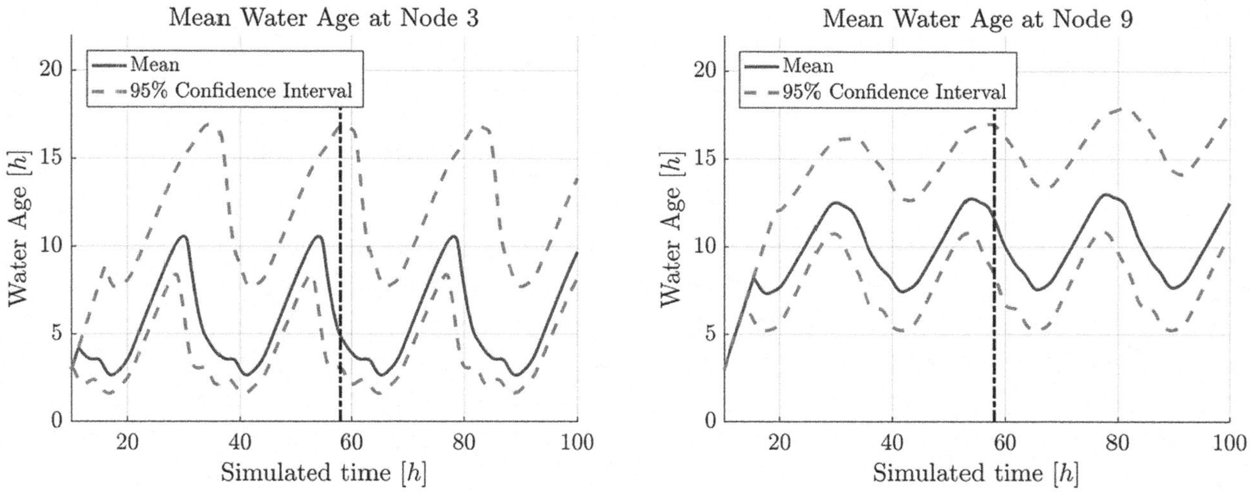

Figure 2 | Temporal development of mean water age at nodes 3 and 9 in the highly aggregated looped network with the 95% confidence interval.

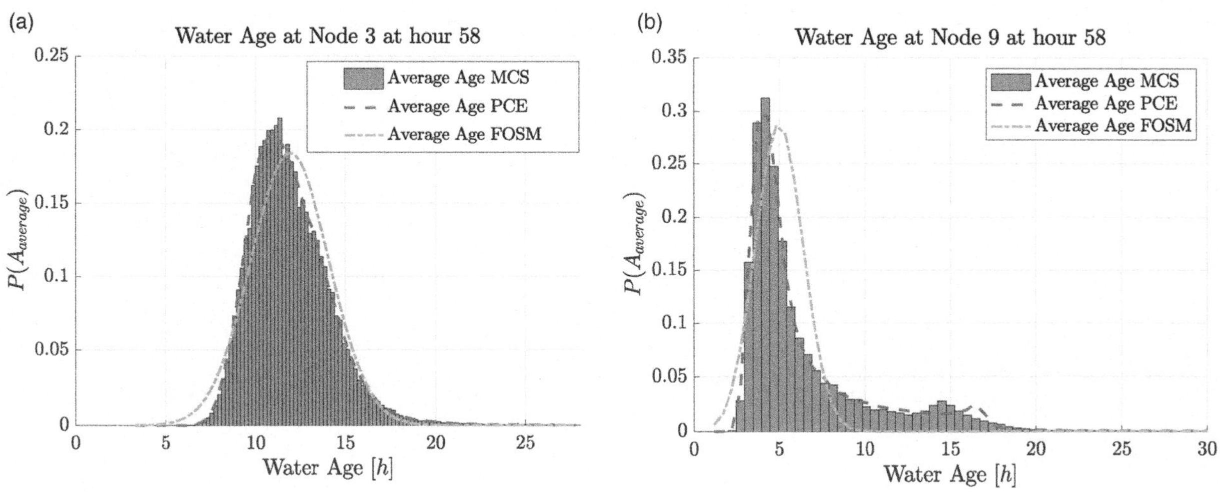

Figure 3 | Water age distributions for small network: (a) node 3 in the meshed part of the system; (b) at node 9 at a branch of the network.

Table 2 | Stochastic moments and PCE order for the probability distribution function of average water age at nodes 3 and 9 of the small network

		Avg water age A_{avg} [h]
Node 3	Mean	$\mu_3 = 11.9$
	Standard deviation	$\sigma_3 = 4.7$
	Expansion order	$O_3 = 7$
Node 9	Mean	$\mu_9 = 6.5$
	Standard deviation	$\sigma_9 = 7.8$
	Expansion order	$O_9 = 9$

than the one in the tree node 9; however, due to the lower standard deviation the level of uncertainty is lower at node 3. The development of the two peaks in Figure 3(b) suggests that, depending on the demand multiplier, the flow structure changes and water arrives from a different node, takes a longer route or flows more slowly. Remixing of old water is unlikely due to the structure of the network with three reservoirs that are connected to the loop. A detailed analysis of the origin reservoir has not been performed as a part of this study but may offer better insight. The results from the Monte Carlo approach and the PCE are in good agreement. For the sensitivity-based FOSM the limitation in capturing the non-linear behaviour becomes obvious. While it gives a good approximation in the mean and variance for node 3, it is unable to approximate the bimodal distribution in node 9.

Figure 4 shows the probability distribution of the water age at node 15Nf79 of the medium-sized network which is indicated by a circle in Figure 1(b). Figure 4(a) displays the

Table 3 | The expansion coefficients for the PCEs of average water age at nodes 3 and 9 of the small network

Order *N*	a_0	a_1	a_2	a_3	a_4	a_5	a_6	a_7	a_8	a_9
Node 3	11.90	−2.12	0.40	0.011	0.05	0.09	0.03	0.04		
Node 9	6.52	−3.67	2.10	−1.52	−0.03	−1.70	0.00	−1.96	0.06	−0.590

distribution after a simulated duration of 116 hours whereas Figure 4(b) gives the distribution after 130 hours. For both times, the water quality has reached a periodicity. However, after a simulated time of 116 hours the water age is at the peak of the water age cycle while at 130 hours it reaches a valley. In contrast to the small network, there is a considerable difference in the three quality indicators of minimum, average and maximum water age and for this reason they are illustrated separately. The first two stochastic moments are given in Table 4 along with the expansion order *N*.

From the stochastic data a few trends can be observed. First, even though the simulation is periodic, it is not cyclic since the mean of all indicators rises with the simulation duration. Second, the distributions of the mean and average water age are very close in the stochastic moments as well as in the expansion coefficients up to the order $N=5$, which is shown in Table 5. From this it may be inferred that the average water age is dominated by the minimum water age and the amount of older water in the network is relatively low. Third, due to the fact that the mean and the variance of the maximum water age are considerably

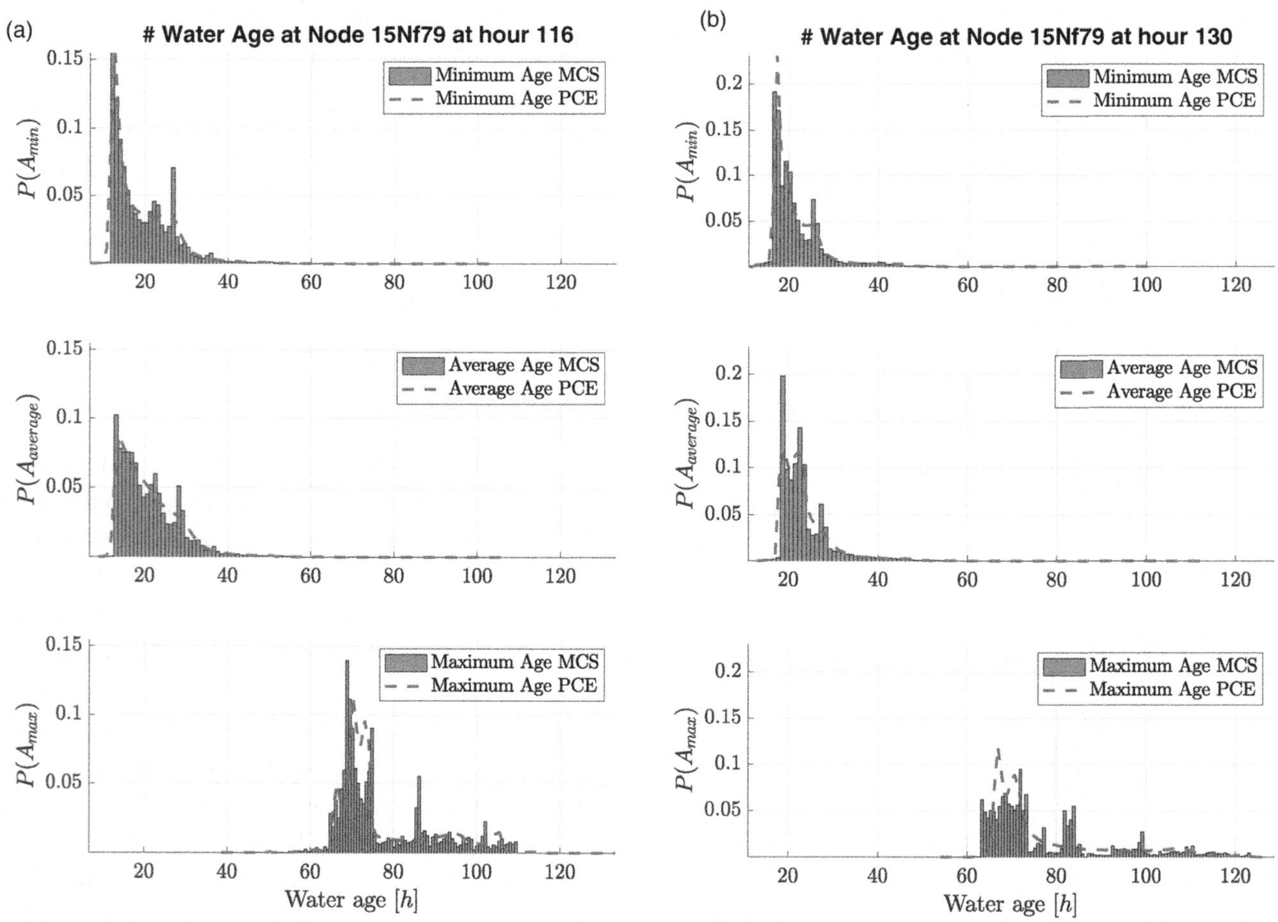

Figure 4 | Water age distributions for the large network: (a) meshed part of the system after 116 hours' simulation time; (b) meshed part of the system after 130 hours' simulation time.

Table 4 | Stochastic moments for the probability distribution function and PCE order of minimum, average and maximum water age at node 15Nf79 for 116 and 130 hours of simulated time in the medium network

Node 15*Nf79*		Min water age A_{min} [h]	Avg water age A_{avg} [h]	Max water age A_{max} [h]
116 hours	Mean	$\mu = 19.2$	$\mu = 21.1$	$\mu = 77.4$
	Standard deviation	$\sigma = 53.7$	$\sigma = 53.1$	$\sigma = 133.0$
	Expansion order	$O=20$	$O=17$	$O=20$
130 hours	Mean	$\mu = 21.6$	$\mu = 23.6$	$\mu = 77.5$
	Standard deviation	$\sigma = 36.3$	$\sigma = 36.7$	$\sigma = 178.3$
	Expansion order	$O=19$	$O=16$	$O=14$

Table 5 | The first ten expansion coefficients for the PCEs of minimal, average and maximal water age at node 15Nf79 for 116 and 130 hours of simulated time in the medium network

	116 hours			130 hours		
Order *N*	A_{min}	A_{avg}	A_{max}	A_{min}	A_{avg}	A_{max}
a_0	19.15	21.11	77.41	21.61	23.60	77.51
a_1	−6.67	−6.38	−10.11	−5.12	−4.67	−11.76
a_2	2.81	3.25	3.63	2.86	3.47	5.57
a_3	−0.24	−0.02	0.00	−1.18	−1.18	0.43
a_4	0.30	0.63	−2.82	0.21	−0.04	−2.33
a_5	−0.86	−1.12	0.82	−0.30	−0.76	0.56
a_6	0.26	−0.13	0.93	0.09	0.25	0.86
a_7	0.24	−0.06	−0.82	0.10	0.36	−0.92
a_8	−0.14	0.13	−0.31	0.24	0.43	−0.19
a_9	−0.20	0.12	1.42	−0.31	−0.42	0.79

higher, it is likely that at least a portion of the water stays in the network by mixing with the new water at pipe junctions. As a result of the more complex structure of the medium-sized network it is not possible to identify changing peaks as an effect of the changing demand multiplier; however, a similar effect may be responsible for the skewed distributions in Figure 3. For the minimum and average water age the PCE is able to approximate the basic shape of the probability density function; however, small fluctuations were not resolved as a high expansion order would be needed in that case. A similar effect can be seen for the maximum water age. The basic shape of the PDF is well approximated by the PCE; however, the histogram from the Monte Carlo approach shows a bifurcation in the distribution with a second peak for a higher water age. These observations are backed by the development of the expansion coefficient in Table 5. In contrast to the application in the strongly aggregated network, the coefficients have not yet converged to zero, implying that a higher order expansion is needed for a more accurate approximation.

Comparing the computational effort, it has to be taken into account that the PCE uses a tenth of the function evaluations used for the Monte Carlo simulation. In comparable applications it has been shown that the PCE converges far quicker in the mean and variance of the probability distribution.

CONCLUSION

The PCE has been applied to a small and a realistic medium-sized water distribution system to evaluate the effect of demand uncertainties on the resident time of water in the network. It has been shown that the PCE gives comparable results to Monte Carlo simulations while reducing the number of evaluations of the full system by an order of magnitude. This means that for a fixed number of simulations the PCE performs considerably better. Even for the relatively small normally distributed perturbation of the demand used in this paper the probability distribution for the water age is far from normal. This means that perturbation methods are not suitable for this task. Further, it has been shown that the size of the network has a considerable influence on the maximum water age.

Additionally, uncertainty quantification was tested on three different kinds of quality indicators. For each of them probability density function may be multi-modal, which has an interest for the WDN management in terms of water safety. The effect of the different water ages is much more distinct in the application of the bigger network. The ensuing difference for the multimodal peaks often correspond to different water paths.

This presented work opens a new direction of research in water distribution networks. It is possible to extend it to applications in a higher dimensional parameter-space; however, the sampling base approach of the MC and non-intrusive PCE is limited to application with lower numbers of parameters, as the computational cost increases exponentially. This effect can, in part, be mitigated through the application of sparse-grid sampling methods or Latin hypercube sampling.

ACKNOWLEDGEMENTS

The work presented in the paper is part of the French–German collaborative research project ResiWater that is funded by the French National Research Agency (ANR; project: ANR-14-PICS-0003) and the German Federal Ministry of Education and Research (BMBF; project: BMBF-13N13690).

REFERENCES

Blatman, G. & Sudret, B. 2011 Adaptive sparse polynomial chaos expansion based on least angle regression. *Journal of Computational Physics* **230** (6), 2345–2367.

Cacuci, D. G., Ionescu-Bujor, M. & Navon, I. M. 2005 *Sensitivity and Uncertainty Analysis, Vol. II: Applications to Large-Scale Systems*, Vol. 2. CRC Press, Boca Raton, FL, USA.

Cross, H. 1936 *Analysis of Flow in Networks of Conduits or Conductors*. University of Illinois at Urbana Champaign, College of Engineering. Engineering Experiment Station, IL, USA.

Dandy, G., Mantopoulos, A., Qiu, M., Shizas, G. & Yao, Z. 2013 Advantages and limitations of the use of water age as a surrogate for water quality in water distribution systems. In: *EWRI 2013*, ASCE, Cincinnati, OH, USA.

Der Kiureghian, A. & & Ditlevsen, O. 2009 Aleatory or epistemic? Does it matter? *Structural Safety* **31** (2), 105–112.

Di Cristo, C., Leopardi, A. & de Marinis, G. 2015 Assessing measurements uncertainty on trihalomethanes prediction through kinetic models in water supply systems. *Journal of Water Supply: Research and Technology-AQUA* **64** (5), 516–528.

Fabrie, P., Gancel, G., Mortazavi, I. & Piller, O. 2010 Quality modeling of water distribution systems using sensitivity equations. *Journal of Hydraulic Engineering-ASCE* **136** (1), 34–44.

Fu, G. & Kapelan, Z. 2011 Fuzzy probabilistic design of water distribution networks. *Water Resources Research* **47**, W05538. doi:10.1029/2010WR009739.

Hart, D., Rodriguez, J. S., Burkhardt, J., Borchers, B., Laird, C., Murray, R., Klise, K. & Haxton, T. 2019 Quantifying hydraulic and water quality uncertainty to inform sampling of drinking water distribution systems. *Journal of Water Resources Planning and Management* **145** (1), 13.

Helton, J. C. & Davis, F. J. 2003 Latin hypercube sampling and the propagation of uncertainty in analyses of complex systems. *Reliability Engineering & System Safety* **81** (1), 23–69.

Jungnickel, D. 2005 *Graphs, Networks and Algorithms*. Springer, Berlin, Germany.

Le Maître, O. & Knio, O. M. 2010 *Spectral Methods for Uncertainty Quantification: With Applications to Computational Fluid Dynamics*. Springer Science & Business Media, Dordrecht, The Netherlands.

Piller, O. & Tavard, L. 2014 Modeling the transport of physicochemical parameters for water network security. *Procedia Engineering* **70**, 1344–1352.

Piller, O., Gilbert, D. & Van Zyl, J. E. 2010 Dual calibration for coupled flow and transport models of water distribution systems. In: ASCE (ed.), *Water Distribution Systems Analysis* WDSA, Tucson, AZ, pp. 722–731.

Piller, O., Gilbert, D., Haddane, K. & Sabatie, S. 2011 Porteau: An object-oriented programming hydraulic toolkit for water distribution system analysis. In: *Eleventh International Conference on Computing and Control for the Water Industry (CCWI 2011)*, 5–7 September 2011, Exeter, UK.

Raei, E., Shafiee, M. E., Nikoo, M. R. & Berglund, E. 2018 Placing an ensemble of pressure sensors for leak detection in water distribution networks under measurement uncertainty. *Journal of Hydroinformatics* **21** (2), 223–239.

ResiWater 2019 *ANR and BMBF Project, Innovative Secure Sensor Networks and Model-Based Assessment Tools for Increased Resilience of Water Infrastructures*. http://www.resiwater.eu/ (accessed 15 January 2019).

Rossman, L. A. & Boulos, P. F. 1996 Numerical methods for modeling water quality in distribution systems: a comparison. *Journal of Water Resources Planning and Management* **122** (2), 137–146.

Savic, D. A., Kapelan, Z. S. & Jonkergouw, P. M. 2009 Quo vadis water distribution model calibration? *Urban Water Journal* **6** (1), 3–22.

Smith, R. C. 2013 *Uncertainty Quantification: Theory, Implementation, and Applications*, Vol. 12. Society for Industrial and Applied Mathematics, Philadelphia, PA, USA.

Sumer, D. & Lansey, K. 2009 Effect of uncertainty on water distribution system model design decisions. *Journal of Water Resources Planning and Management* **135** (1), 38–47.

Tsakiris, G. & Splilotis, M. 2017 Uncertainty in the analysis of urban water supply and distribution systems. *Journal of Hydroinformatics* **19** (6), 823–837.

Xiong, F., Xiong, Y., Greene, S., Chen, W. & Yang, S. 2009 A new sparse grid based method for uncertainty propagation. In: *ASME 2009 International Design Engineering Technical Conferences and Computers and Information in Engineering Conference*, San Diego, CA, USA. American Society of Mechanical Engineers, pp. 1205–1215.

Xiu, D. 2010 *Numerical Methods for Stochastic Computations: A Spectral Method Approach*. Princeton University Press, Princeton, NJ, USA.

Xiu, D. & Karniadakis, G. E. 2002 The Wiener–Askey polynomial chaos for stochastic differential equations. *SIAM Journal on Scientific Computing* **24** (2), 619–644.

Xu, T. Y. & Qin, X. 2014 Integrating decision analysis with fuzzy programming application in urban water distribution system operation. *Journal of Water Resources Planning and Management* **140** (5), 638–648.

First received 15 January 2019; accepted in revised form 4 May 2019. Available online 3 June 2019

Edge betweenness for water distribution networks domain analysis

Antonietta Simone, Francesco G. Ciliberti, Daniele B. Laucelli, Luigi Berardi and Orazio Giustolisi

ABSTRACT

Complex network theory (CNT) studies the relevance of elements in networks using centrality metrics. From the CNT standpoint, water distribution networks (WDNs) are infrastructure networks composed by vertices, named nodes, connected to each other by edges, named pipes, that transfer water to customers following a transfer process based on shortest paths. The present paper proposes the domain analysis of several real WDNs using the edge betweenness in order to capture the hydraulic behaviour based on network structure, i.e., for understanding the role of topological features in the emergent hydraulic behaviour. The strategy is obtained by tailoring CNT studies and tools in order to (i) embed the different hydraulic roles of sources and demand nodes, (ii) move the classic concept of centrality from the nodes to the pipes, i.e., the technically relevant components for WDNs and (iii) include information related to the directional devices, because they constrain flow directions. Results show the usefulness of the novel WDN-tailored edge betweenness for the WDN domain analysis. Therefore, the metric can represent a useful tool for supporting WDNs analysis, design and management tasks.

Key words | centrality metrics, complex network theory, domain analysis, edge betweenness, water distribution networks

Antonietta Simone (corresponding author)
Francesco G. Ciliberti
Daniele B. Laucelli
Orazio Giustolisi
Politecnico di Bari,
Via E. Orabona 4, 70125 Bari,
Italy
E-mail: *antonietta.simone@poliba.it*

Luigi Berardi
University 'G. d'Annunzio' of Chieti Pescara,
Viale Pindaro 42, 65127 Pescara,
Italy

NOTATION

CNT	Complex network theory
WDN	water distribution network
i	generic vertex (node) in the network
V	set of vertices in the network
l	generic edge (link) in the network
E	set of edges in the network
σ_{st}	number of shortest paths from node s to node t
$\sigma_{st}(l)$	number of shortest paths from node s to node t passing along the edge l
C_l^B	edge betweenness of link l
hub	component of a network with a high-degree node
N_f	number of connected fictitious nodes in the network
M	number of reservoirs in the network
N_d	number of demand nodes in the network
V_T	maximum water volume supplied during the emptying process
V_D	average water volume supplied during the operating cycle to the demand nodes

doi: 10.2166/hydro.2019.030

INTRODUCTION

Many centrality metrics have been proposed over the years in order to clarify the concept of importance, in terms of centrality, detailing components and interrelationships among various elements of networks. Freeman (1979) was among the first to propose a definition of centrality by referring to the graph theory, even if the concept of centrality was introduced by Bavelas (1950) for studying communication between small groups. Several centrality metrics have been proposed since by many researchers in order to evaluate the most central element in the network with respect to different physical phenomena: (1) Katz centrality (Katz 1953) computes the relative influence of a node within a network by measuring the number of the immediate and secondary neighbours, i.e., the centrality is based on the number of walks deriving from a node *i*. (2) Eigenvector (Bonacich 1972) measures the influence of a node in a network, based on the idea that the influence of node *i* is related to the influence of its neighbours, i.e., a node with high eigenvector is connected to many nodes also with high eigenvector. (3) Degree (Freeman 1977) is based on the number of connections of each node where the node with the highest degree is the most central. (4) Closeness (Freeman 1977) measures the centrality of a node considering how it is relatively close to all other nodes. (5) Betweenness (Freeman 1977) measures the importance of an element quantifying how many times the element is traversed by shortest paths between two vertices. (6) PageRank (Page *et al.* 1998) is an adjustment of Katz centrality that smooths the centrality of a node when it is only related to the centrality of a connected node. (7) Cross-clique centrality (Everett & Borgatti 1998; Faghani 2013) determines the connectivity of a node to different cliques, i.e., the number of cliques to which it belongs. A node with high cross-clique connectivity facilitates the propagation of information in a graph. (8) Hub and Authorities (Kleinberg 1999) quantify the centrality of nodes as a function of the centrality of the connection nodes. The metric refers to the creation of Web pages, especially hubs. These Web pages, even if not authoritative, represent large directories that lead users directly to the authoritative pages, i.e., a good hub represents a page that points to many other pages while good authority represents a page linked by many different hubs. (9) Decay centrality (Jackson 2008) measures the proximity between a chosen vertex and every other vertex weighted by the decay. (10) Percolation (Piraveenan *et al.* 2013) quantifies the relative impact of nodes based on their percolation states. For a given node, at a given time, it represents the proportion of 'percolated paths' that go through that node, i.e., shortest path between a pair of nodes, where the source node is percolated (e.g., infected). (11) Neighbourhood degree (Giustolisi *et al.* 2017) corresponds to tailoring the degree centrality concept for infrastructure networks by involving adjacent nodes belonging to neighbourhoods at different topological distance.

These centrality metrics aim at ranking the nodal importance in a network, i.e., how much a node is relevant with respect to other nodes in the network. However, the ranking only orders nodes by importance, without quantifying the difference in importance between different levels of the ranking. To overcome this problem, Freeman (1979) proposed the concept of centralization, i.e., the centralization of any network is a measure of how central its most central node is in relation to how central all the other nodes are.

It is important to note that each metric identifies different elements as the most relevant (central) with respect to different applications/physical phenomena (Benzi & Klymko 2015). To this purpose, Borgatti (2005) associated specific metrics to different types of network according to the flow process that characterizes the network. He stated that the use of a metric must be appropriate to the kind of flow process that characterizes the network, i.e., not all metrics can be unconditionally applied to all types of networks. Then, he defined three typologies of flow processes and, for each, suggested the more suitable centrality metrics. Accordingly, the flow process for spatial networks, such as water distribution networks (WDNs), was defined as 'package delivery' like, i.e., a process based on the concept of shortest paths with a fixed target. Finally, Borgatti (2005) stated that for this type of process the more suitable centrality metrics to apply are betweenness and closeness centrality. Later, Demšar *et al.* (2007) proposed an approach towards computing three centrality metrics, i.e., degree, closeness and betweenness, for identifying critical locations in spatial networks by dual graph modelling (Evans & Lambiotte 2009). They claimed that the betweenness is the most suitable for identification of critical locations, especially for structural analysis useful for modelling vulnerability and survivability of spatial networks. From this general picture it emerges that specific Complex network theory (CNT) centrality metrics can be potentially useful for WDNs analysis, planning and management, e.g., in terms of reliability and vulnerability.

The vulnerability analysis of WDNs using centrality metrics was first addressed by Yazdani & Jeffrey (2010, 2011, 2012a, 2012b). They proposed a variety of strategies for understanding the structure, efficiency, robustness and vulnerability of WDNs using centrality metrics and concluded that CNT metrics are useful and necessary but perhaps insufficient for analysing the structural reliability or vulnerability of spatial networks because of spatial constraints. Furthermore, they proposed a novel metric based on the connectivity features for the analysis of weighted and directed WDNs. From here on, CNT centrality metrics were adopted to assess vulnerability analysis (Hawick 2012; Diao *et al.* 2014; Shuang *et al.* 2014), reliability assessment (Ostfeld 2012) and water distribution network (WDN) resilience (Pandit & Crittenden 2016). Later on, Giustolisi *et al.* (2017) proposed the neighbourhood degree, i.e., an extension of the classic nodal degree, in order to demonstrate that the WDN connectivity structure follows a Poisson-like distribution because of spatial constraints, meaning that the domain of these spatial systems, on average (i.e., not considering reservoirs, tanks and directional devices), has a good structural resistance to random failures and intentional threats (Giustolisi *et al.* 2017).

It has to be noted that all proposed works on centrality metrics focus on node relevance, while, in the case of WDNs, pipes are the most important components (Simone *et al.* 2018). This means the focus should move from nodes to pipes, because pipes, with their asset features, are the most important network components for hydraulic systems. To this end, Giustolisi *et al.* (2019) proposed the use of the edge betweenness (Girvan & Newman 2002) for analysing the domain of WDNs. In fact, they proposed to tailor the network topology to embed the different hydraulic roles of nodes (source nodes, tanks and demand nodes) including information related to the directional devices and different weights for pipes, in order to consider pipe properties (e.g., hydraulic resistance). This way, the analysis of the domain can be useful to quantify, for example, the feasibility of management interventions in advance with respect to hydraulic modelling.

The aim of the present work is to show the effectiveness of domain analysis for several real WDNs using the WDN-tailored edge betweenness proposed by Giustolisi *et al.* (2019), in order to demonstrate how it can represent a useful tool to support the hydraulic analysis of a WDN. To this end, this study takes WDNs with different size, asset features and hydraulic schemes in real decision support contexts.

The paper is organized as follows. The next section presents the main concepts of the strategy. This is followed by a section that reports several case studies, a brief discussion of the strategy and an interpretation of the main results. Conclusions are given in the final section.

EDGE BETWEENNESS AND TOPOLOGY TAILORING

This study aims at assuming the edge betweenness proposed by Girvan & Newman (2002) as centrality metric suitable for WDN domain analysis in order to consider the relevance of pipes and not of nodes (Giustolisi *et al.* 2019). In the case of WDNs, nodes are generally elements to transfer water (information) and water is delivered to customers at pipe level. In fact, in WDN models, nodes where water demand is assumed to be delivered are not significant for the connectivity structure of the WDNs because the elements that can fail are pipes while connection nodes generally do not fail.

The edge betweenness is similar to the betweenness centrality, but it refers to the generic edge l. Assume that $\sigma_{st}(l)$ is the number of shortest paths from node s to node t passing along the edge l and σ_{st} is the number of all shortest paths from node s to node t.

The edge betweenness C_l^B of edge l is the sum of the fractions $\sigma_{st}(l)/\sigma_{st}$ for all the couples of nodes s, t in the network:

$$C_l^B = \sum_{\substack{s \neq t \in V \\ l \in E}} \frac{\sigma_{st}(l)}{\sigma_{st}} \tag{1}$$

where V and E are the set of vertices and edges belonging to the network. The edge with the highest edge betweenness centrality is the most relevant.

Therefore, even if the edge and the nodal metrics are performed with the same procedure and algorithm (Dijkstra 1959), the edge betweenness is more meaningful for WDNs, accounting for flow path disruption.

The water source nodes (i.e., reservoirs and tanks) represent the only exception, because they play a different hydraulic role with respect to other nodes and, therefore, they need a specific study in order to define a hydraulics-based topology. In fact, the strategy proposed by Giustolisi *et al.* (2019) considers the source node as a kind of hub with respect to the hydraulic system behaviour. In absence of source nodes, the hydraulics does not exist, and the connectivity structure just embeds the topology. For the water source analysis, let us assume N_f connected fictitious nodes, M reservoirs and N_d demand nodes. This assumption arises from the fact that the reservoirs are generally linked to the networks with only one (suburban) pipe. This topological characteristic gives it a low degree of connection even if, from a hydraulic point of view, it represents a hub from which all water paths towards demand nodes start. This way, the relevance of the position of original source nodes is preserved with respect to the entire network structure although amplified by N_f. Accordingly, the relevance of the pipes close to the reservoir increases, while the other pipes' relevance does not change considerably. For each reservoir there exists a star of N_f connected fictitious nodes whose number is equal to N_d/M. The repartition of the demand nodes is related to the real reservoir influence area, i.e., a different value of N_f for each reservoir exists depending on its area of influence.

The next step is inserting the domain information about tanks, i.e., nodal sources characterized by an emptying/filling process. The information is related to the maximum water volume supplied during the emptying process (V_T) by the tank. Let us assume N_f connected fictitious nodes already attributed to reservoirs and N_d demand nodes. The number of fictious nodes (usually definable as 'star of fictitious nodes') attributed to each tank equals the minimum integer value between $N_d \cdot V_T/V_D$ and N_f, where V_D is the average water volume supplied during the operating cycle to the demand nodes. It is important to note that increasing the volume V_T with respect to V_D, the tank tends to the reservoir and decreasing the volume V_T the tank node tends to a demand or connectivity node. However, tanks are always less relevant hubs than reservoirs.

Directional devices (e.g., pumps, pressure reduction valves, check valves, etc.) and flow direction for pipes next to reservoirs are also considered to enhance the domain analysis. Indeed, the presence of directional devices allows water to flow only in one direction. In this way, for example, the shortest paths between two pipes, one upstream and the other downstream of a specific device, can only be travelled in one direction, i.e., that of the device. The consequence is that the edge betweenness of these pipes is automatically reduced compared to that computed on the network without devices.

The last step is to further refine the WDN-tailored edge betweenness metric attributing weights to the pipes as, for example, the pipes' hydraulic resistance. These weights allow moving the network domain analysis closer to the expected hydraulic behaviour of the system; pipe hydraulic resistances, for example, drive the water fluxes.

CASE STUDIES

The WDN-tailored edge betweenness is here applied and discussed using BBLAWN and other six Apulian real WDNs, in order to show the effectiveness of the analysis for different typology of WDNs. The relevant characteristics of analysed networks are reported in Table 1. Table 1 also reports the Spearman correlation index (Spearman 1904) between two metrics considering several weights, i.e., connectivity, length and resistance, for the edge betweenness. This way the analysis includes the domain characteristics of the networks.

Figure 1 (left panel) reports the pipe with the highest 5% of flow rate values averaged on 1-day hydraulic simulation; (right panel) the pipes with highest 5% values of WDN-tailored edge betweenness for the BBLAWN network. The most relevant pipes correspond for the two metrics. Therefore, the WDN-tailored edge betweenness is able to identify almost all the most important pipes, as ranked by pipe flow rates, only on the basis of the domain structure of the system. The correlation index for BBLAWN in Table 1, i.e., weighting the WDN-tailored edge betweenness with pipe resistance, is equal to 0.74. This result confirms the usefulness of the edge betweenness for WDN domain analysis. It can be argued that information about reservoirs, tanks and directional devices enhances the domain analysis. Furthermore, Figure 1 shows that the analysis defines a main path into the network that starts from the only reservoir and reaches the tank on the left. From this main conduct, two other paths delineate the remaining part of the network

Table 1 | Topological characteristics and correlation index for the seven analysed WDNs

WDN name	Node #	Pipe #	Reservoirs #	Tanks #	Demand nodes #	Inhabitants [×1000] #	Corr. index resistance	Corr. index connectivity	Corr. index length
BBLAWN	390	439	1	7	334	–	0.74	0.75	0.75
Apulia 1	986	1,105	1	0	663	12	0.64	0.71	0.73
Apulia 2	7,716	8,496	3	0	7,011	12	0.60	0.63	0.65
Apulia 3	1,270	1,472	3	0	992	199	0.54	0.50	0.53
Apulia 4	3,166	3,483	2	0	1,582	26	0.66	0.75	0.79
Apulia 5	7,164	7,895	1	0	2,918	70	0.62	0.70	0.75
Apulia 6	1,111	1,307	1	0	770	13	0.54	0.59	0.60

'skeleton', i.e., the main structure of the network. Both two metrics indicate the most relevant pipe as those close to the only reservoir, i.e., the most relevant hub.

Figure 2 reports the flow (left panel) and the WDN-tailored edge betweenness (right panel) for Apulia 1 network. The network is composed of a single reservoir and a pressure control valve, for controlling pressure and reducing water losses. The presence of a single reservoir uniquely defines the main path of the pipes, which starts from the water source, i.e., the only hub, and reaches the centre of the network (see the darker circles). Most of the pipes with slightly lower metrics values (see the lightest circles) act as links between the main path and the rest of the network. The correlation index between two metrics, considering the pipe resistance as weight for the edge betweenness, is equal to 0.64, confirming that the information contained in the domain is relevant for the WDN hydraulics. Table 1 shows that for Apulia 1 network the correlation assumes higher values considering the length as weight for the pipes (i.e., 0.73). This occurs for six of the seven analysed networks, although graphically the pipe hydraulic resistance better corresponds to the flow distribution as expected from an engineering perspective. For this reason, pipe hydraulic resistance was chosen as weight to discuss the metric. However, it is important noting that, regardless of weight, the correlation index is always greater than 0.50, i.e., the edge betweenness well identifies the hydraulic behaviour of the networks.

Figure 3 reports the flow rates (left panel) and the WDN-tailored edge betweenness (right panel) for the Apulia 2

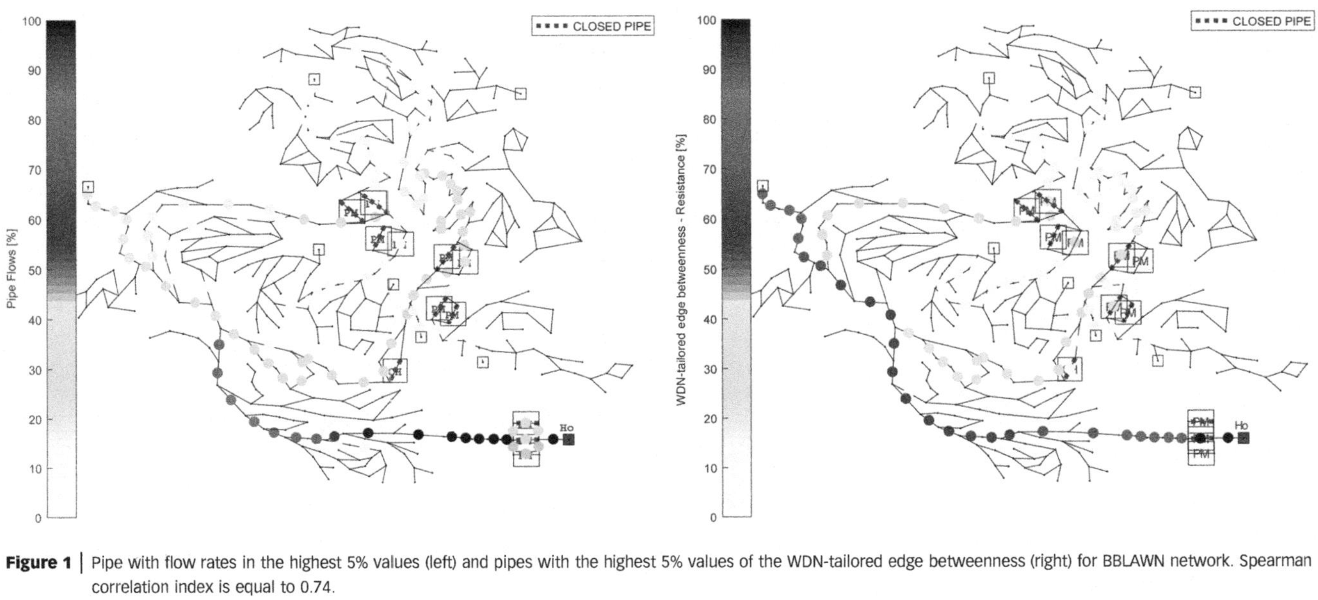

Figure 1 | Pipe with flow rates in the highest 5% values (left) and pipes with the highest 5% values of the WDN-tailored edge betweenness (right) for BBLAWN network. Spearman correlation index is equal to 0.74.

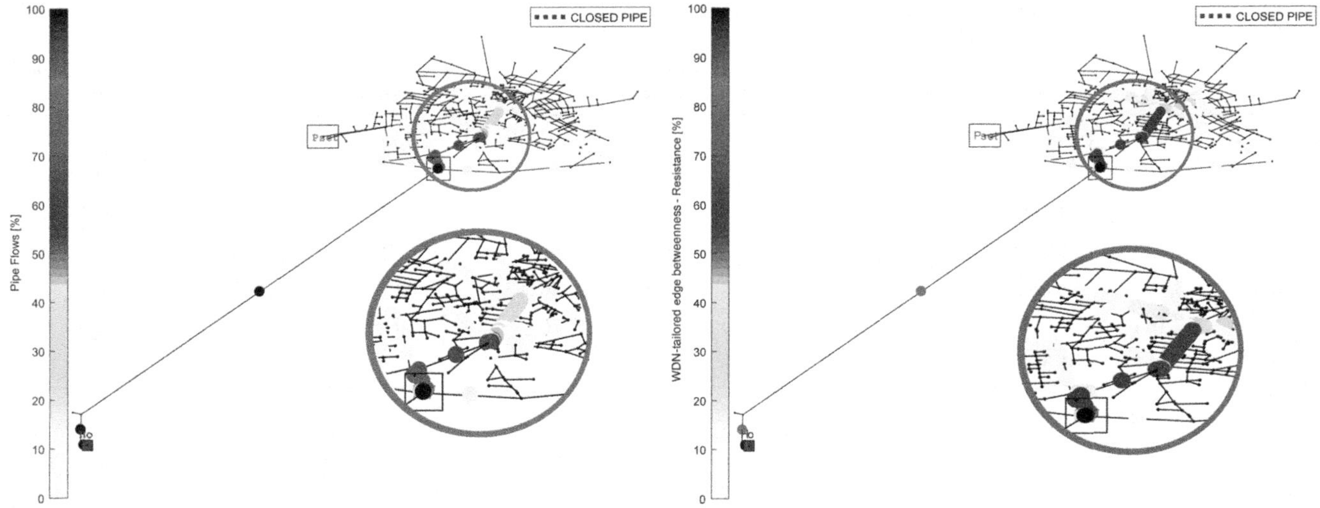

Figure 2 | Pipe with flow rates in the highest 5% values (left) and pipes with the highest 5% values of the WDN-tailored edge betweenness (right) for Apulia 1 network. Spearman correlation index is equal to 0.64.

network. Apulia 2 is a large looped network with three reservoirs, five controlled entries and a wide range of diameters. Comparing the two panels of Figure 3 and looking at the correlation indices, it can be stated that also for such a large network, the proposed tailored-metric identifies quite well all the most important pipes in terms of pipe flow rates. This is also confirmed by correlation indices equal to 0.60, 0.63 and 0.65 in Table 1. It is important to note that, in addition to the five suburban pipes, some of the most important paths are positioned into the looped part of the network, corresponding to the city centre. The complexity of the network, also due to the presence of many reservoirs and pressure control valves, influences the correlation index values. This means that, for Apulia 2, the values obtained considering devices are lower than those without devices. The high correspondence between the two panels of Figure 3 confirms this consideration.

Figure 4 reports pipe flow rates (left panel) and the WDN-tailored edge betweenness (right panel) for the Apulia 3 network. The presence of three reservoirs leads to identifying a higher number of main paths in the network. Nevertheless, since each reservoir has its own area of influence in terms of

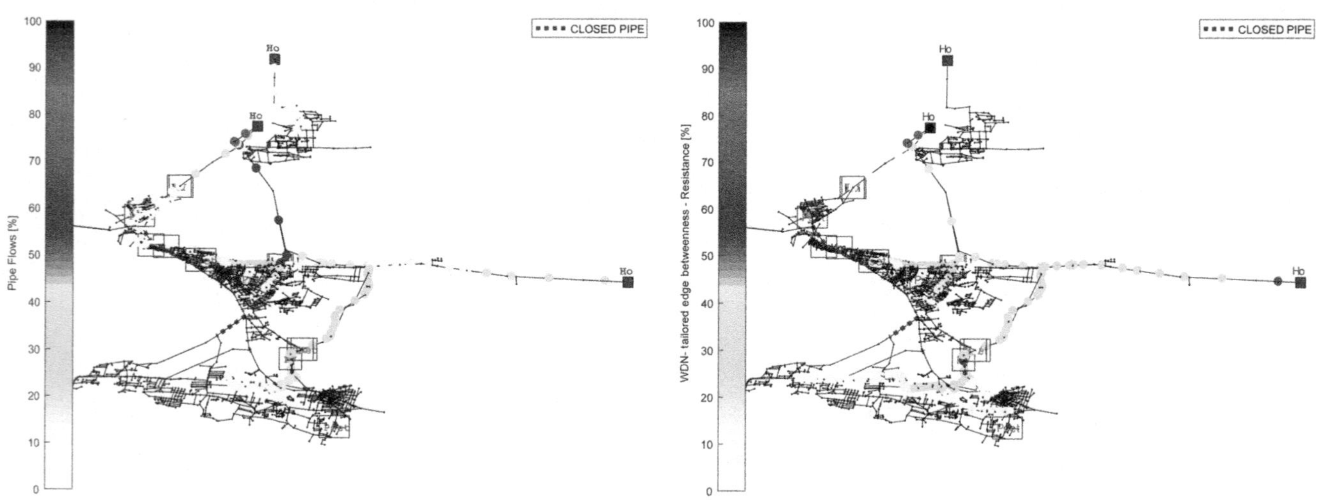

Figure 3 | Pipe with flow rates in the highest 5% values (left) and pipes with the highest 5% values of the WDN-tailored edge betweenness (right) for the Apulia 2 network. Spearman correlation index is equal to 0.60.

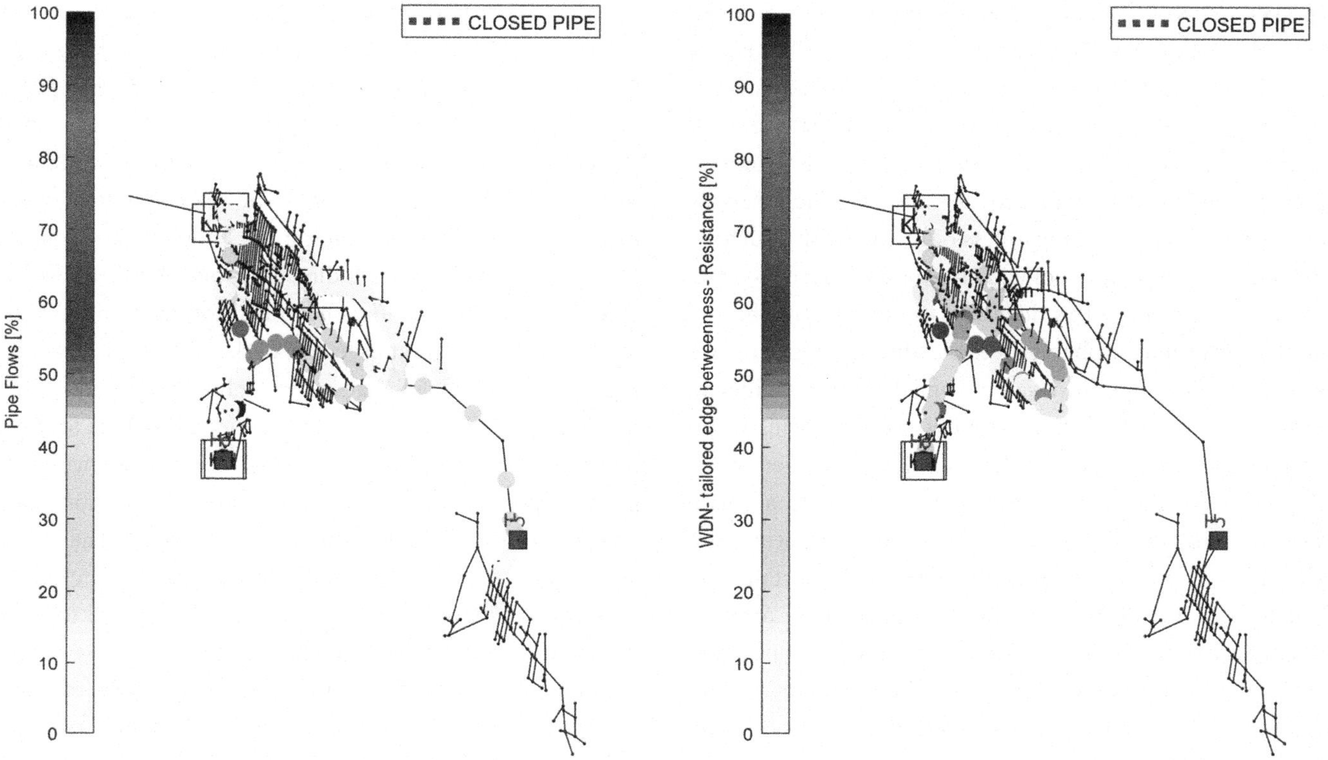

Figure 4 | Pipe with flow rates in the highest 5% values (left) and pipes with the highest 5% values of the WDN-tailored edge betweenness (right) for the Apulia 3 network. Spearman correlation index is equal to 0.54.

demand nodes, each main path has a different relevance. The number of reservoirs influences the correlation index, obtaining a value equal to 0.54 considering the pipe resistance as weight, while for connectivity and length the value decreases to 0.53 and 0.50, respectively, i.e., values lower than those evaluated for less complex networks.

Figure 5 reports pipe flow rates (left panel) and the WDN-tailored edge betweenness (right panel) for the

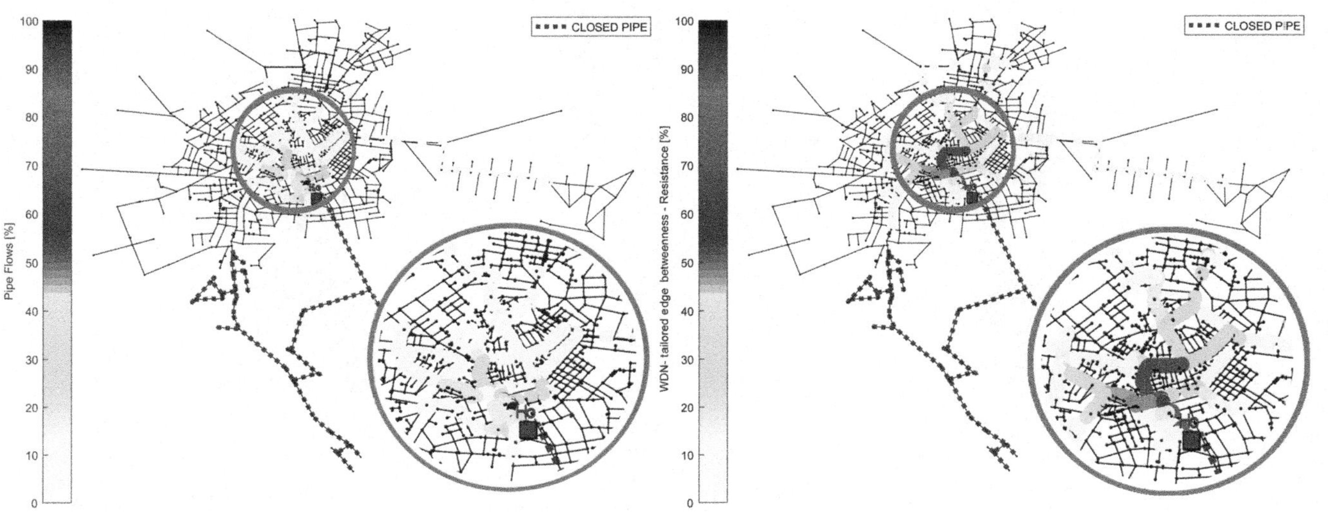

Figure 5 | Pipe with flow rates in the highest 5% values (left) and pipes with the highest 5% values of the WDN-tailored edge betweenness (right) for the Apulia 4 network. Spearman correlation index is equal to 0.66.

Apulia 4 network. The dotted lines indicate a portion of the network no longer in use. The network is strongly looped, i.e., it has a very high redundancy, and this aspect favours the presence of several pipes/paths with similar relevance instead of a few pipes/paths with much higher relevance than the others. It is important to note that the reservoir always represents a hub regardless of the network structure. The most relevant pipes coincide for both the metrics, even if the edge betweenness identifies a greater number of paths as relevant, probably because the topologic metric weighted with pipe resistance considers the diameters regardless of the flow that actually crosses them, while the hydraulic metric defines them uniquely. The very regular nature of the network and the presence of a single reservoir provides a good correlation index between the two metrics, i.e., 0.66 considering the pipe resistance as weight, 0.75 and 0.79 assuming the connectivity and the length, respectively. Once again, the correlation index assumes a higher value considering the length as weight. However, this result makes the analysis very close to the hydraulics of the system.

Figure 6 reports pipe flow rates (left panel) and the WDN-tailored edge betweenness (right panel) for the Apulia 5 network. Apulia 5 is a large network composed of a single reservoir and a pressure control valve downstream of the water source in order to control and manage the pressure into the network. The size of the network means that the hydraulic metric (i.e., flow rates) identifies several main paths, not all directly connected to the reservoir. The edge betweenness identified the same main paths as the most important (see the darker circles). However, Figure 6 (right) shows that in the highest 5% values of the topologic metric there are other pipes/paths that are less relevant (see the lightest circles) from the hydraulic perspective, i.e., the flow has no direct correspondence with the resistance/diameters for all the pipes. The correlation index shows once again that the analysis is close to the hydraulic of the system, obtaining values equal to 0.62, 0.70 and 0.75 in Table 1 weighting the edge betweenness with pipe resistance, connectivity and length, respectively.

Figure 7 reports pipe flow rates (left panel) and the WDN-tailored edge betweenness (right panel) for the

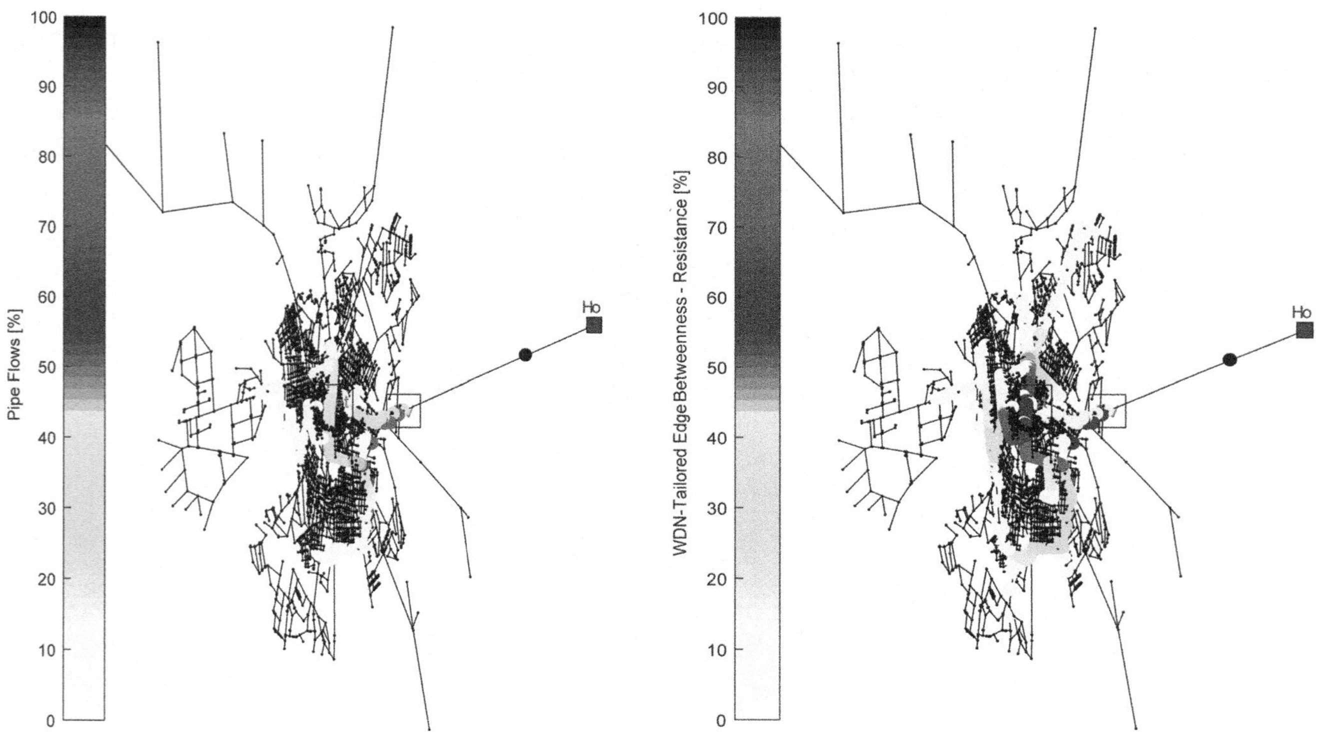

Figure 6 | Pipe with flow rates in the highest 5% values (left) and pipes with the highest 5% values of the WDN-tailored edge betweenness (right) for the Apulia 5 network. Spearman correlation index is equal to 0.62.

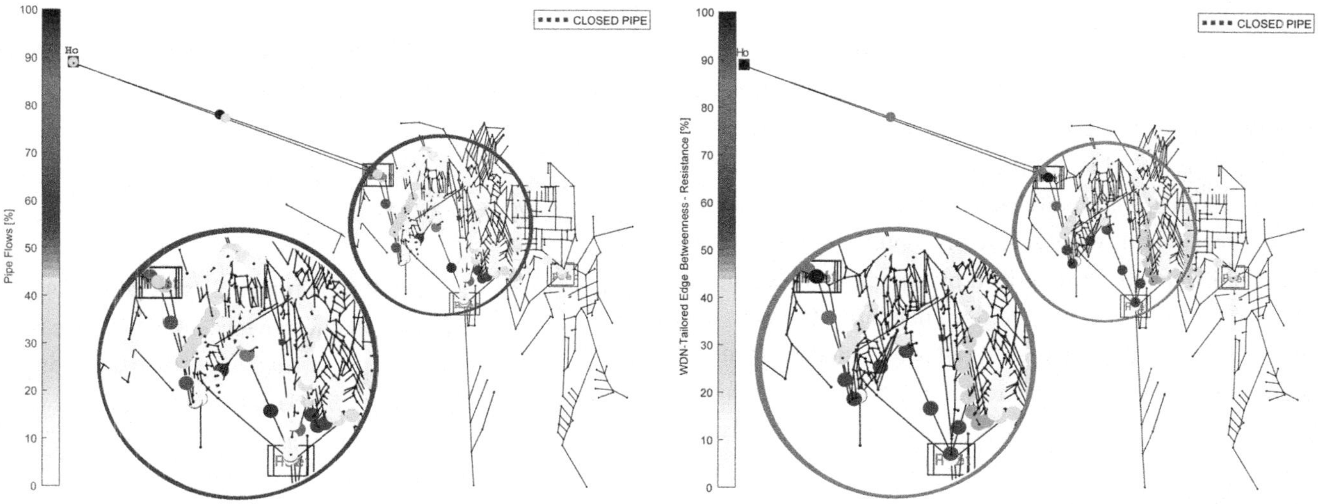

Figure 7 | Pipe with flow rates in the highest 5% values (left) and pipes with the highest 5% values of the WDN-tailored edge betweenness (right) for the Apulia 6 network. Spearman correlation index is equal to 0.54.

Apulia 6 network. The left panel shows that the pipe flow rates do not define properly a main path, but rather highlight the most relevant pipes, i.e., where the value falls in the highest 5% flow rate value, with circles that range from dark to light. The right panel shows the same pipes as the most relevant for the topological metric, confirming that the pipes close to the reservoir are the most relevant. The correlation index between two metrics, considering the pipe resistance as weight, is equal to 0.54, and as for the previous case, Table 1 reports a higher correlation index if considering the length as weight (i.e., 0.60). The presence of four pressure control valves, i.e., four directional devices into the network, probably influenced the evaluation of the correlation indices, because of the directional devices.

The comparison between topological and hydraulic metrics is really positive, also considering that the pipe flows depend on the boundary conditions and on the momentum and continuity equations while the proposed metric focuses only on the connectivity structure. Such results confirm that the hydraulic behaviour is strongly dependent on the domain features and that the proposed metric can be useful to predict it and support further decisions on network operation and management.

In order to show that the hydraulic behaviour is strongly dependent on the domain features embedded in the edge betweenness, Figure 8 shows a diagram that compares the distribution of the pipe flow rates (sorted in descending order) with the corresponding distribution of the tailored edge betweenness for Apulia 6. The comparison confirms

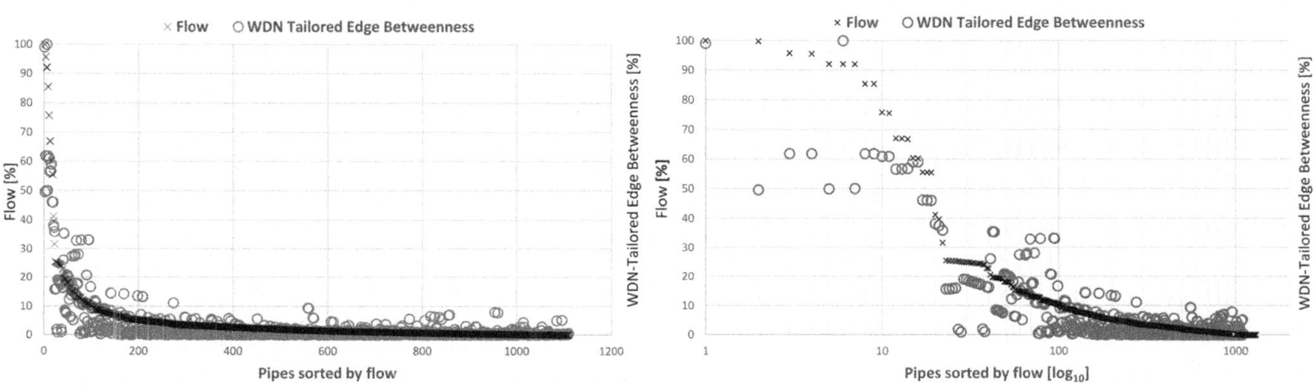

Figure 8 | Comparison between the pipe flows and the tailored edge betweenness.

that the correspondence between the two metrics is very good. The same trend has been obtained for the other WDNs reported herein. Please note that the diagram on the right of Figure 8 has the x-axis in logarithmic scale.

Therefore, looking at the results obtained with the domain analysis, it is possible to state that the edge betweenness, tailored and weighted, can capture the emergent hydraulic behaviour as described by pipe flow rates, due to the connectivity structure of the network. In fact, the domain analysis for seven real WDNs provided correlation index values ranging in the interval [0.50; 0.79], confirming the usefulness of the topological metric for WDN analysis. Furthermore, a further advantage is the possibility to get effective information on the network without the need of many hydraulic simulations and with a low computational cost. For instance, with respect to the largest network herein (Apulia 2), the calculation takes about 60 seconds on a standard laptop equipped with a CPU Intel i7, 2.6 GHz with 16 GB of RAM. Although the time taken is larger than a single hydraulic simulation, it pertains to the main topological domain of the system and needs to be performed only once, unless major connectivity changes happen.

CONCLUSIONS

This contribution proves the effectiveness of the domain analysis of WDNs using the edge betweenness, tailored for accounting the WDNs constraints, as proposed by Giustolisi *et al.* (2019), on seven real networks. The strategy embeds the different hydraulic roles of nodes, moves the concept of centrality from the nodes to the pipes and includes information related to the directional devices. Furthermore, the tailored edge betweenness is weighted with domain information (e.g., pipe resistance). The Spearman correlation index between hydraulic and topological metrics has been evaluated for the seven WDNs considering as weights the connectivity, the length and the resistance. The results confirm the usefulness of the edge betweenness for the WDN domain analysis, i.e., the WDN-tailored edge betweenness can represent a useful tool for supporting WDN analysis, design and management tasks. Indeed, although the WDN-tailored edge betweenness cannot replace hydraulic simulation, it might provide useful indications to drive model validation, calibration, maintenance works as well as relevance of pipes to plan operations.

ACKNOWLEDGEMENTS

This research was part of the Project 'SUstaiNable WATER supply networks in Mediterranean touristic areas – SUNWATER' – Interreg V-A Greece – Italy Programme 2014–2020 (MIS code 5003132), co-funded by the European Union, European Regional Development Funds (E.R.D.F.) and by National Funds of Greece and Italy. The authors wish to thank AQP (Acquedotto Pugliese) for real network data.

REFERENCES

Bavelas, A. 1950 Communication patterns in task-oriented groups. *Journal of the Acoustical Society of America* **22**, 725–730. https://doi.org/10.1121/1.1906679.

Benzi, M. & Klymko, C. 2015 On the limiting behavior of parameter-dependent network centrality measures. *SIAM Journal on Matrix Analysis and Applications* **36** (2), 686–706.

Bonacich, P. 1972 Factoring and weighting approaches to status scores and clique identification. *Journal of Mathematical Sociology* **2** (1), 113–120.

Borgatti, S. P. 2005 Centrality and network flow. *Social Networks* **27** (1), 55–71.

Demsar, U., Spatenkova, O. & Virrantaus, K. 2007 Centrality measures and vulnerability of spatial networks. In: *4th International Conference on Information Systems for Crisis Response and Management, ISCRAM 2007*, Delft, The Netherlands.

Diao, K., Farmani, R., Fu, G., Astaraie-Imani, M., Ward, S. & Butler, D. 2014 Clustering analysis of water distribution systems: identifying critical components and community impacts. *Water Science & Technology* **70** (11), 1764–1773.

Dijkstra, E. 1959 A note on two problems in connexion with graphs. *Numerische Mathematik* **1** (1), 269–271.

Evans, T. S. & Lambiotte, R. 2009 Line graphs, link partitions, and overlapping communities. *Physical Review E* **80**, 016105. doi: 10.1103/PhysRevE.80.016105.

Everett, M. G. & Borgatti, S. P. 1998 Analyzing clique overlap. *Connections* **21** (1), 49–61.

Faghani, M. R. 2013 A study of XSS worm propagation and detection mechanisms in online social networks. *IEEE Transactions on Information Forensics and Security* **8** (11), 1815–1826. doi: 10.1109/TIFS.2013.2280884.

Freeman, L. C. 1977 A set of measures of centrality based on betweenness. *Sociometry* **40** (1), 35–41.
Freeman, L. C. 1979 Centrality in social networks: conceptual clarification. *Social Networks* **1** (3), 215–239.
Girvan, M. & Newman, M. 2002 Community structure in social and biological networks. *Proceedings of the National Academy of Sciences* **99** (12), 7821–7826.
Giustolisi, O., Simone, A. & Ridolfi, L. 2017 Network structure classification and features of water distribution systems. *Water Resources Research* **53** (4), 3407–3423.
Giustolisi, O., Ridolfi, L. & Simone, A. 2019 Tailoring centrality metrics for water distribution networks. *Water Resources Research* **55** (3), 2348–2369.
Hawick, K. A. 2012 Water distribution network robustness and fragmentation using graph metrics. In: *Proceedings of the International Conference on Water Resources Management AfricaWRM 2012*, Gaborone, Botswana, pp. 304–310. doi: 10.2316/P.2012.762-037.
Jackson, M. O. 2008 *Social and Economic Networks*. Princeton University Press, Princeton, NJ, USA.
Katz, L. 1953 A new status index derived from socio metric analysis. *Psychometrika* **18** (1), 39–43.
Kleinberg, J. M. 1999 Authoritative sources in a hyperlinked environment. *Journal of the ACM* **46** (5), 604–632.
Ostfeld, A. 2012 Optimal reliable design and operation of water distribution systems through decomposition. *Water Resource Research* **48** (10), W10521. doi:10.1029/2011WR011651.
Page, L., Brin, S., Motwani, R. & Winograd, T. 1998 The PageRank Citation Ranking: Bringing Order to the Web. In: *Proceedings of the 7th International World Wide Web Conference*, Brisbane, Australia, pp. 161–172.
Pandit, A. & Crittenden, J. C. 2016 Index of network resilience for urban water distribution systems. *International Journal of Critical Infrastructures* **12** (1–2), 120–142.
Piraveenan, M., Prokopenko, M. & Hossain, L. 2013 Percolation centrality: quantifying graph. Theoretic impact of nodes during percolation in networks. *PLoS ONE* **8** (1), e53095. https://doi.org/10.1371/journal.pone.0053095.
Shuang, Q., Zhang, M. & Yuan, Y. 2014 Node vulnerability of water distribution networks under cascading failures. *Reliability Engineering and System Safety* **124**, 132–141.
Simone, A., Ridolfi, L., Laucelli, D. B., Berardi, L. & Giustolisi, O. 2018 Centrality Metrics for Water Distribution Networks. In: *HIC 2018-13th International Conference on Hydroinformatics*, Palermo, Sicily.
Spearman, C. 1904 The proof and measurement of association between two things. *American Journal of Psychology* **15**, 72–101.
Yazdani, A. & Jeffrey, P. 2010 A complex network approach to robustness and vulnerability of spatially organized water distribution networks. *arXiv1008.1770*, 18.
Yazdani, A. & Jeffrey, P. 2011 Complex network analysis of water distribution systems. *Chaos* **21** (1), 016111.
Yazdani, A. & Jeffrey, P. 2012a Robustness and vulnerability analysis of water distribution networks using graph theoretic and complex network principles. In: *12th Annual Conference on Water Distribution Systems Analysis (WDSA)*, Tucson, Arizona, USA, pp. 933–945. https://doi.org/10.1061/41203(425)85.
Yazdani, A. & Jeffrey, P. 2012b Applying network theory to quantify the redundancy and structural robustness of water distribution systems. *Journal of Water Resources Planning and Management* **138** (2). https://doi.org/10.1061/(ASCE)WR.1943-5452.0000159.

First received 28 January 2019; accepted in revised form 10 July 2019. Available online 16 August 2019

Fault detection in level and flow rate sensors for safe and performant remote-control in a water supply system

Sofia Fellini IWA, Riccardo Vesipa, Fulvio Boano and Luca Ridolfi

ABSTRACT

This work presents an algorithm for real-time fault detection in the SCADA system of a modern water supply system (WSS) in an Italian alpine valley. By means of both hardware and analytical redundancy, the proposed algorithm compares data and isolates faults on sensors through analysis of residuals. Moreover, the algorithm performs a real-time selection of the most reliable measurements for the automated control of the WSS operations. A coupled model of the hydraulic and remote-control system is developed to test the performance of the WSS when the proposed algorithm is applied or not. Simulations show that the occurrence of errors in the sensors causes significant worsening in the economic, energy and mechanical performance of the infrastructure. In many cases, the operations of the WSS are seriously compromised. The error detection and measurement assessment performed by the proposed algorithm proves to be crucial for the safe control of the WSS.

Key words | analysis of residuals, fault detection, optimal control, SCADA, water supply

Sofia Fellini IWA (corresponding author)
Riccardo Vesipa
Fulvio Boano
Luca Ridolfi
Department of Environment, Land and Infrastructure Engineering,
Politecnico di Torino,
Corso Duca degli Abruzzi 24, 10129 Turin,
Italy
E-mail: *sofia.fellini@polito.it*

Sofia Fellini
Laboratoire de Mécanique des Fluides et d'Acoustique, UMR CNRS 5509,
Université de Lyon, Ecole Centrale de Lyon, INSA Lyon,
Université Claude Bernard Lyon I, 36, avenue Guy de Collongue, 69134 Ecully,
France

INTRODUCTION

In north Italian mountainous and hilly areas, water supply (WS) to local communities is usually provided by municipal water supply systems (WSSs) that rely on local sources and operate independently from each other. In these regions, in fact, the availability of water resources is generally not an issue, and the high degree of territorial dispersion favors a decentralized water management. In the event of unexpected breakdowns or droughts (Carrera *et al.* 2013), however, this fragmentation results in inefficiencies and water crises. In order to increase the resilience of the WS service, a growing trend is the creation of inter-municipal water networks that connect multiple local WSSs over large areas (Massarutto 2000). Coordination in the operations of multiple WSSs and diversification of water sources entail economic, environmental, and water quality advantages (Anghileri *et al.* 2012; Bel & Warner 2015), and they are in line with the principle of integrated water resources management pursued by the European Water Framework Directive (EC 2000).

doi: 10.2166/hydro.2019.002

Large-scale water infrastructures require an automated regulation aimed at controlling the operations of the entire system, according to a centralized control philosophy. For this reason, they are managed by SCADA (Supervisory Control and Data Acquisition) systems (e.g., Coelho & Andrade-Campos 2014; Meseguer & Quevedo 2017) that (i) acquire and analyze real-time pressure and flow rate measurements provided by sensors in the key points of the network and (ii) remotely control the operations of the regulation devices (e.g., valves, pumps, turbines), according to predefined management rules and as a function of the data measured in real-time throughout the system.

In this framework, faulty sensors in the system may induce the SCADA to perform wrong regulations and, thus, result in faulty water supply system (WSS) operations. To ensure the robustness of the control operations, control systems have to be integrated with procedures that detect errors in sensors and activate in response actions for a safe management. This is crucial for large hydraulic infrastructures that transfer and supply water over large and steep areas. These systems are particularly vulnerable to failures, as the large volumes of water and the high pressures involved can have dangerous consequences when the system regulation is not optimal.

Fault detection and isolation (FDI) methods have been proposed in different fields of engineering and can be roughly classified into signal-based and method-based techniques (Gao *et al.* 2015).

Basic signal-based methods (e.g., Mourad & Bertrand-Krajewski 2002; Burnell 2003) evaluate data as valid or faulty by assessing the value or the variation of the signal. Advanced signal-based methods (e.g., Frank & Köppen-Seliger 1997; Maki & Loparo 1997) include techniques from machine learning, like neural networks. Signal-based methods are generally used when measurement redundancy is not available in the system, when the number of sensors under analysis is huge, and when a model of the system cannot be developed.

On the other hand, model-based methods validate data by using dependencies between different measurable signals. To this end, both hardware redundancy or analytical redundancy (Hwang *et al.* 2010; Gertler 2015) can be exploited. In hardware redundancy, measurements of the same physical variable are acquired by redundant sensors and compared. Conversely, analytical redundancy uses a mathematical model of the system as a comparison term (Frank 1996; Isermann 2005). Hardware redundancy is commonly adopted in safety-critical systems (Goupil 2011), but it involves significant extra costs. Moreover, sensors tend to have a similar lifetime and thus sensors that were installed at the same time in the infrastructure often fail simultaneously. In large-scale systems, analytical redundancy represents a convenient alternative due to the number of sensors, and the high cost of their installation, interconnection, and maintenance (Boukhris *et al.* 2001). Fault detection in model-based methods involves two steps: (i) residual generation and (ii) residual evaluation. A residual is obtained as the difference between two redundant measurements. Its signal is ideally zero when the system is operating correctly and non-zero when faults are present. Residual evaluation is the set of techniques that correctly identify the occurrence of an error starting from the analysis of the residuals (Frank & Ding 1997).

FDI techniques have been widely studied in the field of control engineering, such as in industrial plants (e.g., Gertler 1988; Özyurt & Pike 2004) and in automotive and aerospace engineering (Chen & Patton 1999). In the hydraulic and hydrological field, applications of FDI involved the analysis and reconstruction of hydrological time-series (e.g., Quevedo *et al.* 2010), leak/burst detection in water distribution networks (e.g., Krause *et al.* 2008; Casillas *et al.* 2013), and problems related to water quality monitoring (e.g., Eliades & Polycarpou 2010). A few works instead dealt with the identification of faulty sensors for a safe remote control of WSSs. Among these, Gabrys & Bargiela (1996) determined confidence intervals for flow rate measurements by means of a neural network approach. Ragot & Maquin (2006) proposed a method based on fuzzy logic for measurement fault detection in an urban WSS. Izquierdo *et al.* (2007) applied a FDI hybrid method based on neural networks and fuzzy theory. Meseguer *et al.* (2010) and Cugueró-Escofet *et al.* (2016) used fault signature matrices to identify a faulty sensor by means of the effect that the wrong measurement transmitted by this sensor potentially has on the other variables measured in the system. These methods proved to be efficient but their implementation in real hydraulic systems requires a certain degree of control theory expertise and a relevant computational cost.

Starting from the work of Fellini *et al.* (2018), the goal of this paper is to present an easy-to-implement and robust algorithm for fault detection in the SCADA system of a modern WSS. To increase the safety and the reliability of the control system, measurements from sensors are assessed in real time. By the analysis of residuals, redundant measurements are compared, and faults are automatically detected. Moreover, in case of errors, the developed algorithm ensures continuity in the control operations and prevents interruptions in the water supply. In addition to the introduction of a new FDI method, tools for the application and validation of the algorithm in typical WSSs are also provided in this work. In this way, the application of the method to real WSSs is facilitated.

The work is organized as follows. After this introduction, the 'Materials and methods' section presents first the logic and the structure of the proposed algorithm for fault detection. Then, the application of the algorithm to a modern and multipurpose WSS is illustrated. To conclude the 'Materials and methods' section, the developed tools for the validation of the proposed algorithm are described. In the 'Results' section, the performances of the WSS before and after the application of the fault detection method are assessed. Finally, the main conclusions obtained from the present work are drawn.

MATERIALS AND METHODS

Algorithm for fault detection in sensors

An easy-to-implement and robust method for real-time fault detection in the SCADA system of a WSS has been developed. This method is based on the redundancy concept and thus can be applied when different measurements of the same physical quantity are available.

Usually, critical sensors in a SCADA system are duplicated. However, double redundancy provides error detection but not recovery. In fact, if two different measurements are read by two redundant sensors, the supervisory system can easily detect the presence of an error but cannot automatically recognize which one is the faulty device. In this case, a technician's on-site intervention is the only solution to identify the problematic sensor, and this may lead to an interruption of the control operations of the WSS. In large-scale water infrastructures, this interruption is extremely unsafe.

To guarantee continuity in real-time control operations, a procedure for the automatic detection and correction of faulty measurements is thus required. This is achieved with an error detection system based on triple redundancy. In this case, if one of the three sensors fails, the other two sensors can isolate the faulty sensor and provide a useful and reliable measurement.

The algorithm we present detects faults by comparing three values of the same physical variable, provided by two redundant gauges (hardware redundancy) and by a hydraulic equation (analytical redundancy). In this way, resilience is achieved and the expensive installation of three redundant sensors is avoided.

In particular, the fault detection algorithm (Figure 1) receives as input three values (A, B, and C) of the same

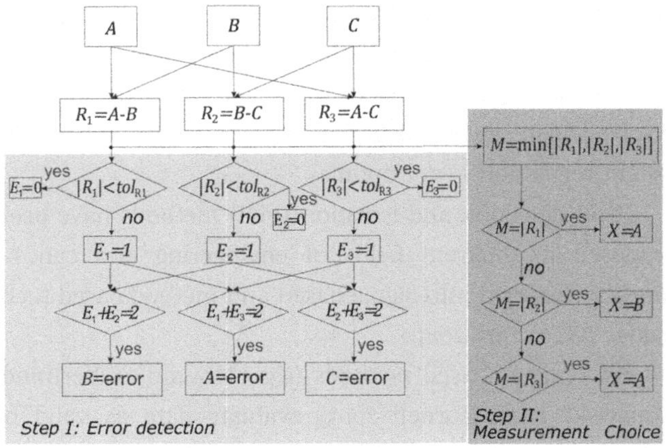

Figure 1 | Algorithm for fault detection (Step I) in a WSS sensors and for the choice of the most reliable measurement (Step II) to be used in the control operations.

observed variable (e.g., water level in a tank or flow rate along a pipe). A and B are transmitted by two redundant sensors, while C is obtained from a hydraulic balance. Residuals R_1, R_2, and R_3 are generated by calculating the difference between measurements two by two:

$$R_1 = A - B, \quad R_2 = B - C, \quad R_3 = A - C \tag{1}$$

Due to instrumental precision, random errors (see the 'Error modeling' section) are intrinsic in instrumental readings and therefore residuals are generally non-zero even if no errors are present. For this reason, faults are detected by comparing each residual to its tolerance. The tolerance intervals are defined in order to include precision errors, as:

$$tol_{R1} = tol_A + tol_B, \; tol_{R2} = tol_B + tol_C, \; tol_{R3} = tol_A + tol_C, \tag{2}$$

where tol_A, tol_B, and tol_C are the maximum instrumental errors expected for measurements A, B, and C, respectively. If the *i-th* residual exceeds its tolerance, the *i-th* error variable, E_i, switches to 1 (e.g., if $|R_1| > tol_{R1}$, $E_1 = 1$). When a sensor fails, two of the three error variables (E_1, E_2, and E_3 in Figure 1) assume value equal to 1, since the error associated with a single measurement appears in the calculation of two residuals. As a consequence, the algorithm can automatically detect which one is the faulty sensor (e.g., if R_1 and R_3 are out of tolerance, the sensor providing measurement A is identified as faulty).

Besides real-time error detection, the developed algorithm automatically selects the most accurate data to be used in the control operations of the SCADA system (step II in Figure 1). For each set of three measurements (A, B, C), the minimum residual is identified (M in Figure 1). The two measurements involved in the calculation of M are the closest to each other and therefore considered the most representative of the real physical value. The algorithm selects one (X in Figure 1) of these two measurements as input for the control operations. This last choice is based on technical considerations. Generally, priority is given to direct measurements compared to indirect ones (i.e., those calculated from quantities different from the measured one). For example, in Figure 1, when $M = R_1$ the algorithm selects A as the input for the control operations. Measurement B would have been an equally correct datum but for technical reasons A is considered more reliable (e.g., A is directly given by a sensor while B is obtained from a hydraulic balance or from a sensor with lower instrumental precision).

The here-presented algorithm is designed for real-time applications. In remote-controlled WSSs, programmable logic controllers (PLC), networked to the SCADA systems, receive measurements from the sensors. The fault detection algorithm is implemented in the PLCs and evaluates the reliability of these measurements. Reliable measurements are thus selected by the algorithm and adopted for the management of the control devices (e.g., valves, turbines). The frequency for this measurement check is case specific. It can be lower than the data acquisition frequency but has to be higher than the frequency of the adjustment operations, and therefore of the rate of variation of the hydraulic properties in the system.

Application of the algorithm to a real WSS

The proposed algorithm for fault detection and measurement assessment is non-specific and applies to WSSs with three basic requirements: (i) a SCADA communication network collects measurements from sensors throughout the system; (ii) programmable logic controllers (PLCs) operate the control devices (e.g., valves, turbines) according to the dynamics of a number of measured variables (e.g., flow rates, tank levels); (iii) three independent measurements of each control variable can be obtained from redundant sensors or models. A case study is introduced to improve the understanding of the method, to present a real application, and to show some useful tools for its validation.

Case study

The case study is described in detail by Fellini *et al.* (2017) and involves a modern WSS (Figure 2) located in an Alpine valley in the northwest of Italy. The WSS consists of an 80-km-long water main connecting 20 municipal WSSs. The water main collects high quality water from an alpine reservoir and conveys a maximum flow rate of 500 l/s. The municipal WSSs (see the inset in Figure 2) are self-sufficient under ordinary conditions as they are supplied by local sources (springs and wells). However, they receive water from the water main when the local sources fail, their quality is low, or water treatment in local plants is expensive. Needle valves with electronic actuators regulate the flow delivered from the water main to the municipal tanks. In the upper valley, four inline tanks (S_1, S_2, S_3, and S_4) split the water main in order to limit the static water pressure in the pipes. A Pelton turbine with electronically controlled Doble needles adjusts the flow entering in each inline tank. In this way, hydropower generation is also performed.

Valves and turbines are the active elements that regulate the flow rate in the entire WSS. Local PLCs control these devices according to (i) predefined management rules, (ii) flow rate and level data measured by sensors installed near the regulation devices, and (iii) data received from the distant sensors networked in the SCADA system. Data transmission in the SCADA system is guaranteed by an optical fiber network.

The management rules of the active devices were developed to optimize water distribution and to maximize the hydropower generation and the energy saving in the entire system (Fellini *et al.* 2017). In particular, flow adjustment through the turbines is aimed at (i) maintaining the water level in the inline tanks between two predefined level thresholds and (ii) minimizing the number of turbine operations. To this end, turbines T_1 and T_2 are regulated according to the water level in the upstream tanks S_1 and S_2, respectively. An upper and a lower level threshold are predefined for each one of these tanks. When the water level crosses one of these thresholds, the status of the

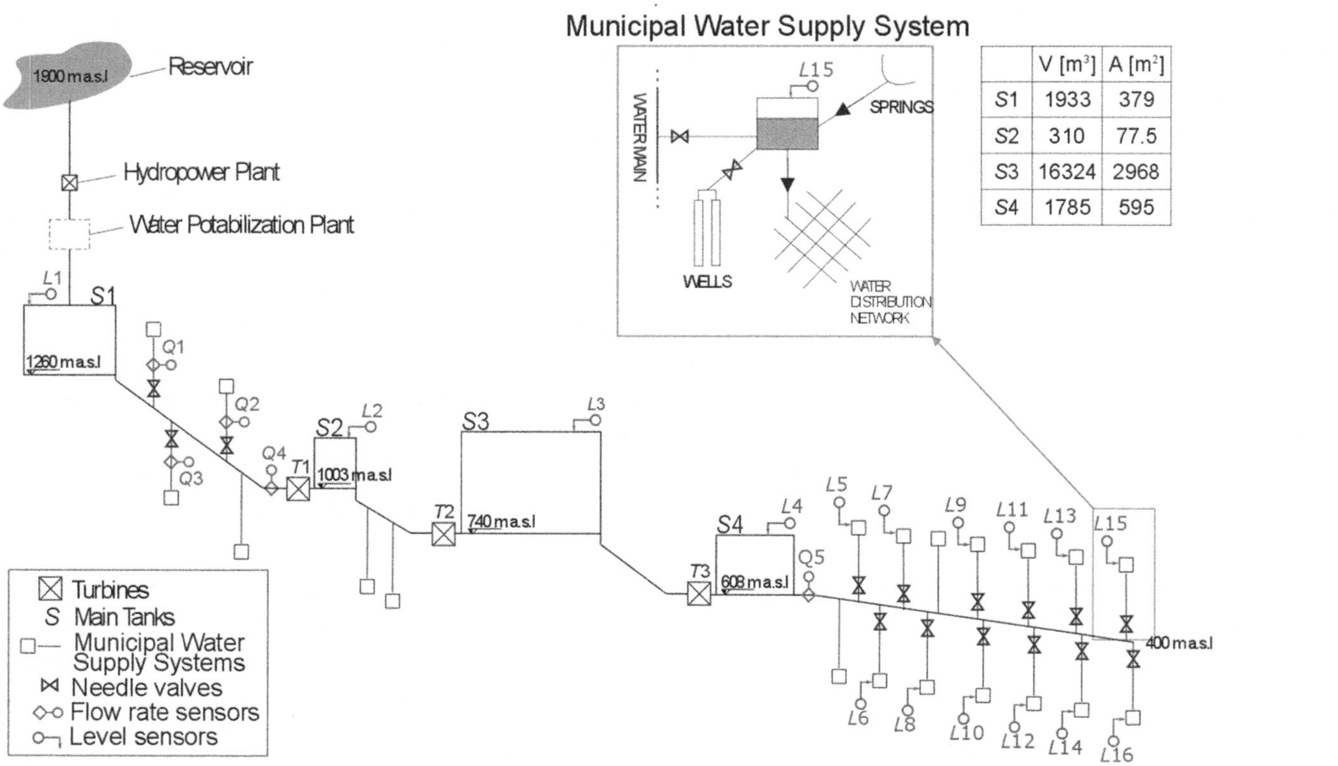

Figure 2 | Scheme of the WSS adopted as case study. The water main is split by four main tanks (the rectangles denominated S_1–S_4) and it supplies 20 municipal tanks (the small white squares). The inset shows a typical municipal water system with the storage tank supplied by mountain springs, local wells, and by the new water main. The flow rate and level sensors involved in the regulation operations are respectively represented as small diamonds and arrows overlaid by a circle. The control devices, i.e., needle valves and turbines, are depicted as 'bow tie' symbols and square with cross symbols.

downstream turbine is updated. The updated value (Q_T) of the flow rate through the turbine is calculated by the SCADA system in order to close the balance:

$$Q_T = Q_{IN} - \sum_i Q_{EX,i} \tag{3}$$

where Q_{IN} is the flow rate entering the tank and $Q_{EX,i}$ is the flow rate supplied to the *i-th* municipal WSS along the pipeline segment between the tank and the turbine. Q_T value is rounded up when the crossed threshold is the higher one. It is rounded down when the crossed threshold is the lower one. In this way, the water level slowly varies between the two thresholds and the number of turbine adjustments is minimized. An additional minimum threshold is set to avoid the emptying of the tank (i.e., to prevent air from entering the water main) in emergency conditions. When the water level falls below this threshold the downstream turbine is completely closed. The same rationale is used to control the flow rate through turbine T_3. However, the regulation is performed according to the water level in the downstream tank S_4. As a consequence, in Equation (3), the flow rate out of the tank (Q_5 in Figure 2) appears instead of Q_{IN}, and when the emergency threshold in tank S_4 is crossed, turbine T_3 is fully opened. The involved measurements for the control of the turbines are thus given by the level sensors in the inline tanks (L_1 to L_4) and by the flow rate sensors installed along the water main (e.g., Q_4 and Q_5) and along the pipes that supply the municipal tanks of the upper valley (e.g., Q_1 to Q_3).

The water level in the inline tank S_3 is instead used to control the flow rate towards the municipal tanks in order to optimize the distribution of water in the whole system. Tank S_3 has the largest capacity, and thus its level is an indicator of water availability in the system. According to L_3 dynamics and to a predefined priority list (based on technical and economic criteria), the municipal WSSs to be supplied by the water main are selected. Alternatively, the municipal tanks are supplied by the local wells. In both cases, the flow rate to the municipal tank is proportional to the tank water level. As for the inline tanks, an emergency level threshold is set for the municipal tanks. When the water level falls below this threshold, the water main supplies the tank with a predefined constant flow rate. The involved measurements for the control of the needle valves are thus given by level sensors in the municipal tanks (e.g., L_5 to L_{16}) and by L_3.

Notice that all the municipal WSSs are actually provided with meters. However, the level and flow rate sensors of four of them are not shown in Figure 2. These sensors are not considered in the simulations discussed in the following sections because the management rules in which they are involved are rather specific.

Assessment of hardware and analytical redundancy in the system

The above-presented fault isolation procedure is effective if three measurements A, B, and C of the same physical quantity are available. As introduced in the 'Case study' section, the key quantities for the control operations are: (i) water levels in the inline tanks (L_1 to L_4 in Figure 2) and in the municipal tanks (L_5 to L_{16}); and (ii) flow rates along the water main (Q_4 and Q_5) and towards the municipal WSSs in the upper valley (Q_1 to Q_3).

Regarding level measurements, each tank is equipped with two redundant level sensors (MH_1 and MH_2 in Figure 3(a)). Moreover, the tank level can be evaluated with a mass balance equation that estimates the actual tank level (C in Figure 3(a)) as a function of the tank level in the previous time step (Δt), the tank area (A_T), and the flow rates into and out of the tank (MQ_{IN} and MQ_{OUT} in Figure 3(a)):

$$C(t) = C(t - \Delta t) + [MQ_{IN}(t - \Delta t) - MQ_{OUT}(t - \Delta t)]\frac{\Delta t}{A_T} \quad (4)$$

Notice that this balance is independent from the level measurement sent by sensors MH_1 and MH_2. Otherwise, any error in one of the two level sensors would also propagate within C and the error identification method would fail.

Regarding flow rate measurements towards the municipalities, two flow meters are installed along the pipes that supply each municipal tank. Generally, one is installed near the tank and the other at the connection to the water main (MQ_1 and MQ_2 in Figure 3(b)). Furthermore, some flow meters intercept the water main (MQ_A and MQ_B in Figure 3(b)). Thus, a third value (C) of the flow rate towards each one of the local WSSs can be evaluated with the following flow rate balance equation:

$$C(t) = MQ_A(t) - MQ_B(t) \quad (5)$$

Finally, for the flow rates along the water main (Q_4 and Q_5 in Figure 2), redundant flow rate sensors at the entrance and at the exit of the inline tanks are installed (Figure 3(c)). The flow rate balance (Equation (5)) is used to compute the third redundant measurement.

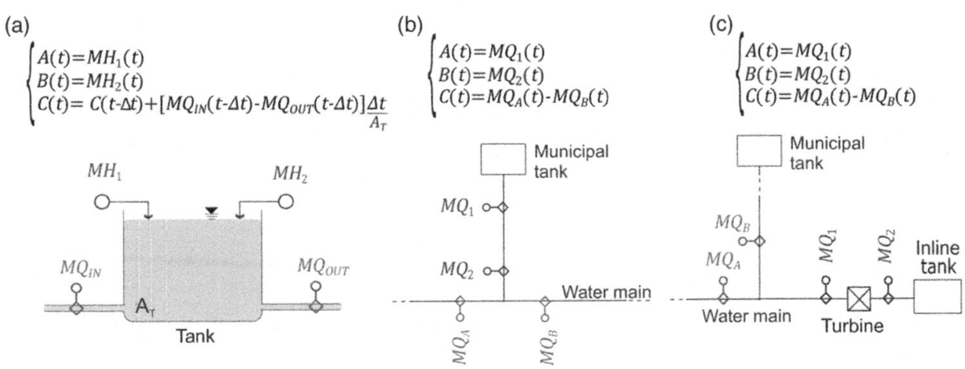

Figure 3 | (a) Level (MH_1 and MH_2) and flow rate (MQ_{IN} and MQ_{OUT}) sensors installed in the tanks (with area A_T) of the WSS. The system of equations provides three redundant values (A, B, and C) of the tank level. (b) Flow rate sensors installed along the pipe that supply the municipal tanks (MQ_1 and MQ_2) and along the water main (MQ_A and MQ_B). The system of equations provides three redundant values (A, B, and C) of the flow rate towards the municipal tank. (c) Flow rate sensors installed at the entrance of the inline tanks (MQ_1 and MQ_2), along the water main (MQ_A), and along the pipes that supply the municipal tanks (MQ_B). The system of equations provides three redundant values (A, B, and C) of the flow rate into the inline tank.

Tolerance intervals for the residuals

In the proposed error detection algorithm, a tolerance interval for each residual has to be defined. Within this interval the residual is accepted. The tolerance interval is obtained from the sum of the maximum precision errors expected for the two measurements involved in the calculation of the residual, i.e., a combination of two between tol_A, tol_B, and tol_C (see Equation (2)).

For flow rate measurements (Figure 3(b) and 3(c)), tol_A and tol_B are set as the maximum instrumental error (E_{max}) of the installed flow meters. tol_C is the maximum precision error expected for the indirect estimation (C) of the flow rate. Equation (5) defines C as a linear combination of flow rate measurements MQ_A and MQ_B. For this kind of indirect measurement, the uncertainty is determined by the law of propagation of uncertainty (BIPM *et al.* 2008). In the case of linear models, the law of propagation states that the variance of an indirect measurement is given by a quadratic combination of the variance of the directly measured quantities. Referring to Equation (5), the variance of C reads:

$$\sigma_C^2 = \sigma_{MQ_A}^2 + \sigma_{MQ_B}^2 \tag{6}$$

Assuming a Gaussian distribution for the measurement results, the maximum precision error can be estimated as three times the standard deviation. As a consequence, tol_C becomes:

$$tol_C = 3\sigma_C = 3\sqrt{\sigma_{MQ_A}^2 + \sigma_{MQ_B}^2} \tag{7}$$

where the variance of the flow rate measurements MQ_A and MQ_B is obtained as a function of their maximum instrumental error:

$$\sigma_{MQ_A}^2 = \left(\frac{E_{max,MQ_A}}{3}\right)^2, \ \sigma_{MQ_B}^2 = \left(\frac{E_{max,MQ_B}}{3}\right)^2 \tag{8}$$

Similarly, for level measurements (Figure 3(a)), tol_A and tol_B are set as the maximum instrumental error stated for the installed level sensors (MH_1 and MH_2). tol_C is the maximum precision error expected for the indirect estimation of the tank level, obtained from the mass balance in Equation (4). Considering the law of propagation of uncertainty, tol_C for the level measurement at time t reads:

$$tol_{C,t} = 3\sigma_{C,t} = 3\sqrt{\left(\frac{tol_{C,t-\Delta t}}{3}\right)^2 + (E_{max,MQ_{IN}}^2 + E_{max,MQ_{OUT}}^2)\left(\frac{\Delta t}{3\,A_T}\right)^2} \tag{9}$$

In Equation (9), tol_C increases with each time step. To avoid this progressive growth of the tolerance range and to keep it significant in the validation process, tol_C is forced equal to tol_A when it exceeds a threshold value equal to twice the tolerance of the installed level sensors (e.g., tol_A).

Kalman filters (e.g., Piatyszek *et al.* 2000; Ciavatta *et al.* 2004) could be used alternatively to Equations (4) and (9) to produce an estimate of the variable C, with its uncertainty σ_C. However, to reduce the uncertainty related to the modeled variable C, Kalman filters require a recursive update of the estimated variable based on the real value measured by a sensor. In the method proposed in this work, however, C has to be independent from the measured values (i.e., A and B). For this reason, the implementation of Kalman filters is here avoided.

Figure 4 shows the residuals R_1, R_2, and R_3 and their tolerance intervals (see Equations (1) and (2)) for the assessment of the level measurement in tank S_1. The tolerance

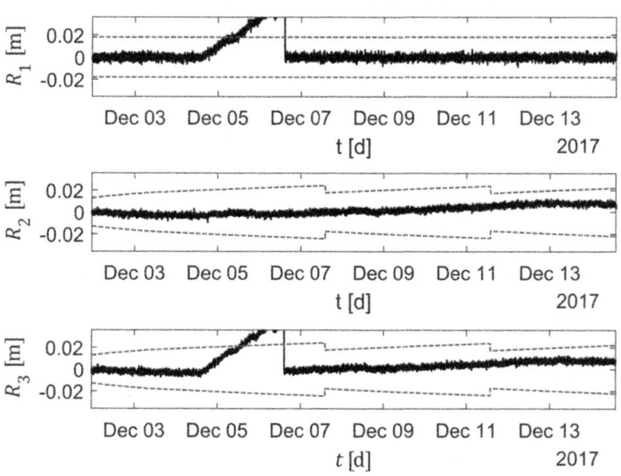

Figure 4 | Residuals (continuous lines) and their tolerance intervals (dashed lines) for level measurement in tank S_1. Level sensor MH_1 is affected by drift. Consequently, residuals R_1 and R_3 present a drifted signal and exceed their respective tolerance ranges. The tolerance intervals of residuals R_2 and R_3 are not constant as they are a function of tol_C (see Equation (9)).

interval of residuals R_2 and R_3 (tol_{R2} and tol_{R3}) present a trend of interrupted growth as R_2 and R_3 are a function of tol_C. In the proposed example, the level sensor MH_1 is affected by drift (see section 'Error modeling'). As a consequence, residuals R_1 and R_3 exceed their tolerance intervals and the sensor malfunction is detected.

In the considered WSS, flow rate is measured by electromagnetic flow meters with a maximum instrumental error (E_{max}) equal to 0.25% of the flow rate in transit, while levels are measured by ultrasonic level sensors with E_{max} equal to 0.15% of the full-scale (this latter is equal to 6 meters).

Tools for the validation of the fault detection algorithm

The efficacy of the proposed algorithm in detecting errors in sensors is assessed through numerical simulations. To perform these simulations, a numerical model of the hydraulic and control operations is developed, and realistic errors in sensors are simulated. To assess in a quantitative way the performance of the WSS under the different simulated scenarios, custom performance indexes are introduced.

Hydraulic and control model

A simulation model was developed to analyze the performances of the WSS in different scenarios (Fellini *et al.* 2017). This model consists of a coupled hydraulic and control model.

The hydraulic model is a system of non-linear equations describing (i) the flow-head loss relation in pipes, at valves and at turbines, (ii) the flow continuity at nodes, and (iii) the boundary conditions at tanks. Time evolution of the system is modeled by a succession of steady states with duration Δt. At each time step, the flow in the pipes and the pressure at nodes are computed. Moreover, the water level in tanks is updated using a mass balance equation.

The control model simulates the supervision and control operations of the SCADA system. A measurement from each one of the installed sensors is modeled as $m = M + \epsilon$, where M is the flow or level computed by the hydraulic model at the exact point where the sensor is located. This value is perturbed with the error ε to model different kinds of sensor faults. These measurements are used as input data for the decision-making algorithms that simulate the control actions of the SCADA system, that is, the governance of valves and turbines.

Error modeling

Measurement errors are of different types and originate from different causes. In this work, we simulate errors that usually affect meters in monitoring systems (Balaban *et al.* 2009; Sharma *et al.* 2010). These are: random errors, instrumental drifts, sensor breakdowns and interruption of signal transmission (Table 1).

Random errors (also called statistical errors) are intrinsic errors that depend on the precision limitation of the measurement devices (e.g., Fuller 2009). In the simulated control system, random errors are modeled as realizations of a Gaussian distribution with mean equal to zero and with a standard deviation that depends on the instrumental precision. In particular, according to the three-sigma rule (Pukelsheim 1994), we estimate the standard deviation as one-third the maximum instrumental error (E_{max}).

Instrumental drift is a progressive bias in the measurement output which increases slowly in time (e.g., Webster

Table 1 | Types of considered errors in sensors and their simulation in the control model of the WSS

Type of error	Subtype	Error simulation
Random error	–	Gaussian distribution with $\mu = 0$ and $\sigma = E_{max}/3$
Drift	*Zero drift*	Instrument calibration curve with intercept increasing over time from 0 to 20% of the instrument full-scale
	Sensitivity drift	Instrument calibration curve with angular coefficient increasing over time from 1 to 1.2
Breakdown	*Minimum constant value*	Zero constant value
	Maximum constant value	Constant value stuck at the instrument full-scale
	Abrupt oscillations	Binomial distribution taking value 0 or the instrument full-scale with equal probability 0.5
Loss of signal	–	NaN

1998). It is caused by various environmental issues and by mechanical wear. In unaltered meters, the relation between the input (i.e., the value of the physical variable) and the output (i.e., the measurement) of the measurement process is a bisector. When this ideal bisector moves vertically, i.e., the instrument output has an offset with respect to the input, the meter is affected by 'zero drift'. This offset may increase over time. When the angular coefficient of the bisector changes over time, the sensor is affected by 'sensitivity drift'. In the simulated control model, the calibration curve of the instrument is disturbed by (i) an increase of the offset from zero to 20% of the instrument full-scale occurring over 10 days and (ii) by a 20% increment of the angular coefficient over the same period.

Sensor breakdowns are usually caused by mechanical damage or electrical issues in internal connections. In the event of a breakdown, the operation of the meter is compromised, and the measured quantity deviates completely from the measured physical variable, i.e., the sensor output is usually stuck at a constant value equal to the full-scale of the instrument or to the minimum detectable value. Alternatively, the measure oscillates unstably between these two values. Stuck sensors are simulated in the control model by setting the measured value on the minimum detectable value or on the full scale of the instrument. Unstable oscillations are generated through a random extraction over time between these two extremes.

Loss of signal is a complete loss of sensor data, due to a failure in the sensor or in the transmission network. In this case, the sensor output is simulated as a NaN (i.e., Not a Number).

Performance indexes

To analyze the operations of the hydraulic system in scenarios with different failures of the sensors, we introduce four performance indices. The first one is a normalized index of the time to breakdown, calculated as:

$$IT = \frac{T_{fin}}{T_{sim}} \tag{10}$$

where T_{fin} is the time in which the WSS operates without criticalities (i.e., empty conditions in the tanks), while T_{sim} is the total simulation time (1 month). If $IT < 1$, errors in sensors are responsible for the failure of the WSS.

For the simulations with $IT = 1$, three additional normalized indexes are calculated. As detailed in the 'Case study' section, the aim of the studied WSS is to supply high quality water and to generate hydropower electricity. Thus, to analyze the performance of the system we calculate the performance quantities

$$E = \text{Net energy} = E_T - E_P \tag{11}$$

$$W = \text{Lost water} = Vol_{over} \tag{12}$$

$$M = \text{Number of turbine operations} = n_{T1} + n_{T2} + n_{T3} \tag{13}$$

E_T and E_P are respectively the energy generated by the three turbines and the energy absorbed by the pumping plants that supply the local tanks. The difference between these two quantities corresponds to the green net energy produced by the system. Vol_{over} is the total volume of overflown water at tanks. When water flows are not properly governed, this parameter is greater than zero and quantifies the lost water in the system. n_{T1}, n_{T2}, and n_{T3} are the number of flow adjustments in the three turbines. For the hydraulic stability of the control devices (Vesipa & Fellini 2019) and for mechanical weariness, this parameter is optimal when minimized.

These metrics are compared with those obtained for the benchmark simulation, that is, the simulation of an ideal behavior of the system, with optimal measurement acquisition by the sensors. This simulation is a reference for the other scenarios and is hereinafter referred to as the benchmark ('B'). The following normalized indexes can thus be defined:

$$IE = \frac{E}{E_B}, \; IW = \frac{W_{max} - W}{W_{max}}, \; IM = \frac{M_B}{M} \tag{14}$$

where W_{max} is the maximum lost water among all the simulations. These indexes are equal to 1 when the performance of the disturbed system is the same as in the benchmark case. Indexes tend to zero as the system performance worsens due to faulty sensors. The water index (IW) is always between 0 and 1 since the lost water in the benchmark simulation is zero. Index $IE(IM)$ can take values greater than 1, if the net energy produced (the number of turbine operations) is higher (lower) than in the benchmark simulation. However, even if one of the indices is greater than 1, the overall performance of the WSS is always expected to be worse than in the benchmark case.

RESULTS

Numerical simulations of the WSS operations are performed to (i) analyze the effects of the transmission of faulty measurements from the meters to the telecontrol system and (ii) verify the efficiency of the error identification algorithm.

First, we simulate the operation of the WSS in the case of errors in the sensors and complete lack of measurement redundancy. Then, the error identification algorithm is introduced into the control model, the same error scenarios are reproduced, and the resilience of the system is investigated.

In these analyses, only the measurements that are essential for the WSS control operations are considered. These measurements are: the water level in the inline and municipal tanks (L_1–L_{16}) and the flow rate along the water main and towards the municipal tanks of the upper valley (Q_1–Q_5). In Figure 2 and in the 'Case study' section, details are given about the location of these measurements and how they are involved in the control operations.

For each of these measurements, three different kinds of errors are simulated: (i) drift, (ii) constant minimum value, and (iii) constant full-scale value. Moreover, in all the simulations, the measurements are disturbed by random errors due to instrumental precision. Interruption of signal transmission and abrupt oscillations in the instrument output were also tested. Differently from measurement errors, the lack of signal is easily detected by the control system even without a specific error detection algorithm. Abrupt oscillations are instead quite rare. For these reasons, and for the sake of conciseness, the results of these simulations are not discussed in the following.

Performances of the WSS without application of the validation algorithm

In Figure 5, the performances of the WSS, without the application of the algorithm for fault detection, are shown. In these simulations, one sensor at a time is affected by errors. Three different types of errors are analyzed: (i) drift, (ii) minimum constant value, and (iii) maximum full-scale value. Only the sensors that are essential for the WSS control operations are considered. In Figure 5(a), the *IT* index for the different scenarios is reported. Level and flow rate meters are ranked on the *x* axis according to the *IT* value. For the simulations with $IT = 1$, the sensors are

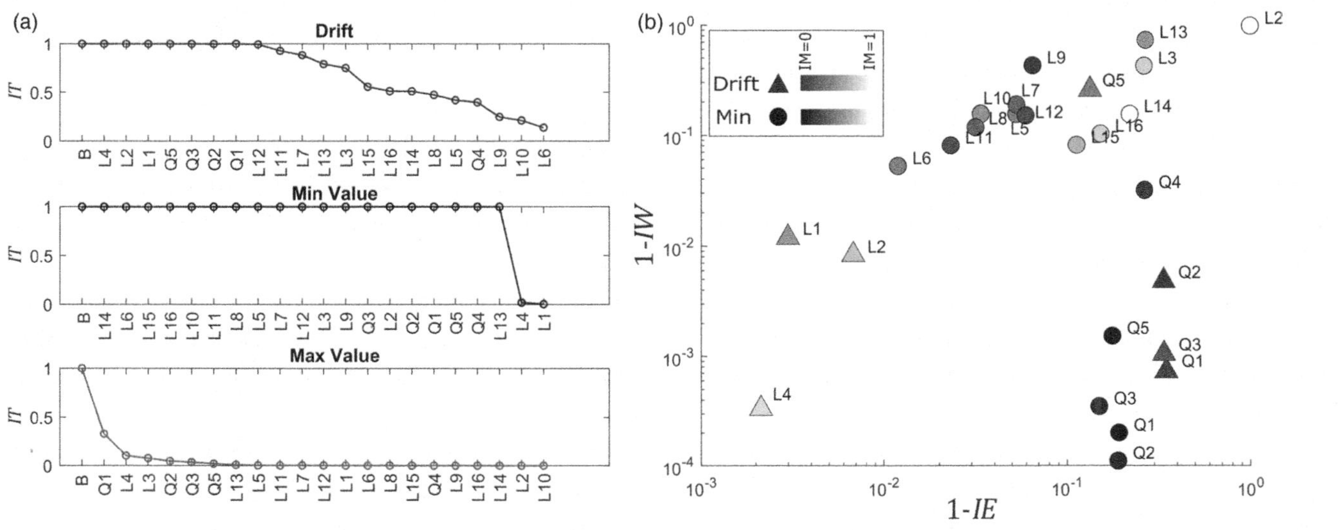

Figure 5 | (a) Normalized time to breakdown (index *IT*) for the three different types of errors. (b) Visualization of the three performance indexes (*IW*, *IE*, and *IM*) for the simulations with $IT = 1$. Circle and triangle markers represent the results for simulations with drift and constant minimum value errors, indexes *IW* and *IE* are reported in the log axes of the chart, while a different shade of the marker fill is assigned according to the mechanical index value (*IM*).

ranked according to the sum of the indexes *IE*, *IW*, and *IM*. In Figure 5(b), the simulations with $IT = 1$ are reported as a function of the three performance indexes *IE*, *IW*, and *IM*. Since the index values are highly concentrated around 1, a better readability is obtained using logarithmic axes of the quantities $1 - IE$ and $1 - IW$, and using a logarithmic scale for the shade of the marker fill that represents the quantity $1 - IM$. In this way, the best scenarios are located near the origin of the axes and present blank markers. The results of the benchmark simulation cannot be visualized on this graph since zero values cannot be represented on a logarithmic scale.

Drift error (first panel in Figure 5(a) and triangle markers in Figure 5(b)) mainly affects level sensors in the municipal tanks. In fact, level sensors L_1 to L_{16} are characterized by $IT < 1$. Based on the drifted level signal, the water level in the tank is overestimated by the SCADA and thus the water flow to the tank from both the water main or the local wells is under-rated and insufficient to match local water consumptions. Drift errors in the level sensors of the four inline tanks (L_1 to L_4) are less critical and jeopardize the system operations only in the long term. In particular, sensors L_1, L_2, and L_4 are rather resilient and their performances are comparable with the benchmark case (see Figure 5(b)). Regarding flow rate sensors, no criticalities are evident except for Q_4, that is, the flow meter involved in the regulation of the T_2 turbine and thus in the regulation of the water level in S_2. This tank is characterized by the smallest volume compared to the other inline tanks, so it is more vulnerable to errors in flow rate measurements. Considering the performances of the WSS under the effect of drifted flow rate measurements to the municipal tanks of the upper valley (Q_1, Q_2, and Q_3), an overestimated flow rate induces a reduction of the flow discharged by the turbines (see Equation (3)), a lower hydroelectric production and thus a decrease of the *IE* index. Moreover, the flow rate balances that govern the opening and closing operations of the turbines are altered and thus a decrease of *IM* index is also observed.

When the faulty sensor settles to the minimum constant value (second panel in Figure 5(a) and circle markers in Figure 5(b)), the WSS operation is generally not compromised. A wrong minimum level in the municipal water tanks (L_5 to L_{16}) entails an excessive water inflow from the local wells and thus a greater energy consumption by the pumping plants and water loss due to water overflowing from the tank. For these reasons, both the *IW* and *IE* indexes greatly decrease (Figure 5(b)). However, critical emptying of the municipal tanks does not occur. On the other hand, inline tanks are more vulnerable to this kind of error since emergency water supply from local wells is not provided. Based on a wrong minimum water level in tanks $S_1(S_2)$, the SCADA system forces a complete closure (see the 'Case study' section) of turbine $T_1(T_2)$, thus impeding the water supply of the downstream tank $S_2(S_3)$. Given its size, tank S_2 is emptied when this error arises in sensor L_1. Similarly, when a minimum level is measured in tank S_4, turbine T_3 is completely opened and the upstream tank S_3 is critically emptied. For these reasons, L_1 and L_4 show the highest vulnerability to this kind of error. A decrease of *IE* and *IM* indexes is observed for errors in the flow rate measurements that disturb the flow rate balances for the regulation of turbine operations (Q_1, Q_2, and Q_3 in Figure 5(b)).

Errors inducing wrong full-scale measurements of level or flow rate in the devices are the most critical for the remote-control system. In municipal tanks, if the level meter indicates that the tank is full, then the tank is not supplied either from the WSS water main or from local sources, and it quickly empties. In inline tanks, based on a wrong maximum level in S_1 and $S_2(S_4)$, the SCADA forces the complete opening(closure) of turbine T_1 and $T_2(T_3)$. Consequently, tanks S_1, S_2, and S_4 are rapidly emptied. When a faulty maximum level is measured in S_3, the water availability in the WSS is overestimated, the municipal tanks are excessively supplied, and thus, the water level in S_3 rapidly decreases. Flow rate sensors are involved in the regulation of the turbine operations and they also exhibit a high vulnerability to this kind of error.

The simulation results shown in Figure 5 evidence the vulnerability of level and flow rate gauges to the most common types of errors. Moreover, the analysis of the performance indexes in Figure 5(b) underlines that even if the system does not reach breakdown, errors still lead to significant worsening in the economic, energy, and mechanical performance of the infrastructure. In particular, it is interesting to observe two main trends in the graph. The simulations related to errors in the level sensors are distributed along the bisector, indicating a similar deterioration both for the *IW* index and for the *IE* index. The simulations of errors in flow rate sensors, on the other hand, are distributed along a vertical line, indicating that the main consequence is a decrease in energy (and mechanical, if the intensity of the marker fill is observed) performances.

Application of the validation algorithm

After the analysis of the WSS vulnerability to faults in sensors, the above-presented algorithm for real-time fault detection and measurement selection was integrated in the control model of the WSS. First, the available gauges in the different districts of the WSS were checked (hardware redundancy). Second, the balance equations that provide an additional value of the measured physical variables were written (analytical redundancy). Third, for each one of the considered sensors, the algorithm reported in Figure 1 was coded. Finally, different error scenarios were simulated to evaluate the efficiency of the integration in the control model of the error detection algorithm.

To show in detail how the control system operates, Figure 6 reports the results for the assessment of level measurement for the inline tank S_1. Figure 6(a) shows the redundant signals of the tank water level. Figure 6(b) depicts the results of error assessment by the fault detection algorithm.

As introduced in Figure 3(a), two level sensors are installed in each inline tank (MH_1 and MH_2 in Figure 3(a)), providing hardware redundancy. A third value of the tank level is obtained using a mass balance equation of the flow rate into and out of the tank, measured by sensors MQ_{IN} and MQ_{OUT} (Figure 3(a)). The combination of hardware and analytical redundancy guarantees three simultaneous measurements of the same physical variable.

In the exemplifying scenario reported in Figure 6, the level sensor MH_1 (continuous line in Figure 6(a), 6(c), 6(d) and 6(e)) is affected by multiple consecutive errors: (1) drifts, (2) minimum and (3) full-scale constant values, (4) abrupt oscillations, and (5) transmission interruption. The level sensor MH_2 (dotted line in Figure 6(a), 6(c), 6(d) and 6(e)) and the flow rate sensors MQ_{IN} and MQ_{OUT} operate correctly instead, and they are only disturbed by slight random errors due to instrumental precision (Figure 6(e)). Note that instrumental errors in the flow rate sensors scarcely affect the precision of the level value (dashed line in Figure 6(e)) obtained from the balance equation.

As shown in Figure 6(a), the level measured by sensor MH_2 is almost equal to the value resulting from the balance equation. On the other hand, the level acquired by MH_1 considerably differs from the previous ones, due to instrumental failures. By means of the residual analysis, the developed algorithm detects with high precision these failures. Figure 6(c) shows in detail the early detection of the drift error. In the first panel, the three curves are very closed one to each other. However, residuals R_1 and R_3, in the second and fourth panels exceed their respective tolerance thresholds represented by the dashed lines. As a consequence, MH_1 is marked as faulty in the last panel. In Figure 6(b), the outcomes of this error detection procedure are highlighted for all the simulation time. Besides error detection, the algorithm selects in real-time the most reliable measurement (Figure 6(b) and 6(c)) to be used in the WSS control operations. Under ordinary conditions, the algorithm alternatively selects the value measured by one of the two level sensors MH_1 and MH_2. In case of error detection for MH_1, the measurement transmitted by MH_2 is the only one to be selected. As shown in Figure 6(b), the level value obtained from the balance equation is never identified as erroneous because the flow meters involved in the balance operate correctly. Furthermore, the value is never selected by the algorithm as the reference datum for the control operations. In fact, the direct measurements from level sensor MH_1 and MH_2 are always favorable to the indirect one. In any case, the level measurement obtained from the balance is fundamental for the operation of the error identification algorithm as it guarantees the triple redundancy on which the algorithm is based.

Performances of the WSS with application of the validation algorithm

The simulations of the response of the system to faulty sensors, discussed above and reported in Figure 5, were performed again. This time, the algorithm for the identification of measurement errors was introduced into the control system. As for the previous analysis, for each one of the considered level and flow rate measurements, three

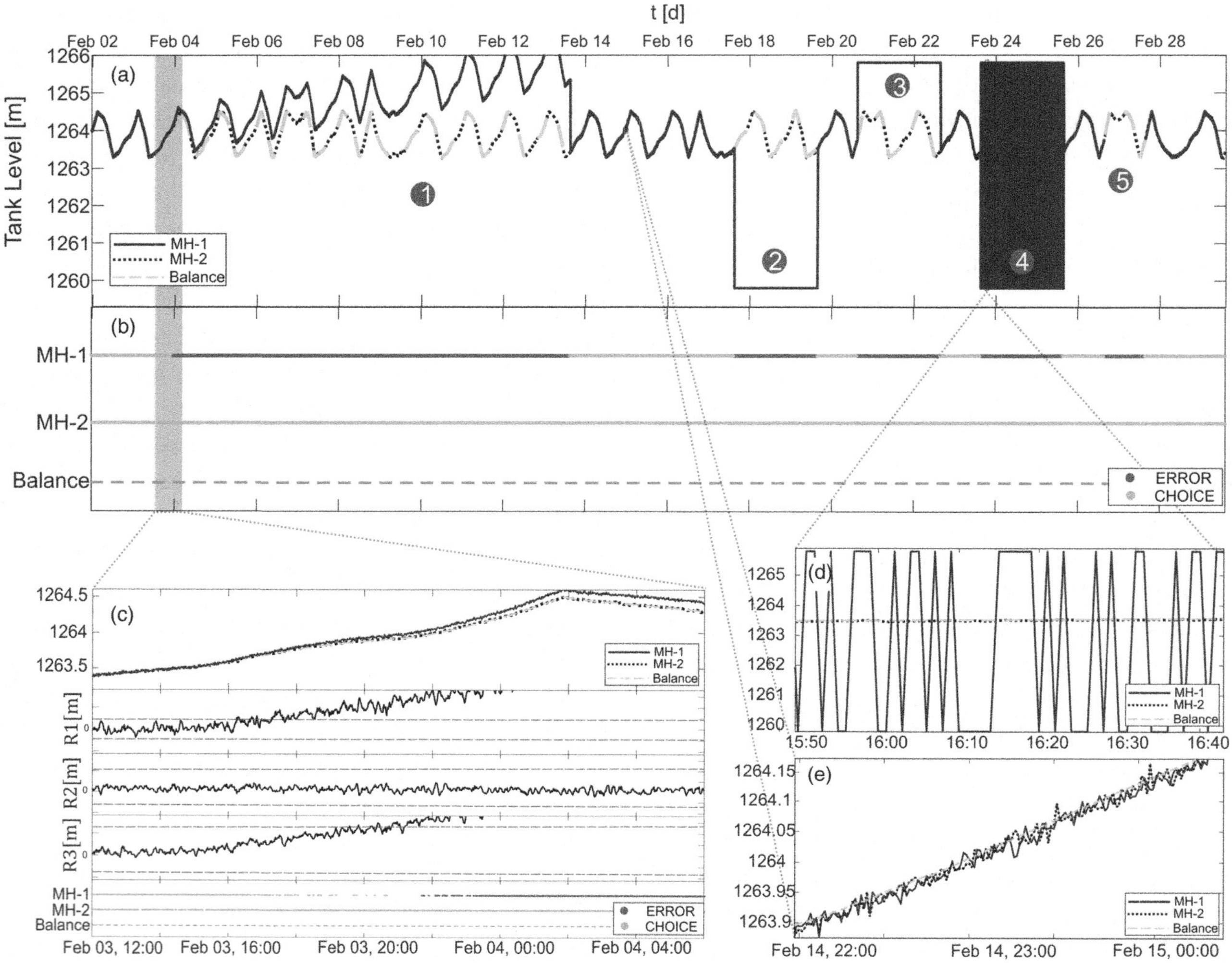

Figure 6 | (a) Level values from level sensors MH_1 (continuous line) and MH_2 (dotted line) and from the balance equation of flow rate into and out of the tank (dashed line). Measurements transmitted by MH_1 experience (1) drifts, (2) minimum and (3) full-scale constant values, (4) abrupt oscillations, and (5) transmission interruptions. (b) Error detection and measurement selection by the fault detection algorithm. (c) Zoom of the early detection of the drift error. As in Figure 4, the three residuals R_1, R_2, and R_3 are reported with the extremes of their tolerance ranges (dashed lines). (d) Zoom of the abrupt oscillations (4). (e) Zoom of the signal showing errors due to instrumental precision.

simulations were conducted. The main sensor was disturbed by (i) drift, (ii) faulty full-scale and (iii) minimum constant values. Precision errors were simulated for the second redundant sensor and for the sensors involved in the indirect quantification of the third redundant measurement. For each scenario, the WSS dynamics were reproduced for a period of 1 month.

Results show that in all the simulated scenarios, the algorithm accurately identifies errors in sensors and transmits reliable measurements to the remote-control system. In this way, adjustments of the control devices (turbines and valves) are regularly managed and critical situations that compromise the safety of the aqueduct are prevented. The positive feedback of the system to the introduction of the fault detection algorithm is at first assessed observing that the *IT* index is equal to 1 for all the error scenarios. This indicates that the WSS operates without criticalities for the entire duration of the simulation. However, the *IT* index is not enough to evaluate the improvement of the system performance for the scenarios in which the WSS proved to be resilient even in the absence of the error identification algorithm (i.e., scenarios with $IT = 1$ in Figure 5(a)). For these cases, we calculated the difference between the *IE*, *IW*, and *IM* performance indexes after and before the introduction of the fault detection procedure (Figure 7).

Results reveal a general increase in the energy index (Δ_{IE}), in particular when errors are detected in the flow rate sensors (Q_1 to Q_5). In fact, the correct regulation of the flow rate through the turbines guarantees the optimization of the hydroelectricity production.

Water saving (Δ_{IW}) is enhanced when the error detection algorithm is applied to level sensors. The water level is the control parameter for the flow rate towards the tanks. When levels are properly measured, the excessive supply of water for the inline and municipal tanks is avoided, thus limiting overflow conditions.

The mechanical index (Δ_{IM}) significantly increases when the signal from the flow rate sensors is verified. As mentioned in Equation (3), these measurements are involved in the balance equations that control the flow rate through the turbines. When the flow rate sensors are faulty, these water balances are only met at the monitoring level but not in the hydraulic system. This results in continuous adjustments of the flow through the turbines. Conversely, when flow rate measurements are assessed, the number of turbine operations is minimized. The implementation of the algorithm induces instead a slight decrease in the mechanical index, when level sensors are stuck on the minimum value (circle markers for level sensors in Figure 7(c)). However, this decrease proves that the WSS is operating properly. In fact, as discussed in the 'Case study' section, when a minimum constant value is erroneously measured in one of the municipal tanks, the control system automatically supplies the tank with an emergency constant flow rate from the water main. Due to this constant flow rate, the natural variation of water supply given by consumptions in the municipalities is reduced, as well as the adjustments of the flow rate through the turbines. When level sensors in the inline tanks are stuck on the minimum constant value, turbines are either completely open or completely closed. Therefore, in these cases, the control operations of the WSS are less dynamic, the number of turbine maneuvers decreases, but also the energy and water saving performances evidently worsen. In fact, a negative Δ_{IM} index is always followed by positive Δ_{IE} and Δ_{IW} indices.

DISCUSSION AND CONCLUSIONS

In this work, an algorithm is proposed for real-time assessment of data measured in the SCADA system of modern and multipurpose WSSs. An automated remote-control system based on reliable flow rate and level measurements is crucial for the safe and optimal operation of these water infrastructures. The developed algorithm compares redundant data provided by both redundant sensors and analytical models. By means of the analysis of residuals, failures and gross errors in sensors are detected. Moreover, the algorithm performs a real-time selection of the most reliable measurements to be used in the control operations.

The effectiveness of the method was assessed through numerical simulations of a coupled hydraulic and control model of an alpine WSS, taken as case study. The considered WSS is currently under construction and thus real datasets from sensors are not available. However, the most common measurement errors have been carefully simulated, based on the technical specifications of the sensors most

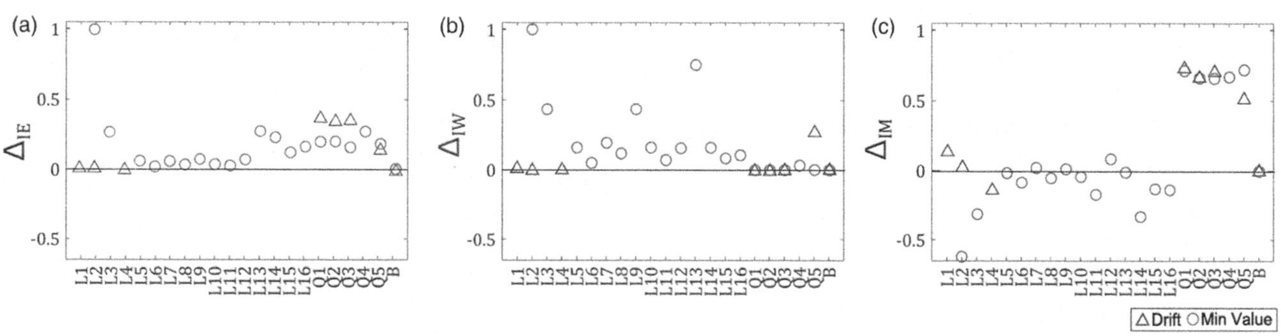

Figure 7 | Increment of the performance indices *IE* (a), *IW* (b), and *IM* (c) after the introduction of the fault detection algorithm in the control system of the WSS. Triangle (circle) markers report the results of simulations with drift (constant minimum value) errors in the sensors.

frequently installed in modern WSSs. By means of these simulations, the system vulnerability to different types of errors in sensors was first analyzed. Four performance indices were defined to assess the achievement of safety, energy, mechanical and water supply targets of the water infrastructure in the different scenarios. Results showed that in most cases, errors in sensors critically undermined the operation of the WSS. Moreover, the most vulnerable sensors were clearly identified. Then, the proposed algorithm was introduced in the control model. The simulations revealed that the algorithm ensures error detection with a high degree of accuracy and guarantees continuity in the system operations. As a result, the performance indices showed a noticeable increase.

The developed algorithm is robust and easy to implement. Moreover, the proposed tools for modeling the hydraulic behavior of the WSS, its control system, and sensor errors are useful to carry out preliminary studies on the performance and safety of a WSS in the design phase. Within this approach, hardware redundancy can be optimized.

The algorithm can verify the reliability of a control variable only in the case of a single faulty sensor among those involved in the estimation of the three redundant estimates of the variable. In the event of multiple, simultaneous errors the algorithm is not effective. Also, water leaks can disturb the hydraulic balances involved in the analytically redundant measurements. In this case, if the corresponding sensors (hardware redundancy) are not faulty, the algorithm identifies the presence of an error in the hydraulic balance and can thus help in localizing water leaks. Once detected, water leaks must be fixed, or the balance equation implemented in the algorithm must be modified to take these outflows into account. If one of the two redundant sensors is out of order, the unexpected occurrence of water leaks precludes the correct functioning of the algorithm. At the state of the art, the method therefore requires that in the event of a sensor malfunction, a maintenance intervention on site is arranged in a short time so as to prevent a second failure being added to the first one. Another aspect to be considered for the application of the proposed FDI in real WSSs is the frequency of sensor data validation. The fault detection algorithm should assess data with a frequency that is in line with the measurement frequency of the sensors, and with the speed of the WSS dynamics. All these aspects, i.e., the frequency of sensor maintenance, on-site human intervention, and data validation, are case-dependent and should be examined during the design of the control and fault detection system.

ACKNOWLEDGMENTS

We gratefully acknowledge SMAT Group for the financial support to this research and for providing valuable information.

REFERENCES

Anghileri, D., Castelletti, A., Pianosi, F., Soncini-Sessa, R. & Weber, E. 2012 Optimizing watershed management by coordinated operation of storing facilities. *Journal of Water Resources Planning and Management* **139** (5), 492–500.

Balaban, E., Saxena, A., Bansal, P., Goebel, K. F. & Curran, S. 2009 Modeling, detection, and disambiguation of sensor faults for aerospace applications. *IEEE Sensors Journal* **9** (12), 1907–1917.

Bel, G. & Warner, M. E. 2015 Inter-municipal cooperation and costs: Expectations and evidence. *Public Administration* **93** (1), 52–67.

BIPM, IEC, IFCC, ILAC, ISO, IUPAC, IUPAP & OIML 2008 *Evaluation of Measurement Data - Guide to the Expression of Uncertainty in Measurement* (GUM 1995 with minor corrections). Joint Committee for Guides in Metrology, JCGM, 100.

Boukhris, A., Giuliani, S. & Mourot, G. 2001 Rainfall–runoff multi-modelling for sensor fault diagnosis. *Control Engineering Practice* **9** (6), 659–671.

Burnell, D. 2003 Auto-validation of district meter data. In: *Proceedings of the CCWI'03 Conference In Advances in Water Supply Management*, 15–17 September, London, UK, pp. 13–22.

Carrera, L., Mysiak, J. & Crimi, J. 2013 Droughts in Northern Italy: Taken by Surprise, Again. Review of Environment, Energy and Economics (Re3). https://papers.ssrn.com/sol3/papers.cfm?abstract_id=2278860 (accessed 21 June 2019).

Casillas, M., Puig, V., Garza-Castañón, L. & Rosich, A. 2013 Optimal sensor placement for leak location in water distribution networks using genetic algorithms. *Sensors* **13** (11), 14984–15005.

Chen, J. & Patton, R. J. 1999 *Robust Model-based Fault Diagnosis for Dynamic Systems*. Springer Science & Business Media 3, New York, USA.

Ciavatta, S., Pastres, R., Lin, Z., Beck, M. B., Badetti, C. & Ferrari, G. 2004 Fault detection in a real time monitoring network for water quality in the lagoon of Venice (Italy). *Water Science and Technology* **50** (11), 51–58.

Coelho, B. & Andrade-Campos, A. 2014 Efficiency achievement in water supply systems-A review. *Renewable and Sustainable Energy Reviews* **30**, 59–84.

Cugueró-Escofet, MÀ, Quevedo, J., Alippi, C., Roveri, M., Puig, V., García, D. & Trovò, F. 2016 Model-vs. data-based approaches applied to fault diagnosis in potable water supply networks. *Journal of Hydroinformatics* **18** (5), 831–850.

EC (European Communities) 2000 *Directive 2000/60/EC of the European Parliament and of the Council Establishing a Framework for Community Action in the Field of Water Policy*. OJ L327, 22.12.2000.

Eliades, D. & Polycarpou, M. 2010 A fault diagnosis and security framework for water systems. *IEEE Transactions on Control Systems Technology* **18** (6), 1254–1265.

Fellini, S., Vesipa, R., Boano, F. & Ridolfi, L. 2017 Multipurpose design of the flow-control system of a steep water main. *Journal of Water Resources Planning and Management* **144** (2), 05017018.

Fellini, S., Vesipa, R., Boano, F. & Ridolfi, L. 2018 Real-time measurement fault detection and remote-control in a mountain water supply system. In: *HIC 2018. 13th International Conference on Hydroinformatics*, Vol. 3 (G. La Loggia, G. Freni, V. Puleo & M. De Marchis, eds), pp. 698–705.

Frank, P. M. 1996 Analytical and qualitative model-based fault diagnosis – a survey and some new results. *European Journal of Control* **2** (1), 6–28.

Frank, P. M. & Ding, X. 1997 Survey of robust residual generation and evaluation methods in observer-based fault detection systems. *Journal of Process Control* **7** (6), 403–424.

Frank, P. M. & Köppen-Seliger, B. 1997 Fuzzy logic and neural network applications to fault diagnosis. *International Journal of Approximate Reasoning* **16** (1), 67–88.

Fuller, W. A. 2009 *Measurement Error Models*. John Wiley & Sons, New York, USA.

Gabrys, B. & Bargiela, A. 1996 Integrated neural based system for state estimation and confidence limit analysis in water networks. In: *Proceedings of European Simulation Symposium ESS'96*, Genoa, pp. 398–402.

Gao, Z., Cecati, C. & Ding, S. X. 2015 A survey of fault diagnosis and fault-tolerant techniques-Part I: Fault diagnosis with model-based and signal-based approaches. *IEEE Transactions on Industrial Electronics* **62** (6), 3757–3767.

Gertler, J. 1988 Survey of model-based failure detection and isolation in complex plants. *IEEE Control Systems Magazine* **8** (6), 3–11.

Gertler, J. 2015 *Fault Detection and Diagnosis*. Springer, London, UK.

Goupil, P. 2011 AIRBUS state of the art and practices on FDI and FTC in flight control system. *Control Engineering Practice* **19** (6), 524–539.

Hwang, I., Kim, S., Kim, Y. & Seah, C. 2010 A survey of fault detection, isolation, and reconfiguration methods. *IEEE Transactions on Control Systems Technology* **18** (3), 636–653.

Isermann, R. 2005 Model-based fault-detection and diagnosis–status and applications. *Annual Reviews in Control* **29** (1), 71–85.

Izquierdo, J., López, P. A., Martínez, F. J. & Pérez, R. 2007 Fault detection in water supply systems using hybrid (theory and data-driven) modelling. *Mathematical and Computer Modelling* **46** (3–4), 341–350.

Krause, A., Leskovec, J., Guestrin, C., Van Briesen, J. & Faloutsos, C. 2008 Efficient sensor placement optimization for securing large water distribution networks. *Journal of Water Resources Planning and Management* **134** (6), 516–526.

Maki, Y. & Loparo, K. A. 1997 A neural-network approach to fault detection and diagnosis in industrial processes. *IEEE Transactions on Control Systems Technology* **5** (6), 529–541.

Massarutto, A. 2000 *Agriculture, water resources and water policies in Italy*. Fondazione Eni Enrico Mattei, Milano, Italy.

Meseguer, J. & Quevedo, J. 2017 Real-time monitoring and control in water systems. In: *Real-time Monitoring and Operational Control of Drinking-Water Systems*. Springer, Cham, Switzerland, pp. 1–19.

Meseguer, J., Puig, V. & Escobet, T. 2010 Fault diagnosis using a timed discrete-event approach based on interval observers: Application to sewer networks. *IEEE Transactions on Systems, Man and Cybernetics: Part A* **40** (5), 900–916.

Mourad, M. & Bertrand-Krajewski, J. L. 2002 A method for automatic validation of long time series of data in urban hydrology. *Water Science and Technology* **45** (4–5), 263–270.

Özyurt, D. B. & Pike, R. W. 2004 Theory and practice of simultaneous data reconciliation and gross error detection for chemical processes. *Computers & Chemical Engineering* **28** (3), 381–402.

Piatyszek, E., Voignier, P. & Graillot, D. 2000 Fault detection on a sewer network by a combination of a Kalman filter and a binary sequential probability ratio test. *Journal of Hydrology* **230** (3–4), 258–268.

Pukelsheim, F. 1994 The three sigma rule. *The American Statistician* **48** (2), 88–91.

Quevedo, J., Puig, V., Cembrano, G., Blanch, J., Aguilar, J., Saporta, D., Benito, G., Hedo, M. & Molina, A. 2010 Validation and reconstruction of flow meter data in the Barcelona water distribution network. *Control Engineering Practice* **18** (6), 640–651.

Ragot, J. & Maquin, D. 2006 Fault measurement detection in an urban water supply network. *Journal of Process Control* **16** (9), 887–902.

Sharma, A. B., Golubchik, L. & Govindan, R. 2010 Sensor faults: Detection methods and prevalence in real-world datasets. *ACM Transactions on Sensor Networks (TOSN)* **6** (3), 23.

Vesipa, R. & Fellini, S. 2019 Instability of the tank-level control system of water mains in mountainous environments. *Journal of Hydraulic Engineering* **145** (7), 04019025.

Webster, J. G. 1998 *The Measurement, Instrumentation and Sensors Handbook*. CRC Press, Boca Raton, FL, USA.

First received 30 December 2018; accepted in revised form 15 May 2019. Available online 19 June 2019

MultiRain: a GIS-based tool for multi-model estimation of regional design rainfall for scientists and practitioners

Susanna Grasso, Andrea Libertino and Pierluigi Claps

ABSTRACT

Extreme rainfall estimation is a long-standing challenge for hydrological hazard assessment and infrastructure design, particularly if considering the need to deal with climate change. Advances in statistical methods and in rainfall data availability allow for frequent updates of regional rainfall frequency analyses. These allow for new estimates that, however, cannot simply replace older ones in the risk management, due to technical, socio-economic and legislative reasons. To preserve compatibility between old and new regional estimates a multi-model approach could be used, where model uncertainties can be combined to help reach a final decision. To make this possible, one has to face the uneasy retrieval of data and results of older analyses and, quite often, non-trivial areal rainfall estimates are needed with all methods. To give an answer to these technical needs, a tool named *MultiRain* has been developed. The tool computes depth–duration–frequency (*DDF*) curves, both related to a point and integrated over an area, from multiple regional statistical analyses. The *MultiRain* procedure is based on Python scripting, geographic information system (GIS) functions and web technologies, and can be performed via web browser or in a desktop GIS environment. A demonstration version has been built using four different regional analyses proposed in a 20-year period for the north-west of Italy.

Key words | areal rainfall, *DDF* curves, design rainfall, hydrologic design tool, multi-model, regional frequency analysis

Susanna Grasso (corresponding author)
Andrea Libertino
Pierluigi Claps
Department of Environment, Land and Infrastructure Engineering,
Politecnico di Torino,
Corso Duca degli Abruzzi, 24, Torino 10129, Italy
E-mail: *susanna.grasso@gmail.com*

INTRODUCTION

Risk assessment related to severe precipitation is a growing priority of hydrological practice, due to the significant consequences of these phenomena on human safety, in a context of apparent increase of the number and severity of extreme meteorological events. Most of the adverse hydro-meteorological events that led to about 25,000 fatalities per year across the globe (Munich Re's NatCatService 2017) relate directly to intense rainstorms, which are difficult to forecast, both in location and in terms of expected effects, particularly in urban areas. Intensity–duration–frequency (*IDF*) or depth–duration–frequency (*DDF*) curves are commonly the starting point for design and operation of flood protection works (Grimaldi *et al.* 2011). They are mathematical relationships that allow the estimation of the average rainfall intensity i (or the total rainfall depth h) over a given timescale d for a given return period T (e.g., Koutsoyiannis *et al.* 1998). From the *DDF* one can obtain a design hyetograph, i.e., a reference rainfall sequence to be used in rainfall–runoff modelling with various methods (e.g., Alfieri *et al.* 2008). For the evaluation of the *DDF* curves in ungauged sites, or averaged over an area, regional rainfall frequency analysis (*RFA*) is required, and many studies have been carried out to predict rainfall depths with a certain probability of occurrence in a generic location within large areas (see, e.g., Reed *et al.* 1999; Svensson & Jones 2010). The estimation is usually carried out by interpolating spatially the data or the distribution parameters estimated at the station points (e.g., Myers 1994; Libertino *et al.* 2018), or by pooling

doi: 10.2166/hydro.2019.016

the data to form statistically homogeneous regions (see, e.g., Hosking & Wallis 1997).

It is not uncommon that different methods of *RFA* of extreme precipitation exist for the same regions. Quite often they adopt different techniques for identifying homogeneous regions and use different probability distributions or different methods for estimating distribution parameters. Most of the time, a new *RFA* is based on the availability of new data, even only for a limited number of years or of stations. The rationale for not totally replacing the older results with the new ones is that the hydrologic design figures used to assess the reliability of infrastructure need to be consistent over time. That is, if the design of a given infrastructure is based on estimates obtained by an 'old' study, the design of related and successive works should comply with the previous one, even though data and knowledge have increased and improved over time. At the same time, the institutions in charge of the risk management of infrastructure are very interested in the results of new methodologies, that use advanced geo-statistical techniques and data with a higher spatio-temporal resolution, to face the challenges of climate change. In fact, even if the widespread perception of an increase in the frequency of severe rainstorms does not yet find clear confirmation in the scientific literature, local correlations between the increase in temperature and the intensification of extreme rainfall are being consolidated (e.g., Westra *et al.* 2013).

To help preserve the consistency in design rainfall estimates over time, some scientific studies have promoted the simultaneous use and association of different methodological approaches over the search of a single 'best model', sometimes hard to identify (see, e.g., Shamseldin *et al.* 1997; Di Baldassarre *et al.* 2009; Velázquez *et al.* 2011; Okoli *et al.* 2018). The so-called *model-averaging* is particularly efficient when the goal is the estimation of quantiles for large return periods (Claeskens & Hjort 2008). For these high quantiles, the high estimation variance generally allows different distribution functions to be admissible after the hypothesis testing. The need to comply with this undetermined model selection assumes practical significance, for example, when considering that hydraulic safety of dams in Italy relies on the estimate of a 1,000-year flood, and that for small basins it is very important to check that estimate using regional flood frequency analyses and rainfall–runoff methods.

The above considerations underline the usefulness of merging and harmonizing the information deriving from different methods which have evolved over time. This process, which includes the above-mentioned *model-averaging*, is called here a *multi-model* approach for the selection of the hydrologic design values. Non-trivial technical difficulties arise for the practitioners when managing this merging process, deriving from the different variables required by the methods, the different detail of reproducibility of estimates, the topographical bases used (from paper maps to geographic information systems (GIS)), the different probability distributions, etc.

This paper presents the technical rationale of the development of a tool that allows a fast comparison and combination of the results from different rainfall mapping studies (see, Grasso *et al.* 2018). More specifically, this tool, named *MultiRain*, is a set of GIS-based procedures designed to produce concurrent *DDF* estimates from various regional frequency analyses both at a point and over an area. To better describe the features of the *MultiRain* product, all its building phases are presented, starting with the case study of the north-west of Italy, where four existing *RFA* studies are available. The four *RFA* methodologies are first briefly summarized, along with some details required for the customization of the software tool. The *MultiRain* general features are then described, with a focus on the solutions adopted for harmonizing the presentation of the results of the four methods. An application of the *MultiRain* customized for north-west Italy is finally presented, with regard to the assessment of the rarity of a real rainfall event, where the results can be obtained with the combination of the different approaches demonstrating the tool capabilities.

DATA AND METHODS

Study area and available RF analysis

In this section, the case study will be introduced, and the available extreme rainfall mapping analyses will be described with the details needed to explain the customization required by the software tool. The case considered here refers to the

Piemonte region, an area of about 25,000 km^2 in the north-west of Italy. Four *RFA* procedures are currently available for the area of interest, each one providing maps and procedures for the estimation of the *DDF*s for each location in the region. Automatic procedures for *DDF* computation are partly available only for one of the methods, along with some GIS support. Automatic areal rainfall estimates are not possible for all the methods. In some cases, the *RFA* parameters are provided through non-spatial tables or non-properly geo-referenced maps, that poses critical technical challenges for their systematic application.

Three of the considered methodologies are based on the index-rainfall method (Dalrymple 1960), in which the quantile of the extreme rainfall for a duration *d* and a return period *T* is obtained as:

$$h_T(d) \quad = \bar{h}(d) \cdot K_T(d) \tag{1}$$

In Equation (1), $h_T(d)$) is the design rainfall depth, $\bar{h}_d$ is the mean depth–duration curve (the index value for the duration *d*) and K_T is the growth function, i.e., the inverse of the non-dimensional frequency distribution, expressed as a function of T.

For the index rainfall $\bar{h}_d$ all the methods adopt the power law function:

$$\bar{h}(d) = a \cdot d^n \tag{2}$$

with *a* and *n* as coefficients estimated at-station and, in most cases, mapped by spatial interpolation.

Three *RFA*s were built using almost the same database; however, different statistical approaches are adopted, and extreme precipitation quantiles can differ considerably. A brief description of the methodologies is provided below, focusing on the characteristics that can make a direct (manual) comparison quite cumbersome.

PAI-Po design rainfall map (Po River basin)

The *PAI-Po* procedure has been developed by the Po River Authority on the 60,000 km^2 area covering the Po drainage basin, in the context of the Po River Hydrogeological Master Plan (*PAI* – Piano Stralcio per l'Assetto Idrogeologico, AdbPo 2001). This latter organization represents the framework behind the hydrologic and hydraulic risk assessment plan, providing rules, limitations and guidelines to be used in land and urban planning in the Po River basin.

The analysis has been carried out on 277 rain gauges, distributed over the whole basin area using the time series of the Servizio Idrografico e Mareografico Italiano (*SIMN*) of annual maxima of rainfall depth collected for the durations: 1, 3, 6, 12 and 24 hours. Data from 1929 to 1989 were considered.

The methodology is not based on the index rainfall paradigm. *DDF* curves are provided for selected return periods and are estimated by fitting the power law (Equation (2)) separately on the rainfall quantiles estimated on the data for the aforementioned durations. Therefore, independent *DDF* curves are provided for different return periods as:

$$h_T(d) \quad = a_T \cdot d^{n_T} \tag{3}$$

with a_T and n_T as the coefficients fitted by linear regression on the logarithms of the five quantiles estimated for the return period *T* at each station point. All quantiles are estimated using a Gumbel distribution. The return periods considered are: $T = 20$, 100, 200 and 500 years.

The a_T and n_T parameters are then interpolated on a 2 km spatial grid using ordinary kriging, to estimate *DDF* curves at ungauged locations. All the parameters in the grid covering the Po basin are provided in a table in which the location of each cell is obtained by sequential ordering of rows and columns indices. Querying individual cells on an online map is also possible (AdbPo 2001).

As this procedure does not provide results according to the index-rainfall framework, a systematic conversion has been proposed here, to make *PAI-Po* results easily comparable to the other *RFA*s. All details are provided in the Appendix.

VAPI – De Michele & Rosso RFA (NW Italy)

For the whole of north-west Italy, a study was conducted (De Michele & Rosso 2001) in the framework of a research project for the estimation of floods in Italy (*VAPI* project), driven by the National Group for the Prevention of Hydrogeological Disasters (CNR-GNDCI) in the 1990s. The *VAPI* project represents an important pillar, at national level, in hydrological risk assessment, since its main objective was to produce methods for extreme rainfall and flood estimation at the national scale.

In this analysis (De Michele & Rosso 2001), a database of 366 stations was considered: 270 in the Po River basin and 96 in Liguria, out of the Po River basin. Station record length was at least 20 years and the average sample length was 34, ending up with a database not very different from that of the *PAI-Po* study. *DDF* curves were estimated through the index-rainfall method, using a generalized extreme value (GEV) distribution (Jenkinson 1955) to represent the growth curve K_T. Parameters were estimated using the *PWM* method (Hosking *et al.* 1985). Ordinary kriging interpolation was used for mapping the parameters of Equation (2) and of the GEV at ungauged sites. The maps of *a*, *n* and of the GEV parameters (Equations (1) and (2)) for north-west Italy (including Lombardia, Piemonte, Valle d'Aosta, Liguria and Emilia-Romagna regions) were made available as images in pdf format (see Figure 1), so that the numeric information cannot be extracted.

VAPI – Villani RFA (Piemonte region)

The *VAPI – Villani RFA* (Villani 2003) was later developed for the Piemonte region according to the standards suggested by CNR-GNDCI. It uses a hierarchical approach based on the two-component extreme value (TCEV) distribution (Rossi *et al.* 1984), which takes into account separately the estimation of parameters of the ordinary and of the extra-ordinary components of the curve. The analysis was carried out on 517 rain gauges with at least 10 years of observations, adding some rainfall data collected by *Regione Piemonte* and by some neighbouring French institutions during 1913–1986 to the main database of the national hydrologic service (SIMN). The spatial mapping of the TCEV parameters is obtained through a noiseless kriging (Furcolo & Villani 1998). Note that the extraordinary component of the growth curve resulted as not significant in a large proportion of the region. In that case, the TCEV collapses on a Gumbel distribution, with the higher-order parameters set as $\boldsymbol{\theta}_* = 1$ and $\boldsymbol{\Lambda}_* = 0$. Together with the publication of the procedure report (Villani 2003), a support software (named *Va.Pi.*) was developed, aimed at the production of *DDF* curves for single stations and for a given set of Piemonte basins. However, the limited distribution of *Va.Pi.*, which was originally developed for Windows98 and never updated afterwards, discouraged its use. The *Va.Pi.* map grids have a resolution ranging from 1 km to 5 km, depending on the variability of the represented data.

Patched Kriging RFA (Piemonte region)

The *Patched Kriging* procedure (Libertino *et al.* 2018) was undertaken to take advantage of a consistent increment in the data availability that had occurred since the previous

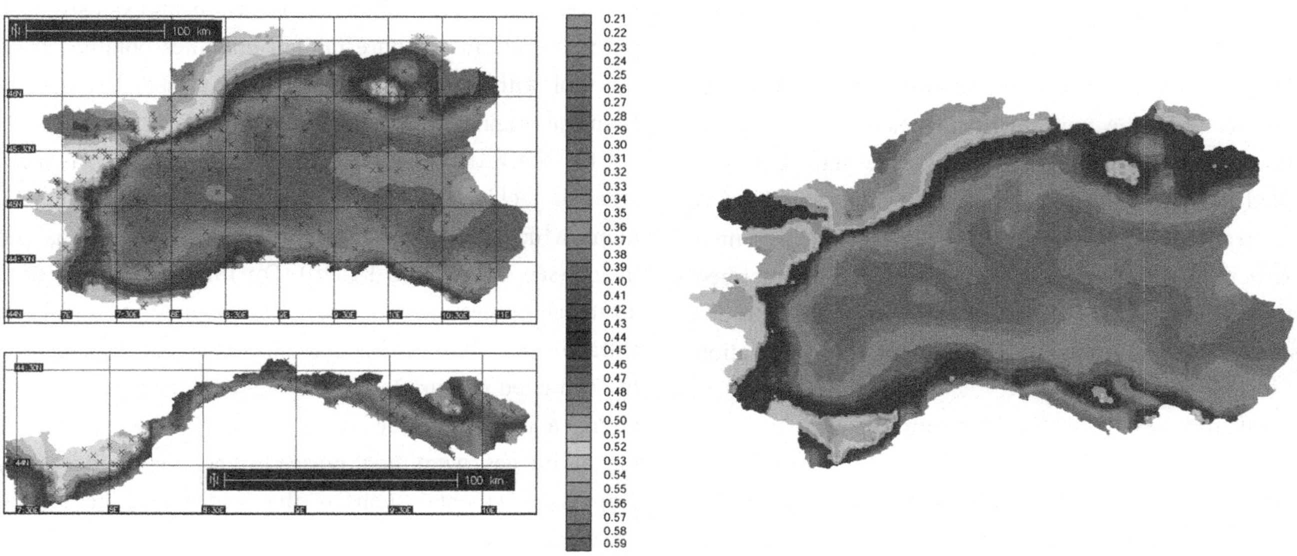

Figure 1 | Maps of the *n* parameter of the *VAPI – De Michele & Rosso* procedure in the original paper format (left) and in the digital version (right). Please refer to the online version of this paper to see this figure in colour: http://dx.doi.org/10.2166/hydro.2019.016.

analyses were published. The study uses annual extreme data from 1913 to 2010, gathered from the *SIMN* and from the regional hydrologic services, for the durations 1, 3, 6, 12 and 24 hours. The methodology involves the sequential application in time of the ordinary kriging equations, producing a homogeneous dataset of synthetic series with uniform lengths. On these series, the index method is adopted and the *IDF* curves are computed according to Equations (1) and (2). The authors decided not to choose a 'best' probability distribution and to provide a map of the a and n parameters of the *DDF* and of the first three sample L-moments. This allows the end-users to choose the most appropriate probability distribution after autonomously checking of the goodness of fit. Results are provided on a 250 m gridded domain covering the areas of the Piemonte and Valle D'Aosta regions. After resampling on a 1 km grid, the results were included in a digital map embedded in a Web-GIS called *Atlante delle piogge intense in Piemonte* (ARPA Piemonte 2013) developed by the Regional Agency for Environmental Protection.

The *MultiRain* tool development

The software tool *MultiRain* described in this work operates in a geographic consistent framework, that facilitates the selection of the location of interest and the comparison of results. This is intended to allow easy comparison of results of different *RFA* methodologies and to compute the areal-averaged *DDF* curves within a basin. In the first part of this section we describe the customization problems we tackled to allow the uniform management of the existing *RFA*s in the tool. In the second part, the main approach adopted and the features that describe the possibilities of generalization for the adoption of *MultiRain* are presented. In particular, data preprocessing techniques are described, along with a brief explanation of the implementation process and of the technological details of the GIS application.

Customization issues for the Piemonte region

The main goal in the customization of the *MultiRain* tool is the building of a unique geospatial database containing all the necessary elements for the application of the different *RFA* methods available. Problems that can arise in this task derive from the heterogeneous criteria for mapping the parameters (e.g., digitized scanned images, tabulated values, ASCII raster data with different formats and resolutions) and by the methodological differences between the various regionalization approaches (e.g., the homogeneous region procedures are based on vector maps while the methods that involve the continuous spatial variation of the parameters refer to raster maps). To exemplify, some details of the customization of the four *RFA* analyses in Piemonte are provided below.

The *PAI-Po* methodology proved to be the most difficult to customize, as it provides the a_T and n_T parameters of Equation (3) on a 2 km grid, for $T = 20$, 100, 200 and 500 years. Index curve (Equation (2)) and growth curve (Equation (3)) are not provided. Starting from the knowledge that the *PAI-Po* procedure adopts a Gumbel distribution, for each cell of the gridded domain, the transformation of the results (provided as in Equation (3)) to a mapping of a and K_T parameters was accomplished by following a procedure reported in a master thesis (Magro 2007) briefly explained in the Appendix.

After the reconstruction described in the Appendix, raster maps of parameters a and n (Equation (2)), and of the two Gumbel parameters θ_1 and θ_2 have been built. The computerized procedure needed to provide the tool's functionalities, compute the *DDF* curves for whatever return period in a single cell, and is applicable to all Gumbel-based *RFA*s. After customization, an extensive testing was done to compare rainfall quantiles obtained from the tool with the ones available, e.g., to selected stations from the original report.

For the *VAPI - De Michele & Rosso* method, the main issue to tackle was the need to digitize the parameters' maps, which are only available in paper format and as .pdf image (see, e.g., Figure 1 (left)). For this purpose, we used an image processing procedure implemented in Matlab for extracting the colour information from the .pdf map, relating the extracted colours to the provided colour bar, and attributing a numeric value to each pixel of the image. The digital images were first noise-cleaned, i.e., unnecessary graphical elements, such as the grid lines, the station indicators, etc. were removed (see Figure 1 (left)). The numeric values derived from the Matlab procedure were then converted into raster format, suitably geo-referenced and then merged, to connect the two separate maps provided for the Po basin and the Liguria catchments. An example of the resulting map obtained for the n coefficient is shown in Figure 1 (right).

For the *VAPI – Villani RFA*, we obtained from the *Va.Pi* software folder the raster files of parameters a and n for the mean, and of θ_*, Λ_*, Λ_1 for the TCEV growth curve. Georeferred parameter maps were obtained converting the raster from a binary grid format (.grd) to the ASCII format (.asc) and referring the cell centroids to the WGS84/UTM Zone 32N coordinate system. In the cells where $\Lambda_* = 0$, the TCEV distribution is known to reduce to a Gumbel distribution, and this was reproduced in the *MultiRain* tool.

The maps related to the *Patched Kriging* methodology (Libertino *et al.* 2018) have been requested from the authors in raster format. Using the provided L-moments raster maps with a 250 m × 250 m resolution, we have computed the θ_1 and θ_2 parameters of the Gumbel distribution and the θ_1, θ_2 and θ_3 parameters of the GEV, allowing evaluation of the growth curve K_T with two probability functions. This procedure has entailed minimum pre-processing customization efforts, as the raster maps of a and n were also available and already georeferenced.

MultiRain technical features

In the early stage of the *MultiRain* tool development, a Python plugin for the QGIS software (i.e., 'QGIS local plugin version') was built. In that version, the digitized regional maps, necessary to infer *DDF* curves at ungauged sites, were inserted into the plugin configuration package, to be used locally. Subsequently, with the aim to more easily distribute GIS data and functionality over the internet through standard internet protocols, specific web services have been implemented according to the OGC Web Processing Service (WPS) specification, accessible both by web browser and by using some desktop GIS programs. The use of server-side scripting allows the users to access calculations independently of the underlying software and avoid the problem of updating the plugin to keep it compatible with the updates of the various GIS programs. Moreover, data do not need to be stored locally (client-side) but are maintained by the hosting entity and can be updated and expanded on the server, without the need for the users to download updates. It should be added that computing times are usually shorter than those of the client scripting. This WPS format appeals to practitioners and planners as it reduces uncertainties and calculation errors and makes it possible to obtain *DDF* curves in an automatic way.

The connection to the WPS services is also available through a Desktop GIS software (i.e., QGIS, ArcGIS). This setting is interesting for several reasons: it allows the use of the geospatial algorithms available in the GIS software and the integration of data stored locally. Furthermore, results can be visualized directly on the map canvas enabling integration and overlay of spatial information from different sources in order to gain a deeper insight into relationships and scenario evaluation.

For sharing these procedures, a web platform has been developed, with free and open-source software, using PyWPS to set up the WPS processes, GRASS GIS as a backend to access all the geoprocessing functionalities, GisClient3 to build the WebGIS (accessible through client browsers) and Apache as web server. GisClient3 is an interesting web authoring tool configurator for PostGIS and MapServer that has the ability both to build Mapfiles and to provide OpenLayers maps. All the spatial data, necessary for *RFA*s implementation, have been registered in the GRASS database.

Despite all the pros deriving from the use of the WPS service, the use of the 'QGIS local plugin version' remains attractive, as it allows researchers to further explore rainfall frequency analysis and to carry out enlarged spatial analysis and also provides to any Python and QGIS skilled user the possibility to add other *RFA* methods.

The *MultiRain* tool, despite differences in scripting, has the same capabilities and provides the same benefits for the user in both the developed versions (the 'QGIS local plugin' and the WPS one). The user can either provide a point (as a shapefile or by inserting its coordinates) or select an area of interest (by uploading a proper shapefile). In the first case, the results are referred to the values of a pixel; in the second case, *DDF* curves of all the cells falling into the basin divide are computed and averaged spatially. Both versions of the tool allow for the following operations for all the available methodologies:

- query of all the available maps, for extraction of the parameters (point or areal);

- computation of the index rainfall curve $\bar{h}(d)$;
- production, in a text format, of the values of the local or areal parameters of the *DDF* curves;
- production of a graphical representation of the *DDF*s for some significant return periods.

Moreover, if the user specifies a return period *T* and a duration, the available scripts:

- compute the growth factor K_T;
- provide the design rainfall for the duration and return period requested;
- provide a graphical representation of the *DDF*s for the considered return period and duration, obtained with the different methodologies.

The WPS procedures implemented can be invoked client-side in two ways: directly from the built web-mapping application, accessible from the project website (soon available at http://www.idrologia.polito.it/MultiRainTool), or from any GIS desktop software that supports the OGC WPS standard (some software may require external plugins). In the first case, the user can either upload via web browser the polygon vector file of the area of interest or proceed, directly online, using an automatic watershed delineation service (performed using GRASS functionalities starting from a digital elevation model map, based on the NASA SRTM, designed by Ganora *et al.* (2016)). In the second case, the user can execute the WPS processes choosing the vector layer that delimitates the basin of interest from those already available into the current GIS session.

RESULTS AND DISCUSSION

The final configuration of the *MultiRain* tool customized for the Piemonte region is shown here with a series of screenshots and related comments.

Figure 2 shows an example of the execution of one of the WPS procedures from the QGIS desktop software using a WPS Client plugin. In particular, the figure

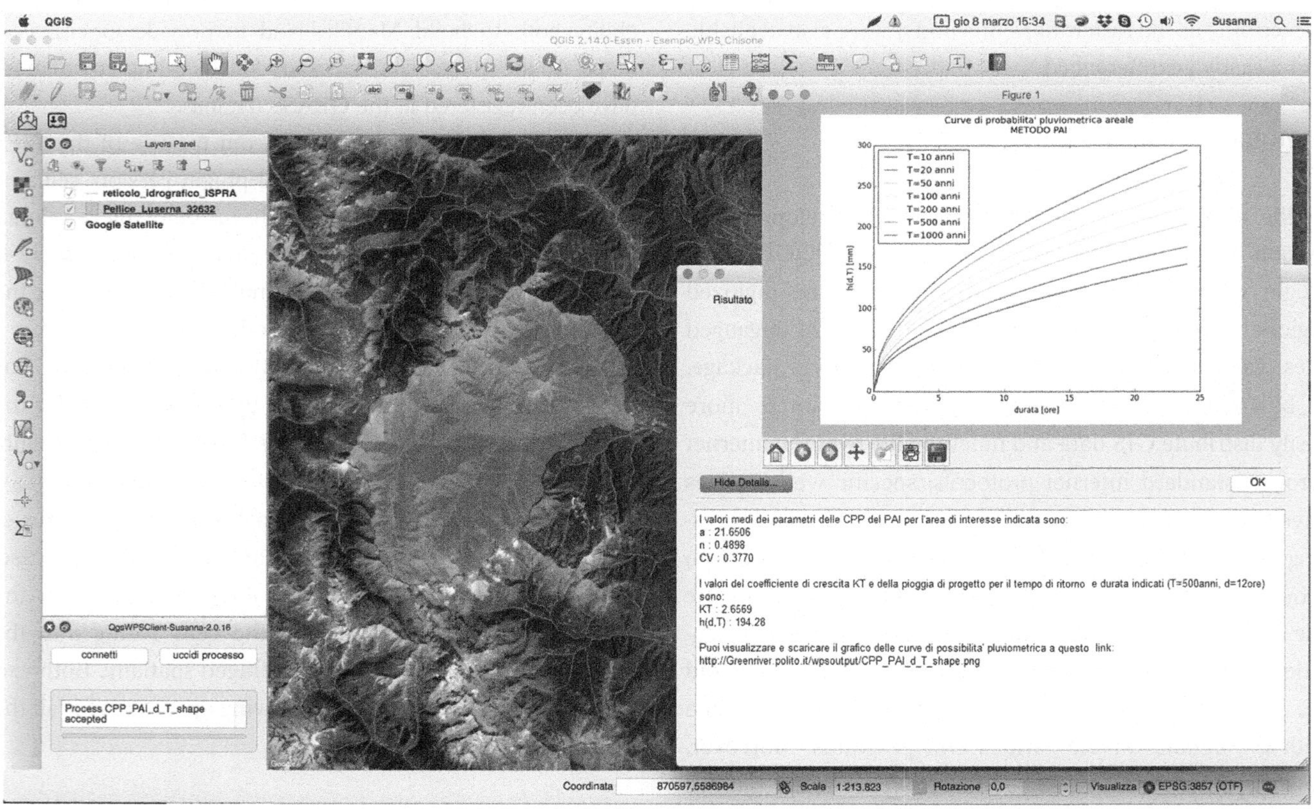

Figure 2 | View of the results of the WPS process, served up to estimate regional depth–duration–frequency curves applying the *PAI-Po* methodology, executed from the QGIS Desktop software using a WPS plugin.

represents the results of the application of the WPS process to estimate *DDF* curves with the *PAI-Po* methodology for a given catchment (the vector layer added into QGIS project and displayed in the map canvas) with return period $T=$ 500 years and duration $d=12$ hours. The WPS procedure provides data outputs as 'text/plain' and 'image/svg'. The values of the areal parameters of the *DDF*s and the rainfall quantile for the duration and return period requested are reported in the textbox, while a graphical representation of the *DDF*s for the considered *T* and some others significant return periods are represented in a svg format.

When the WPS process is executed directly from the web-mapping application, textual results are displayed in a webpage message box, where an URL-endpoint for retrieving the created graph is provided. In case the watershed divide is not available, it can be delimited through a WPS procedure available through the web browser, as depicted in Figure 3, starting from the definition of the outlet's coordinates.

The *MultiRain* tool QGIS plugin version allows for more in-depth analysis not only at the basin scale, but also at regional level. Indeed, the plugin provides not only the graphs or the parameters' value of the *DDF* curves for the area of interest, but also allows maps to be built of the quantiles and of the growth factors for a selected return period. This last functionality is applied to build maps of Figure 4, where the rainfall quantiles for $T=1{,}000$ years and $d=3$ hours are represented for all methods. Consistent differences among the obtained quantiles are apparent, as well visible is how the final grid resolution reflects the different spatial resolution of the original analyses.

Application example and discussion

To demonstrate the capabilities of the *MultiRain* tool, the version we have customized for NW Italy is applied with regard to the assessment of rarity of a real rainfall event, where the results that can be obtained with the different approaches demonstrate a typical experience from the practitioner viewpoint.

The example concerns the Pellice River, a basin with an area of about 216 km^2, which was affected by significant flooding in November 2016 (ARPA Piemonte 2016) generated by a strong and persistent precipitation event. The chosen reference rainfall station is Luserna San Giovanni (361006, 4963859 in the reference system WGS84 – UTM

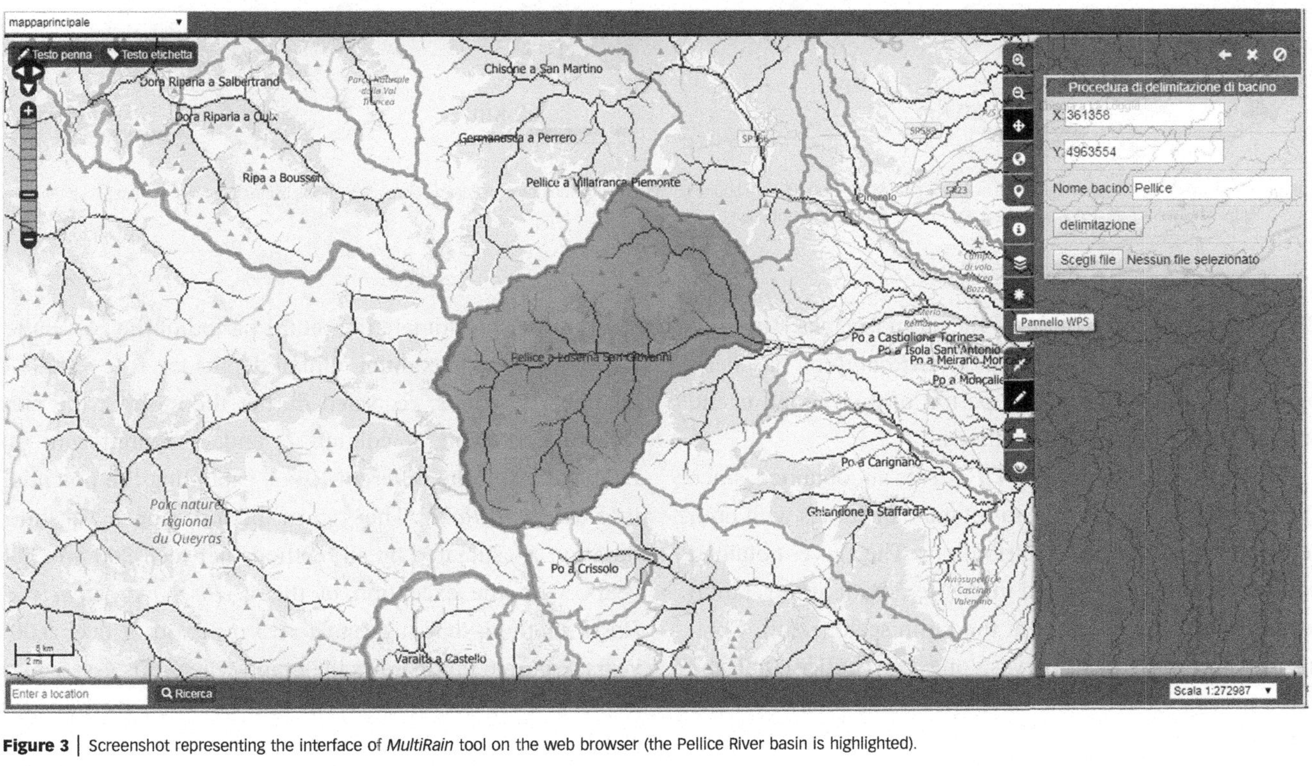

Figure 3 | Screenshot representing the interface of *MultiRain* tool on the web browser (the Pellice River basin is highlighted).

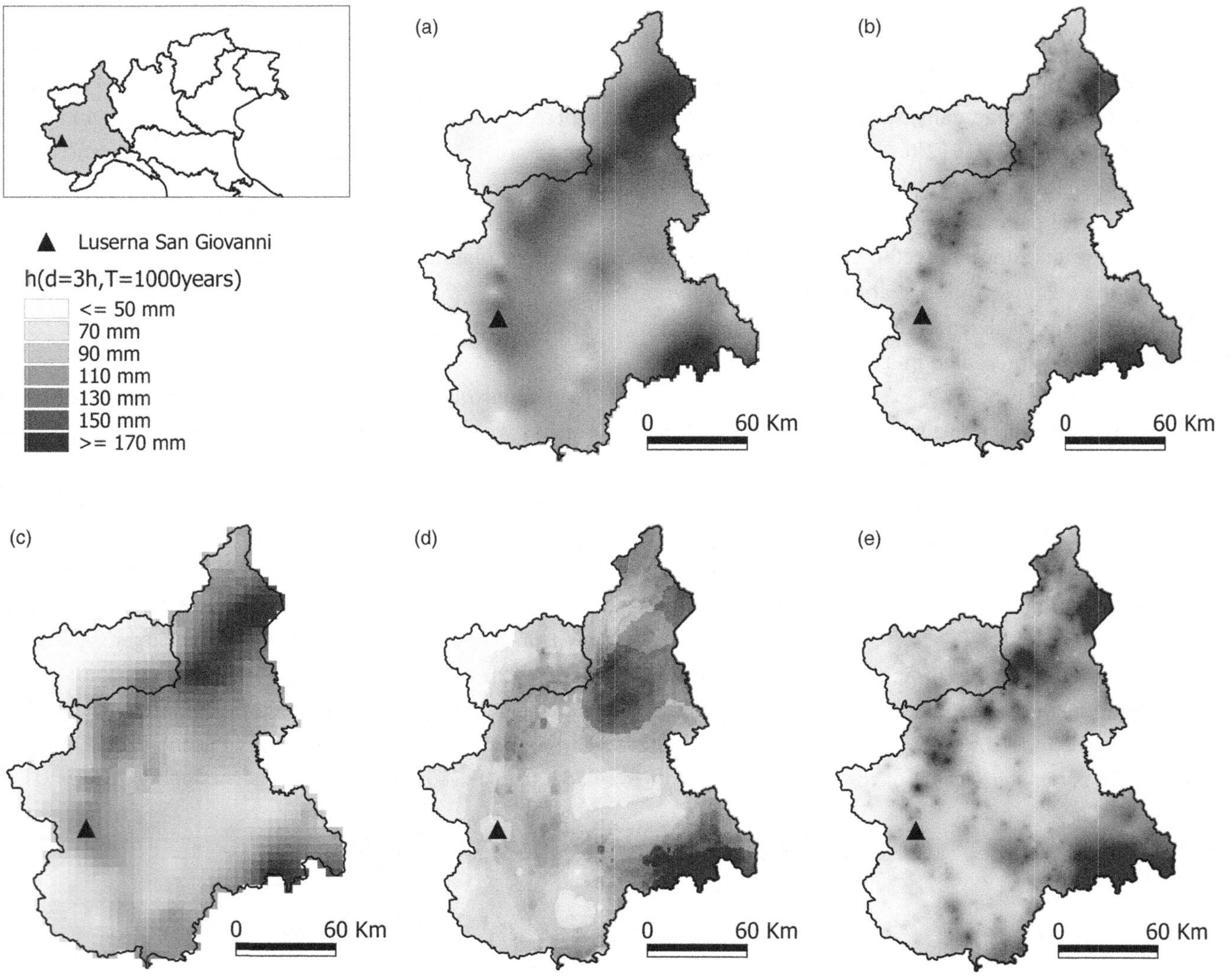

Figure 4 | Comparison of design rainfall depth maps for $T = 1{,}000$ years and $d = 3$ hours, obtained for the methods: (a) *PAI-Po*, (b) *Patched Kriging* using Gumbel distribution, (c) *VAPI Villani*, (d) *VAPI De Michele & Rosso*, (e) *Patched Kriging* using GEV distribution.

32N), positioned at the outlet of the basin (see black triangle in the maps of Figure 4). The *MultiRain* tool has been applied to estimate the return period of the rainfall event, through the four described methods.

Five different families of *DDF* curves are obtained, since the *Patched Kriging* model can estimate quantiles with both the Gumbel and the GEV distributions. The curves obtained with $T = 500$ are represented in Figure 5 both at the station location (left) and averaged at the basin scale (right). Comparing local and areal curves, one can notice that the spatial average reduces the differences that emerge on site estimation.

In order to estimate the return period of the November 2016 event, the maximum rainfall depths for 1, 3, 6, 12 and 24 hour durations, recorded during the storm were considered (see Table 1, column 2) and compared with the *DDF* curves of the different *RFA*s, exploiting the potentialities of *MultiRain*. The return periods for a 24 hour duration is estimated by all methods between 200 and 500 years, with the exception of the *VAPI - De Michele & Rosso* method, that suggests a return period of over 5,000 years. Figure 6 (left) shows the graph where this assessment is made using the most recent and data-rich method (the *Patched Kriging*) using a Gumbel distribution.

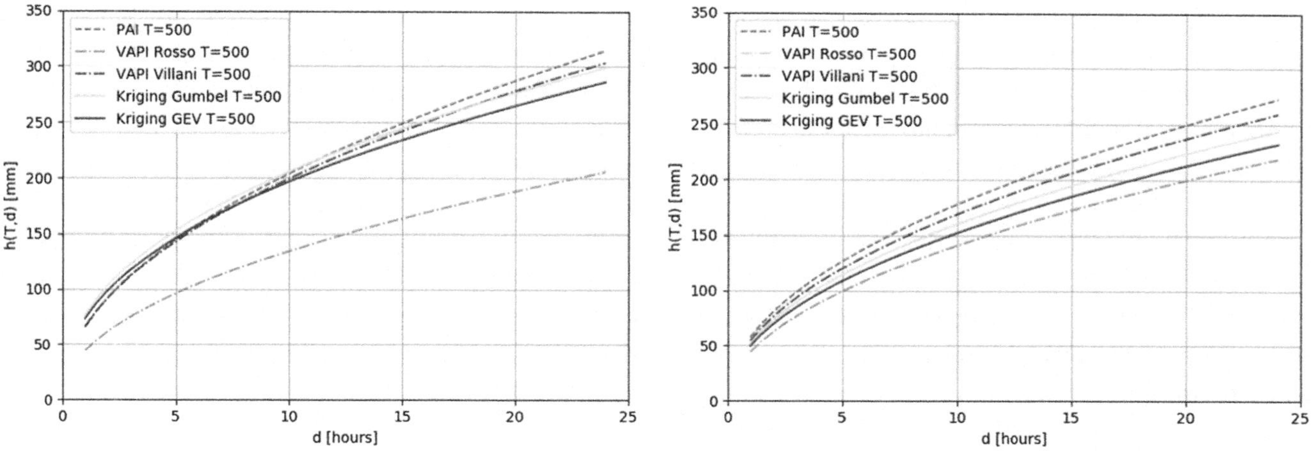

Figure 5 | (Left) *DDF* curves for $T = 500$ years for the Luserna San Giovanni station and (right) areal *DDF* curves for $T = 500$ years for the Pellice basin at Luserna San Giovanni.

Table 1 | Estimated return periods using all the available *RFA*s for the observed extreme precipitation event at Luserna San Giovanni rainfall gauge

		Return time *T* (years) for different models				
Duration (h)	**Cumulated rainfall depth (mm)**	**PAI**	***VAPI Villani***	***VAPI Rosso***	***Patched Kriging* Gumbel**	***Patched Kriging* GEV**
1	17.6	1.20	1.37	1.31	1.06	1.06
3	39.4	1.81	2.17	2.16	1.42	1.42
6	75.6	4.55	5.74	8.95	3.53	3.48
12	140.6	17.00	22.31	210.01	16.14	16.87
24	278.6	175.84	242.23	>5,000	263.27	375.89

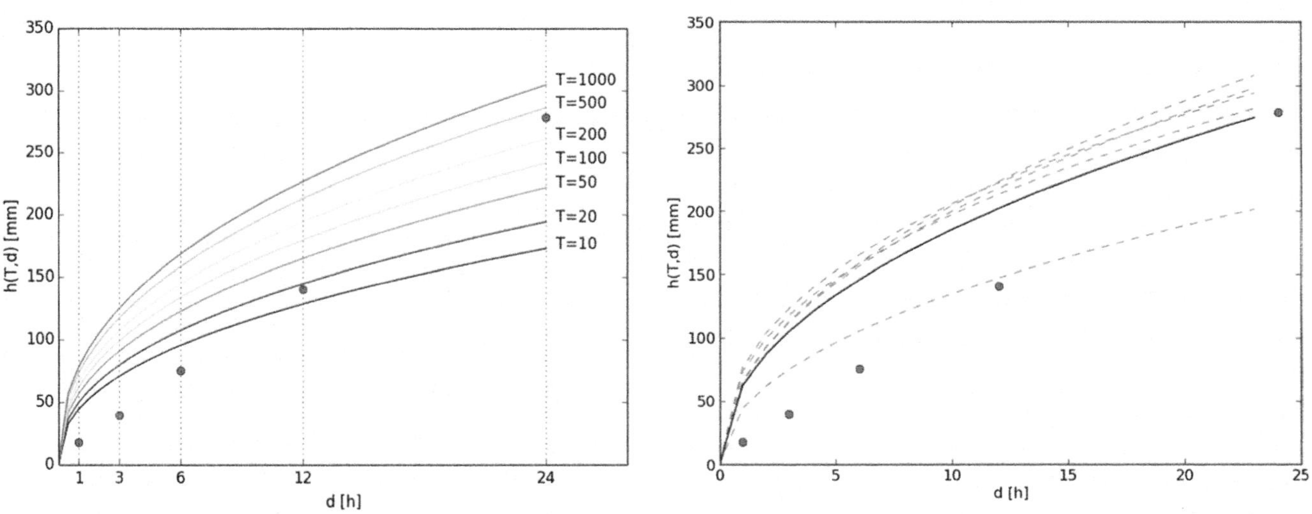

Figure 6 | Comparison of the event-maxima recorded in Luserna San Giovanni in November 2016 with (left) the *DDF* curves estimated with the *Patched Kriging* and with (right) the model-averaged *DDF* curve for $T = 500$ years (in bold) as compared to the individual $T = 500$.

To provide wider understanding of what can happen in similar situations, when various methods provide significantly different results and it is not possible to identify a 'best' model, the *MultiRain* tool helps in adopting a *model-averaging* approach, commonly considered a robust way to manage parameters and model uncertainty in quantile estimation.

In detail, for a certain return period, the averages of the a, n and $K(T)$ among the available methodologies can be computed to obtain an averaged curve. Differential weight can be given to the different estimates in the averaging phase, according to the expertise of the user. As a demonstration of the model-averaged result, the $T = 500$ averaged curve is computed, giving equal weight to the considered *RFAs*, and shown in Figure 6 (right) together with the 2016 event dots. The result of the overlapping of the data with the curve shows that a robust estimation of the return period for the 24 hour duration is closer to $T = 500$ than to $T = 200$ or to $T = 5{,}000$ years.

CONCLUSIONS

When hydrologic design values encounter updates of the estimation procedures, information deriving from different sets of data and of regional statistical estimation procedures needs to be harmonized or merged. In other conditions, e.g., for critical infrastructures, when a 'best model' is difficult to identify, the *multi-model* approach allows for the consideration of a wide range of scenarios and the adoption of robust estimates in the design and control phases.

In this framework, the *MultiRain* tool presented here aims at providing a fast operational aid to apply a *multi-model* approach to the regional rainfall frequency estimation. The tool offers estimates of depth–duration–frequency curves both for a generic location in the region considered, and as a spatial average. The estimate and comparison of spatial averages is given when the user uploads a shapefile of the contour of the area of interest (usually a watershed divide). The *MultiRain* Web Processing Services implemented, accessible both by web browser and by desktop GIS software, allows the user to carry out a thorough comparative analysis based on the results derived from different methodologies, without installing software and downloading data.

Despite the web platform being developed in a modular and reproducible way to allow future integrations, the need for maintaining appropriate control on the security of the data, of the scripts and of the Web server itself has led us to keep the online platform closed to editing. All the extensions and editing of the *MultiRain* WPS Service must be performed by the authors. Skilled users, interested in expanding the potentialities of the tools, can take advantage of the free availability of the 'QGIS local plugin version', that allows the source code to be edited and the integration of further data and methodologies.

Future developments will be addressed first and foremost at integrating the tool with procedures for improving its customization, needed for its implementation in further regions.

Tool availability

Final refinements to the user-interface of the web services are in development and the *MultiRain* tool web page will be available soon at: wwww.idrologia.polito.it/MultiRainTool.

SUPPLEMENTARY MATERIAL

The Supplementary Material for this paper is available online at https://dx.doi.org/10.2166/hydro.2019.016.

REFERENCES

AdbPo – Autorità di bacino distrettuale del fiume Po 2001 *Piano stralcio per l'Assetto Idrogeologico (Po Basin Authority, Hydrogeological Master Plan)*. http://pai.adbpo.it/ (accessed 1 January 2018).

Alfieri, L., Laio, F. & Claps, P. 2008 A simulation experiment for optimal design hyetograph selection. *Hydrological Processes* **22** (6), 813–820.

ARPA Piemonte 2013 *Atlante delle piogge intense in Piemonte (Atlas of Extreme Rainfalls for the Piemonte Region)*. http://webgis.arpa.piemonte.it/atlante_pioggia_webapp/ (accessed 12 December 2018).

ARPA Piemonte 2016 Rapporto d'Evento Novembre (Weather Events Report, November 2016). https://www.arpa.piemonte.it/pubblicazioni-2/relazioni-tecniche/analisi-eventi/eventi-2016/rapporto-preliminare-novembre-2016-def.pdf (accessed 12 December 2018).

Claeskens, G. & Hjort, N. L. 2008 *Model Selection and Model Averaging*. Cambridge University Press, Cambridge, UK.
Dalrymple, T. 1960 Flood-frequency Analyses. Manual of Hydrology: Part 3. Geological Survey Water-Supply Paper, 1543A, U.S. G.P.O., Washington.
De Michele, C. & Rosso, R. 2001 *Rapporto sulla valutazione delle piene Italia Nord Occidentale. Portata al colmo di piena Bacino del Fiume Po e Liguria Tirrenica Estratto dal Rapporto Nazionale VAPI con aggiornamenti (Flood Estimation in North-Western Italy Summary Report. Extract of the final National VAPI Report)*. CNR-GNDCI u.o. 1.8, Milano.
Di Baldassarre, G., Laio, F. & Montanari, A. 2009 Design flood estimation using model selection criteria. *Physics and Chemistry of the Earth, Parts A/B/C* **34** (10–12), 606–611.
Furcolo, P. & Villani, P. 1998 Spatial analysis of uncertain data: an application to extreme rainfall. *Systems Research and Information Science* **8** (2–3), 151–181.
Ganora, D., Laio, F., Masoero, A. & Claps, P. 2016 Spatially-smooth regionalization of flow duration curves in non-pristine basins. In: *The Spatial Dimensions of Water Management – Redistribution of Benefits and Risks*, Vol. 373 (A. H. Schumann, G. Blöschl, A. Castellarin, J. Dietrich, S. Grimaldi, U. Haberlandt, A. Montanari, D. Rosbjerg, A. Viglione & S. Vorogushyn, eds), 7th International Water Resources Management Conference of ICWRS, 18–20 May, Bochum, Germany, pp. 73–80.
Grasso, S., Libertino, A. & Claps, P. 2018 A Web-GIS tool for rainstorm hazard management over large areas. In: *HIC 2018. 13th International Conference on Hydroinformatics*, Vol. 3 (G. La Loggia, G. Freni, V. Puleo & M. De Marchis, eds), EPiC Series in Engineering, pp. 844–851.
Grimaldi, S., Kao, S. C., Castellarin, A., Papalexiou, S. M., Viglione, A., Laio, F., Aksoy, H. & Gedikli, A. 2011 Statistical hydrology. In: *Treatise on Water Science*, Vol. 2 (P. Wilderer, ed.), Elsevier Science, Amsterdam, The Netherlands, pp. 479–517.
Hosking, J. R. M. & Wallis, J. R. 1997 *Regional Frequency Aanalysis: An Approach Based on L-Moments*. Cambridge University Press, Cambridge, UK.
Hosking, J. R. M., Wallis, J. R. & Wood, E. F. 1985 Estimation of the generalized extreme-value distribution by the method of probability-weighted moments. *Technometrics* **27** (3), 251–261.
Jenkinson, A. F. 1955 The frequency distribution of the annual maximum (or minimum) values of meteorological elements. *Q.J.R. Meteorological Society* **81** (348), 158–171.
Koutsoyiannis, D., Kozonis, D. & Manetas, A. 1998 A mathematical framework for studying rainfall intensity-duration-frequency relationships. *Journal of Hydrology* **206** (1), 118–135.
Libertino, A., Allamano, P., Laio, F. & Claps, P. 2018 Regional-scale analysis of extreme precipitation from short and fragmented records. *Advances in Water Resources* **112**, 147–159.
Magro, P. 2007 *Gestione informatizzata del rischio di nubifragio sul bacino del Po (Computer Model for Risk Management of Extreme Rainfall in Po River Basin)*. Master thesis, Environmental Engineering, Politecnico di Torino, Torino, Italy.
Munich Re's NatCatSERVICE (Natural Catastrophe SERVICE) 2017 http://natcatservice.munichre.com (accessed 1 April 2017).
Myers, D. E. 1994 Spatial interpolation: an overview. *Geoderma* **62** (1), 17–28.
Okoli, K., Breinl, K., Brandimarte, L., Botto, A., Volpi, E. & Di Baldassarre, G. 2018 Model averaging versus model selection: estimating design floods with uncertain river flow data. *Hydrological Sciences Journal* **63** (13–14), 1913–1926.
Reed, D. W., Faulkner, D., Robson, A., Houghton-Carr, H. & Bayliss, A. 1999 *Flood Estimation Handbook. Volume II: Rainfall Frequency Estimation*. Institute of Hydrology, Wallingford, UK.
Rossi, F., Fiorentino, M. & Versace, P. 1984 Two-component extreme value distribution for flood frequency analysis. *Water Resources Research* **20** (7), 847–856.
Shamseldin, A. Y., O'Connor, K. M. & Liang, G. C. 1997 Methods for combining the outputs of different rainfall–runoff models. *Journal of Hydrology* **197** (1–4), 203–229.
Svensson, C. & Jones, D. A. 2010 Review of rainfall frequency estimation methods. *Journal of Flood Risk Management* **3** (4), 296–313.
Velázquez, J. A., Anctil, F., Ramos, M. H. & Perrin, C. 2011 Can a multi-model approach improve hydrological ensemble forecasting? A study on 29 French catchments using 16 hydrological model structures. *Advances in Geosciences* **29**, 33–42.
Villani, P. 2003 Rapporto sulla valutazione delle piene in Piemonte (Flood study report for Piedmont Region). In: *Relazione delle attività del CUGRI fino al 2001* (D. Paguro, ed.), Fisciano, Italy, pp. 89–118.
Westra, S., Alexander, L. V. & Zwiers, F. W. 2013 Global increasing trends in annual maximum daily precipitation. *Journal of Climate* **26** (11), 3904–3918.

First received 15 January 2019; accepted in revised form 15 May 2019. Available online 25 June 2019

The mode of the climacogram estimator for a Gaussian Hurst-Kolmogorov process

Panayiotis Dimitriadis and Demetris Koutsoyiannis

ABSTRACT

Geophysical processes are often characterized by long-term persistence. An important characteristic of such behaviour is the induced large statistical bias, i.e. the deviation of a statistical characteristic from its theoretical value. Here, we examine the most probable value (i.e. mode) of the estimator of variance to adjust the model for statistical bias. Particularly, we conduct an extensive Monte Carlo analysis based on the climacogram (i.e. variance of the average process vs. scale) of the simple scaling (Gaussian Hurst-Kolmogorov) process, and we show that its classical estimator is highly skewed especially in large scales. We observe that the mode of the climacogram estimator can be well approximated by its lower quartile (25% quantile). To derive an easy-to-fit empirical expression for the mode, we assume that the climacogram estimator follows a gamma distribution, an assumption strictly valid for Gaussian white noise processes. The results suggest that when a single timeseries is available, it is advantageous to estimate the Hurst parameter using the mode estimator rather than the expected one. Finally, it is discussed that while the proposed model for mode bias works well for Gaussian processes, for higher accuracy and non-Gaussian processes, one should perform a Monte Carlo simulation following an explicit generation algorithm.

Key words | climacogram, long-term persistence, mode estimator, statistical bias, stochastic uncertainty

Panayiotis Dimitriadis (corresponding author)
Demetris Koutsoyiannis
Department of Water Resources and Environmental Engineering, School of Civil Engineering,
National Technical University of Athens,
Heroon Polytechneiou 5, 15880 Zographou,
Greece
E-mail: *pandim@itia.ntua.gr*

INTRODUCTION

An important attribute characterizing geophysical processes is the high spatio-temporal dependence, in the sense that a random variable of such a process at a specific time or location strongly depends on several (even infinite) past, or of different location, random variables of the same process. This type of dependence requires long samples for its identification, which is a rare case in most natural processes, and thus, for the estimation of its parameters, it is advised to use only up to the second-order statistics (Lombardo *et al.* 2014) and only in cases where very long samples are available to expand to higher orders. The above issues are further highlighted in Dimitriadis (2017), where several (overall thirteen) such processes with various lengths and physical properties expanding from small-scale turbulence to large-scale hydrometeorological processes are analyzed in terms of their long-term behaviour using massive databases and unbiased estimators of the second-order dependence structure. Interestingly, all the examined processes exhibited long-term persistence, otherwise known as Hurst-Kolmogorov (HK) behaviour (coined by Koutsoyiannis & Cohn (2008)), i.e. power-law decay of the autocorrelation function with lag (for a literature review on long-term persistent processes in hydrometeorology, see also O'Connell *et al.* (2016)). Additionally, Koutsoyiannis (2011) provided a theoretical justification of the HK behaviour in geophysical processes, showing that it is linked to the second law of

doi: 10.2166/hydro.2019.038

thermodynamics (i.e. entropy extremization), and specifically, the stronger the persistence of the dependence structure of a process, the higher the entropy of the process at large scales.

The identification of the dependence structure of a process can be highly affected by the sample uncertainty and statistical bias where the true statistical properties (mean, variance etc.) of a statistic (e.g. variance) of a stochastic process may differ from the one estimated from a series with finite length. The deviations of the statistical characteristics from their true values should be taken into account not only for the marginal characteristics but also for the dependence structure of the process. Therefore, to correctly adjust the stochastic model to the observed series of the physical process, we should account for the bias effect since all series are of finite (and often short) lengths.

The second-order properties can be similarly assessed by common stochastic tools such as the autocovariance function (a function of lag), power spectrum (a function of frequency), and variation of statistics (e.g. variance) of the averaged process vs. scale, a tool known as climacogram (Koutsoyiannis 2010). It is shown that the latter estimator of the second-order dependence structure, as compared to the other two metrics, encompasses additional advantages in stochastic model building and interpretation from data; for example, it is characterized by smaller statistical uncertainty and easier to handle expressions of the statistical bias (Dimitriadis & Koutsoyiannis 2015). Therefore, it is advisable that the sample uncertainty of the second-order dependence structure be tackled with the estimator with the lower variation, such as the climacogram. When multiple sample realizations (i.e. recorded series) are known, the handling of the statistical bias arising from a selected stochastic model may be based on the unbiased estimator of the expected value of the climacogram (Dimitriadis & Koutsoyiannis 2018). However, when a single data series of observations is available for the model fitting (which is the case when geophysical processes are studied), it would be interesting to examine the mode of the climacogram, instead of the expected value; the two may differ in case of strong HK behaviour. This estimator is equivalent to a maximum-likelihood estimator (e.g. Kendziorski *et al.* 1999) for processes with zero (i.e. white noise) or short-term (e.g. Markov) dependence structure, while here we further extend it for HK processes (see also the work of Tyralis & Koutsoyiannis (2011) for the expectation of the climacogram). It is noted that while the climacogram is often based on the second central moment (i.e. variance), other types of moments (e.g. raw, L-moments or K-moments; Koutsoyiannis 2019) can be used to measure fluctuation in scale, and here, we focus on the central second-order climacogram (i.e. fluctuation measured by variance vs. scale).

METHODS

In this section, we present the applied methods, namely the climacogram estimator, the statistical bias expressions for the mode and expected values of the estimator and the algorithm for the stochastic synthesis of the Gaussian HK process for the Monte Carlo analysis.

The climacogram

The analysis of a process through the variance of the averaged process vs. scale has been thoroughly applied in stochastic processes (e.g. Papoulis 1991; Vanmarcke 2010). However, its importance to the analysis of the second-order dependence structure is highlighted mainly by more recent works (see a historical review in Koutsoyiannis (2018)). Also, the simple name *climacogram* allowed its further understanding through visualization; indeed, the term originates from the Greek *climax* (meaning scale) and *gramma* (meaning written; cf. the terms autocorrelogram for the autocorrelation, scaleogram for the power spectrum and wavelets).

It has been shown that the climacogram, treated as an estimator (rather than just a tool for the identification of long-term behaviour of the second-order dependence structure), has additional advantages from the more widely applied estimators of the autocovariance and power spectrum (Dimitriadis & Koutsoyiannis 2015). Namely, the climacogram could provide a more direct, easy, and accurate means to make diagnoses from data and build stochastic models in comparison to the power spectrum and autocovariance. For example, the climacogram, compared to other tools, has the lowest standardized estimation error for processes with short- and long-term persistence, zero discretization error for averaged processes, simple and analytical expression for the statistical bias, always positive values, well-defined and usually monotonic behaviour, smallest fluctuation of skewness on small scales while closest to zero skewness in larger scales, and mode closest to the expected (i.e. mean) value in large scales. Also, the climacogram is directly linked to the entropy production of a process (Koutsoyiannis 2011, 2016). Furthermore, the climacogram expands the notion of *variance* by making it a function of *time scale* and is *per se* further expandable for statistics different from the central estimators of fluctuation (e.g. second raw moment and second L-moment vs. scale; Koutsoyiannis 2019) for different characteristics of the estimator (e.g. mode and median) and even for moments of higher (e.g. third and fourth) orders (Dimitriadis & Koutsoyiannis 2018). Recently, Koutsoyiannis (2019) extended the notion of climacogram for orders higher than two and showed how to substitute the joint moments of a process, allowing in this manner to tackle some limitations of the latter such as the discretization effect and statistical bias.

Symbolically, the climacogram is:

$$\gamma(k) := \mathrm{var}[\underline{x}(k)] \tag{1}$$

where var[ ] denotes the variance and $\underline{x}(k) := 1/k \int_0^k \underline{x}(t)\mathrm{d}t$ is the continuous-time process at scale k (in dimensions of time), which equals the discrete-one averaged in time intervals Δ, i.e. $\underline{x}_\kappa := 1/\kappa \sum_{i=1}^{\kappa} \underline{x}_i$, in the discrete-time scale $\kappa = k/\Delta$ (dimensionless natural number whereas for real numbers see the adjustment in Koutsoyiannis (2011)).

The Gaussian long-term persistent process and its stochastic synthesis

The most common processes employed in geophysics, and particularly in hydrology, are the white noise process, the Markov process (with an exponential decay of the autocorrelation), and long-term persistent processes, which are characterized by a power-law decay of the climacogram (or equivalently of the autocorrelation) as a function of scale (or lag). A typical representative of the latter processes is the Gaussian HK process defined as follows:

$$\underline{x}(k) - \mu =_{\mathrm{d}} (k)^{(H-1)}(\underline{x}(1) - \mu) \tag{2}$$

where $=_{\mathrm{d}}$ denotes equality in distribution with μ the mean and $\gamma(k) = \gamma(\Delta)/\kappa^{2-2H}$ the variance of the process for each scale k, H is the Hurst parameter ($0 < H < 1$) otherwise defined as (Dimitriadis *et al.* 2016a) $H := 1 + \frac{1}{2}\lim_{k\to\infty} \mathrm{d}\ln(\gamma(k))/\mathrm{d}\ln k$; the quantity in the limit is the derivative of $\ln(\gamma(k))$ with respect to $\ln(k)$.

It is noted that this process has infinite variance at scale zero and thus, it should not be used to model the small scales of a physical process (in spite of the fact that the fractional-Gaussian-noise–fGn–process is widely used to model several processes at small scales; Koutsoyiannis *et al.* 2018). For the stochastic synthesis of the Gaussian HK model, we may use the sum of arbitrarily many independent Markov processes, thus expressing the target climacogram as follows (Dimitriadis & Koutsoyiannis 2015):

$$\gamma(\kappa\Delta) = \sum_{i=1}^{l} \frac{2\lambda_i}{(\kappa\Delta/q_i)^2}\left(\kappa\Delta/q_i + e^{-\kappa\Delta/q_i} - 1\right) \tag{3}$$

where λ_i is the variance, q_i a time scale parameter for each Markov model i, and l the total number of Markov processes. Mandelbrot (1963) has shown that for $l \to \infty$, the above model can adequately describe an fGn (or else Gaussian HK) process for any generated length (see also Mandelbrot & Van Ness 1968; Mandelbrot & Wallis 1968). Koutsoyiannis (2002) has analytically estimated the parameters of three AR(1) models ($l = 3$) to capture the HK process for $n < 10^4$. Dimitriadis & Koutsoyiannis (2015) have expanded this framework to the sum of arbitrarily many AR(1) models (abbreviated as SAR) for the generation of any type of process with an autoregressive dependence structure and up to any number of scales, by using a suitable function with only two parameters, namely p_1 and p_2, that link the lag-1 autocorrelations of each Markov model, e.g. through the expression $q_i = p_1 p_2^{i-1}$, with $i = 1, \ldots, l$ and l often taken equal to the integer part of $\log(n) + 1$. For example, for $n = 10^6$ and $H = 0.8$, we have $l = 7$, $p_1 = 0.394$ and $p_2 = 12.356$ for a maximum standardized error between the true γ_{t} (Equation (2)) and modelled γ_{m} (Equation (3)) climacogram (i.e. $\max|(\gamma_{\mathrm{t}} - \gamma_{\mathrm{m}})/\gamma_{\mathrm{t}}|$ for all scales) equal to 0.009 (Table 1).

Table 1 | Parameters p_1 and p_2 estimated to approximate different types of the $N(0,1)$-HK model (i.e. $\mu = 0$ and $\gamma(\Delta) = 1$) with $l = 7$ and $n \leq 10^6$

H	p_1	p_2	Maximum error (standardized)
0.51	0.022	17.122	0.001
0.60	0.091	12.607	0.006
0.70	0.124	13.317	0.009
0.80	0.394	12.356	0.009
0.90	0.395	14.708	0.005
0.99	0.548	19.465	0.001

The mode of climacogram estimator and its statistical bias

The climacogram can be estimated from a sample through an estimator as similarly done for the estimators of the marginal moments. Here, for the climacogram we use a classical estimator:

$$\underline{\hat{\gamma}}(\kappa\Delta) = \frac{1}{[n/\kappa] - 1} \sum_{i=1}^{[n/\kappa]} (\underline{x}_i^{(\kappa)} - \underline{\bar{x}})^2 \tag{4}$$

where $[n/\kappa]$ is the integer part of n/κ, $\underline{x}_i^{(\kappa)} = \sum_{l=\kappa(i-1)+1}^{\kappa i} \underline{x}_l/\kappa$ is the averaged process at scale $\kappa = k/\Delta$ for $i \in [1, [n/\kappa]]$, $\underline{\bar{x}} = \sum_{l=1}^{n} \underline{x}_l/n = \underline{x}_1^{(n)}$ is the sample average, and n is the series length.

Since the above estimator is a random variable, it has a marginal distribution (see an illustration in Figure 1). The true value of a statistical characteristic (e.g. variance) of a stochastic model may differ from the one estimated from a series with finite length. To correctly adjust the stochastic model to the observed series of the physical process, one should account for the bias effect. An important question is how the statistical bias is generally handled through the second-order dependence structure in case of long-term persistent processes. Particularly, the selected stochastic model should be adjusted for bias before it is fitted to the sample dependence structure. It is noted that neglecting the bias effect in case of a long-term persistent process leads to underestimations of the stochastic model parameters such as the Hurst parameter and to erroneous conclusions. An adjustment of the models for bias is usually done by equating the observed dependence structure to the expected value of the applied estimator. The alternative studied here is the mode, instead of the expected value, of the dependence structure, which represents the most probable value (and thus, the most expected) of the variance estimator at each scale.

The statistical bias of an estimator is the difference of the expected value of the estimator from its true value (e.g. Papoulis 1991). Thus, the bias of the climacogram is shown as follows (e.g. Koutsoyiannis 2011):

$$\mathrm{B_E}[\underline{\hat{\gamma}}(\kappa\Delta)] = \mathrm{E}[\underline{\hat{\gamma}}(\kappa\Delta)] - \gamma(\kappa\Delta) = \frac{(\kappa/n)\gamma(\kappa\Delta) - \gamma(n\Delta)}{1 - \kappa/n} \tag{5}$$

where $\mathrm{B_E}[\,]$ denotes the bias of the expected value of a statistical estimator of a process. Clearly, for the mean value of a process, we have that $\mathrm{B_E}[\underline{\hat{\mu}}] = \mathrm{E}\left[\sum_{l=1}^{n} \underline{x}_l/n\right] - \mu = 0$.

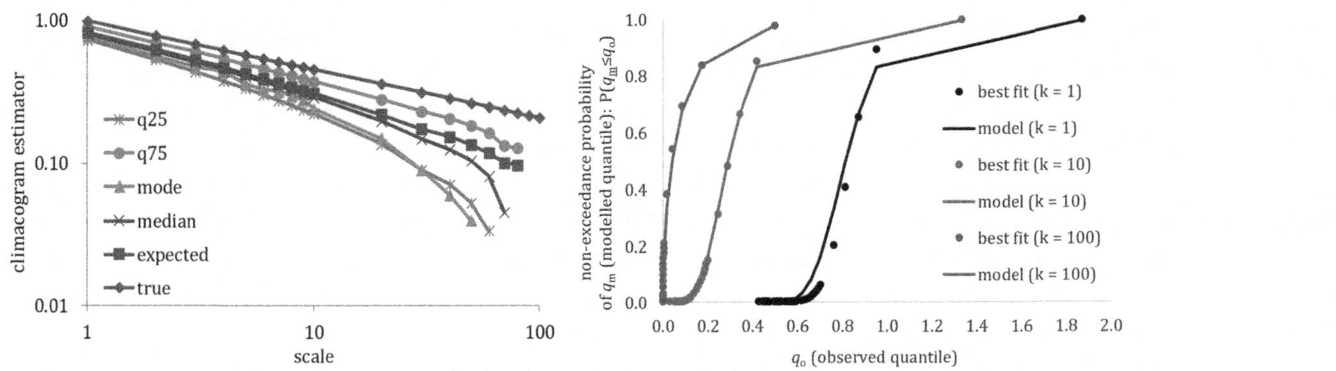

Figure 1 | An illustration for an $N(0,1)$-HK ($H = 0.83$, $n = 200$) process of (left) how several statistical characteristics of the climacogram estimator vary with scale and (right) the observed quantile (q_o) vs. the non-exceedance probability of the modelled quantile $P(q_m \leq q_o)$, showing how the gamma distribution can adequately approximate the distribution of the climacogram estimator especially at large scales.

Following the same rationale, we define an expansion of the notion of bias for the mode of the above estimator of the climacogram, i.e.:

$$\mathrm{B_M}[\underline{\hat{\gamma}}(\kappa\Delta)] = \mathrm{M}[\underline{\hat{\gamma}}(\kappa\Delta)] - \gamma(\kappa\Delta) \quad (6)$$

where $\mathrm{M}[\underline{x}] := \arg\max\,[f(\underline{x})]$ denotes the mode of the variable $\underline{x}$ with density function $f(x)$. We refer to $\mathrm{B_M}[\ ]$ as the mode bias.

For a Gaussian white noise process of length n and variance $\gamma(1)$, the distribution of its sample variance follows the gamma distribution $\Gamma((n-1)/2\,,2\gamma(1)/(n-1))$ (Cochran 1934). The averaged process at scale κ, with a sample length of n/κ and variance $\gamma(k) = \gamma(\Delta)/\kappa$, follows $\Gamma((n/\kappa-1)/2\,,2\gamma(\Delta)/(n-\kappa))$, with $\mathrm{M}[\underline{\hat{\gamma}}(\kappa\Delta)] = \gamma(\Delta)\dfrac{n-3\kappa}{\kappa(n-\kappa)}$ for $n/\kappa \geq 3$, or else 0. Hence, for $n/\kappa \gg 3$, we have that $(n-3\kappa)/(n-\kappa) \approx 1$, and $\mathrm{M}[\underline{\hat{\gamma}}(\kappa\Delta)] \approx \mathrm{E}[\underline{\hat{\gamma}}(\kappa\Delta)] = \gamma(\Delta)/\kappa = \gamma(\kappa\Delta)$, i.e. zero bias. However, for long-term persistent processes, the mode bias is non-zero and its analytical solution is no longer easy to derive.

From the above results, it becomes evident that the statistical bias always depends on the selected model and not on the data as commonly thought. For example, consider the Gaussian HK process in the previous section with an autocorrelation function in discrete time $\rho_\upsilon = 1/2(|\upsilon+1|^{2H} + |\upsilon-1|^{2H}) - |\upsilon|^{2H}$, where υ is the discrete-time lag. The bias of the autocorrelation is similarly defined as $\mathrm{B_E}[\underline{\hat{\rho}}(\upsilon)] = \mathrm{E}[\underline{\hat{\rho}}(\upsilon)] - \rho(\upsilon)$, and thus depends on the model parameter H. It is noted that the above apply even to the so-called non-parametric models, since they also involve estimation from data, and thus, these models should be similarly adjusted for statistical bias to avoid underestimation of the process variability during a Monte Carlo simulation.

For simplicity and without loss of generality, we set $\Delta = 1$ for the rest of the analysis. It is evident that $\mathrm{B_M}[\underline{\hat{\gamma}}(\kappa)] \leq \mathrm{B_E}[\underline{\hat{\gamma}}(\kappa)] \leq 0$ or else $|\mathrm{B_E}[\underline{\hat{\gamma}}(\kappa)]| \leq |\mathrm{B_M}[\underline{\hat{\gamma}}(\kappa)]|$, since the sample variance is positively skewed, i.e. $\mathrm{E}[\underline{\hat{\gamma}}(\kappa)] \geq \mathrm{M}[\underline{\hat{\gamma}}(\kappa)]$ and the equality holds when $n \to \infty$, where the variance of the sample variance is zero for an ergodic process. A preliminary analysis of common HK-type processes has shown that the mode climacogram is close to the low quartile (25% quantile) of the marginal distribution of variance at each scale (Dimitriadis *et al.* 2016c; Gournary 2017). Therefore, when the mode of the variance estimator is of interest, we may use a Monte Carlo technique (as described in the next section) to accurately estimate the mode bias or, in case the marginal distribution of the climacogram is known, to calculate the 25% quantile at each scale to approximate the mode bias.

MONTE CARLO ANALYSIS FOR THE MODE OF THE VARIANCE ESTIMATOR

We perform Monte Carlo experiments over the $N(0,1)$-HK model for a wide range of Hurst parameters H (i.e. 0.5 to 0.95) and for a wide range of series lengths n (i.e. 20–2,000). Specifically, we produce a number (N) of synthetic series through the SAR model described in the section 'The Gaussian long-term persistent process and its stochastic synthesis', where N depends on the sample mean value to reach the expected one at scale $\kappa = n/10$ based on the rule of thumb when using the climacogram as shown in Dimitriadis & Koutsoyiannis (2015). We found that for $N \approx 10^6/n^{2-2H}$, the standardized error between the theoretical expected value and the sample one (Equation (5)) is lower than 1% at scale $\kappa = n/10$. In this way, the mode is expected also to be well preserved with a similar error. However, caution should be given to the selection of the sample mode estimator to ensure that its variance permits a robust estimation of the true value of the mode. Since the distribution function of the estimator of variance is unknown for long-term persistent processes and given that the mode value is the most likely to occur within the sample, we calculate the sample mode from each simulated series by finding the most probable value with an accuracy of two decimal digits. Specifically, we round up each value of the time series, and for each scale, to the second decimal digit, and we estimate the most probable value of the rounded time series (for higher accuracies a larger N was required). Also, other estimators for the sample mode (e.g. Bickel & Fruwirth 2006) could be used and compared to the proposed one in future research to optimize the performance of the analysis.

Here, to derive an easy-to-fit empirical expression to approximate the mode bias, we adopt the assumption that the above distribution is nearly gamma for smaller scales (see also a similar analysis in Gournary (2017) and Dimitriadis *et al.* (2018)). Using the results from the Monte Carlo analysis, we then evaluate the parameter of the gamma distribution for each H, n, and κ, and we build a model for the mode, then later test its performance. Although the true autocorrelation function of the averaged process for a long-term persistent process does not vary with scale, the sample autocorrelation will be also prone to bias (e.g. Dimitriadis & Koutsoyiannis 2015) affecting the distribution function of the sample variance at each scale. To minimize the sample error for the fitting of the two-parameter gamma distribution, we use the theoretical expression for the expected value of the sample climacogram, i.e. $\mathrm{E}[\underline{\hat{\gamma}}(\kappa)]$, and the variance of the sample climacogram, i.e. $\mathrm{Var}[\underline{\hat{\gamma}}(\kappa)]$, as evaluated from the Monte Carlo analysis, which exhibits the lowest variability in estimation among the four central moments (Dimitriadis & Koutsoyiannis 2018; Figure 2). Based on these two measures, we estimate the two parameters of the gamma distribution.

We first set the scale parameter of the gamma distribution such as to simulate the sample ratio of the aforementioned parameters, i.e. $b(H, n, \kappa) = \mathrm{Var}[\underline{\hat{\gamma}}(\kappa)]/\mathrm{E}[\underline{\hat{\gamma}}(\kappa)]$ and so, the shape parameter can be also estimated as $a(H, n, \kappa) = \mathrm{E}[\underline{\hat{\gamma}}(\kappa)]/b(H, n, \kappa)$.

We observe (Figure 2) that for $a(H, n, \kappa) > 1$, the shape parameter $a(H, n, \kappa)$ is approximately proportional, by a function $c(H)$, to the corresponding shape parameter for the white noise process $a(0.5, n, \kappa) = (n/\kappa - 1)/2$ raised to a function $p(H)$, i.e.:

$$a(H, n, \kappa) = c(H)((n/\kappa - 1)/2)^{p(H)} \tag{7}$$

where $a(H, n, \kappa) > 1$ is a function corresponding to the shape parameter of the gamma distribution function, while for $a(H, n, \kappa) \leq 1$ or $\kappa \gtrsim n/3$, the mode is considered close to zero.

The two functions of the above expression are fitted as follows (Figure 2):

$$c(H) = 1.68(H - 0.5)^2 - 0.3025(H - 0.5) + 1 \tag{8}$$

and

$$p(H) = -2.4865(H - 0.5)^2 + 0.1485(H - 0.5) + 1 \tag{9}$$

The above two adjustments allow us to empirically express the mode of the climacogram estimator as a function of H, n, and κ:

$$\begin{aligned} \mathrm{M}[\underline{\hat{\gamma}}(\kappa)] &= (a(H, n, \kappa) - 1)b(H, n, \kappa) \\ &= (1 - 1/a(H, n, \kappa))\mathrm{E}[\underline{\hat{\gamma}}(\kappa)] \end{aligned} \tag{10}$$

It is noted that based on the above assumptions, the standard deviation, and the skewness and excess kurtosis coefficients of the climacogram estimator can be estimated as $b(H, n, \kappa)\sqrt{a(H, n, \kappa)}$, $2/\sqrt{a(H, n, \kappa)}$, and $6/a(H, n, \kappa)$, respectively. Since $a(H, n, \kappa) \leq a(0.5, n, \kappa)$ all the above measures will be larger than those in case of a white noise process.

The above expression can approximate the mode by an absolute difference of 0.005 from the Monte Carlo estimates, while for better approximations it is advised to implement a new Monte Carlo analysis (see also discussion and application in the section 'Applications to annual

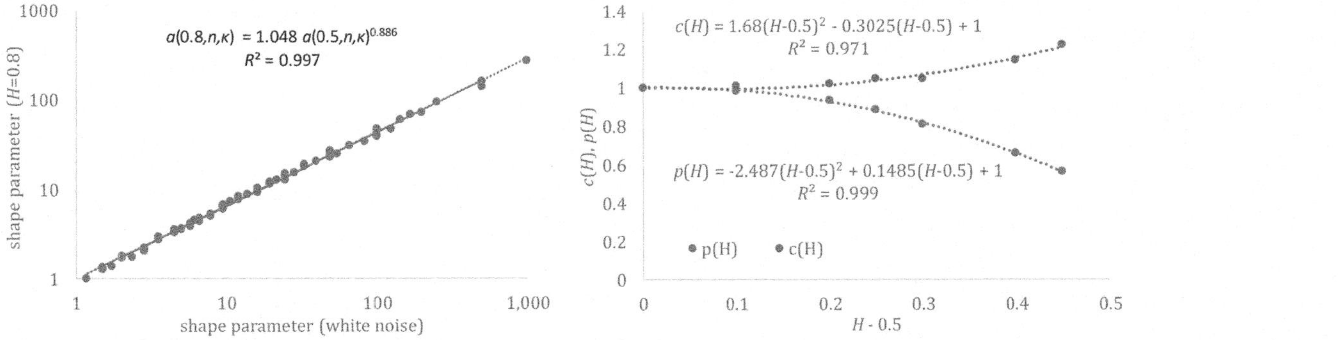

Figure 2 | (left) The shape parameter assuming a gamma distribution for the mode estimator of the climacogram of an $N(0,1)$-HK process (for $H = 0.8$ and for all n and κ simulated in the Monte Carlo analysis) vs. the theoretical shape parameter of the white noise process. (right) Proposed model for the $c(H)$ and $p(H)$ functions for all examined H from the Monte Carlo analysis.

streamflow'). Interestingly, the standardized error between the mode and expected values of the estimator, i.e. $\varepsilon = \left|\mathrm{E}\left[\underline{\hat{\gamma}}(\kappa)\right] - \mathrm{M}\left[\underline{\hat{\gamma}}(\kappa)\right]\right| / \mathrm{E}\left[\underline{\hat{\gamma}}(\kappa)\right]$, is calculated from the Monte Carlo analysis to reach a maximum value of 67% corresponding to cases with $H \geq 0.6$ and $n/\kappa \leq 10$, while for the white noise process it can be theoretically estimated as $\varepsilon = 2/(n/\kappa - 1)$, which for $\kappa = n/10$ is approximately 20%.

APPLICATIONS TO ANNUAL STREAMFLOW

For illustrations of possible implications of the above results, we apply a stochastic analysis based on the expected and the mode values of the climacogram to a streamflow process at the Peneios river (Thessaly, Greece), where a historical streamflow annual time series is available at the upstream station of Ali Efenti with only a 13-year length (for more information on the study area, see Dimitriadis *et al.* (2016b)). For the identification of the stochastic model, we adjust for statistical bias and, in particular, we fit the mode of the estimator rather than its expectation. It is noted that the proposed empirical model for the mode bias (Equation (10)) is derived from a Monte Carlo analysis for sample lengths of $n \geq 20$, and so for this application, we perform a new Monte Carlo analysis to fit the observed climacogram for scales $1 \leq \kappa \leq n/10$ (rule of thumb; Dimitriadis & Koutsoyiannis 2015) and so here, for the first two scales (Figure 3). We find that an HK model can adequately simulate the observed standardized climacogram, i.e. $\hat{\gamma}(\kappa)/\hat{\gamma}(1)$, with $H = 0.9$. We also estimate the Hurst parameter with the expectation of the estimator, and we find $H' \approx 0.8$ and $H'' \approx 0.7$, with or without adjusting for bias, respectively. Evidently, both latter values underestimate the long-term persistence behaviour (Figure 3).

It is noted that the dependence structure of a process (e.g. streamflow) will have a small effect at the risk imposed by the expected number of peaks over threshold (e.g. for the design of a dam or for flood risk mapping) as compared to the effect of the marginal distribution of the process (Volpi *et al.* 2015; Serinaldi & Kilsby 2018). However, the dependence structure will have a great effect (especially for processes with long-term behaviour) at the duration of successive peaks over threshold (e.g. maximum duration of wet/dry periods or of flood inundation), which may highly affect urban as well as agricultural areas and insurance

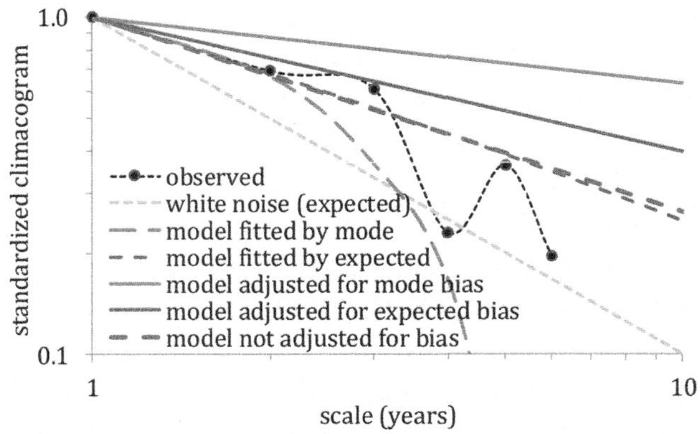

Figure 3 | Standardized climacogram estimations of the observed standardized time series (black line), the white noise model (grey line), and the three fitted $N(0,1)$-HK stochastic processes: (a) adjusting for bias of the mode of the estimator (green line), i.e. $\mathrm{M}[\underline{\hat{\gamma}}(\kappa)]/\mathrm{M}[\underline{\hat{\gamma}}(1)]$, and of its expectation (red line), i.e. $\mathrm{E}[\underline{\hat{\gamma}}(\kappa)]/\mathrm{E}[\underline{\hat{\gamma}}(1)]$, and (b) not adjusting for bias (blue line), i.e. $\underline{\hat{\gamma}}(\kappa) = \gamma(\kappa)$, also corresponding to the non-parametric model configuration. Please refer to the online version of this paper to see this figure in colour: http://dx.doi.org/10.2166/hydro.2019.038.

policies (e.g. Serinaldi & Kilsby 2016; Goulianou *et al.* 2019). To illustrate this, we generate an adequate number N (see the section 'Monte Carlo analysis for the mode of the variance estimator') of HK synthetic timeseries with $H = 0.5$ ($N = 5 \times 10^3$), $H = 0.7$ ($N = 4 \times 10^4$), $H = 0.8$ ($N = 10^5$), and $H = 0.9$ ($N = 3 \times 10^5$). For convenience, we assume an $N(0,1)$ distribution for all processes. We then estimate the expected frequency of the number of peaks over various thresholds (PoT) as well as the expected frequency of the maximum duration of successive peaks over various thresholds (MdT), and we standardize them with the PoT and MdT values of the white noise process (Figure 4).

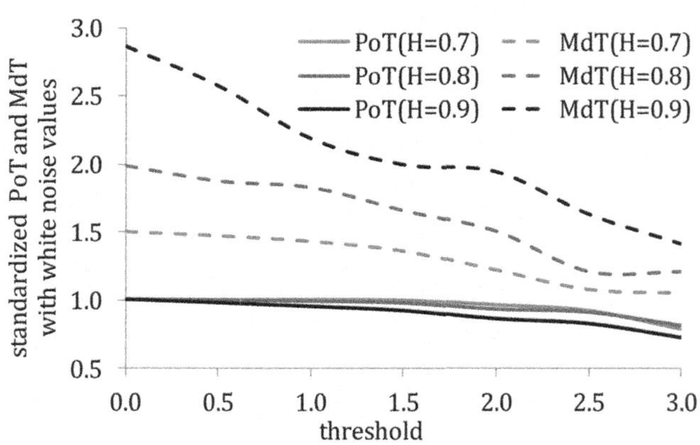

Figure 4 | Expected frequency of peak over threshold (PoT) and expected maximum duration of successive peaks over threshold (MdT) standardized with the PoT and MdT values of the $N(0,1)$ white noise process for various HK-$N(0,1)$ processes.

We find that the MdT varies with threshold and long-term persistence, while the PoT stays almost unaffected by both. Additional analyses and quantifications on the reflection of long-term term persistence in terms of clustering in time can be found in Iliopoulou & Koutsoyiannis (2019).

The results from this study suggest that the sample estimator of the variance can be skewed even for long samples in the presence of long-term persistence behaviour as opposed to the white noise process. Therefore, the mode is different from the expectation and more suitable to use in estimation. We propose that when a single recorded series is available and a Gaussian HK process is fitted with small sample size and relatively high Hurst parameter, it is advantageous to employ the mode of the estimator as calculated from the empirical model of Equation (10), rather than its expectation (Equation (5)), so as to avoid underestimation of the Hurst parameter (and thus, the uncertainty of the process). In case of a non-Gaussian distribution, larger accuracy, or a different estimator of the second-order dependence structure (e.g. other climacogram estimator, autocovariance, power spectrum, variogram etc.), we should employ the Monte Carlo technique and test whether the mode of the estimator used is close enough to its expected value. If this is true, then the expected value can be used to adjust the model for bias, whereas if the two values vary then the model should be adjusted for bias based on the mode estimator. For Monte Carlo analysis of a non-Gaussian-correlated process, an explicit algorithm should be preferred (Dimitriadis & Koutsoyiannis 2018) since the mode value is expected to highly depend on higher-order moments in case of long-term persistent processes.

CONCLUSIONS AND DISCUSSION

Awareness of uncertainty in assessing the dependence structure of a process is of paramount importance as it may critically affect the interpretation of results. Estimation uncertainty may introduce large statistical bias, which can be additionally magnified in the presence of long-term persistence (Dimitriadis & Koutsoyiannis 2015). In addition, if the uncertainty is underestimated, then a regular cluster of events could be erroneously regarded as an extreme cluster. Although the mode of the examined classical estimator for variance is close to its expectation for small Hurst parameters and large lengths, we show that for larger values of the Hurst parameter and small sample lengths, equating the expected climacogram to the observed one may lead to underestimation of the long-term persistence and thus the uncertainty of the process.

We propose that when the available series have short lengths or when the empirical Hurst parameter is estimated larger than 0.5, we should always account for statistical bias. Particularly for the bias adaptation, when information is available on only a single series/realization of the process, it is advantageous to equate the mode instead of the expectation of the climacogram estimator to the sample values. Interestingly, in case of an $N(0,1)$-HK process, the absolute difference between the mode and expected values of the estimator is calculated (from a Monte Carlo analysis performed in this study) to reach a maximum value of 67% of the expected value, corresponding to cases with $H \geq 0.6$ and $n/\kappa \leq 10$, while for the white noise process the value is approximately 20% for $\kappa = n/10$. In cases of different stochastic processes or estimators or when a larger accuracy of the mode bias is of interest, one should employ a Monte Carlo technique through an explicit generation algorithm (Dimitriadis & Koutsoyiannis 2018) to estimate the mode climacogram estimator or use the lower quartile (25% quantile) of the estimator (in case its distribution is known) as an approximation.

From the Monte Carlo analysis performed in this study, it is also observed that for an $N(0,1)$-HK process with variance $\gamma(\kappa) = \gamma(1)/\kappa^{2-2H}$ and for large n and small n/κ, the distribution of the climacogram estimator tends to that of $\Gamma((n/\kappa - 1)/2\,, 2\gamma(\kappa)/(n/\kappa - 1))$, with a mean value equal to $\gamma(\kappa)$, i.e. zero bias. However, given the estimation uncertainty present in records exhibiting persistence, the autocorrelation of the averaged process is independent of the scale, and thus, the above distribution will never be truly reached. The underestimation of the persistence of the parent process also has critical implications for the estimation of the properties of its extremes, since it was shown that the maximum duration of successive peaks over threshold is greatly affected by the degree of dependence. Additional analyses and quantifications on the reflection of long-term term persistence in terms of clustering in time can be found in Iliopoulou & Koutsoyiannis (2019).

A final remark for discussion, considering the etymology of the terms, is that the expected value of a random process is less expected to occur than its mode (i.e. most probable value; a term coined by Pearson (1895, p. 345)), where the two coincide only in symmetrical distributions. Therefore, when only one value is known (here, only one realization of the climacogram estimator), it is more accurate to fit the model and evaluate the Hurst parameter based on the proposed mode estimator rather than the expected one.

ACKNOWLEDGEMENT

The authors would like to thank the editor Luigi Berardi for handling the paper, one anonymous reviewer for useful comments, and Federico Lombardo for his fruitful discussion, comments, and suggestions that helped us improve the paper.

CODE AVAILABILITY

The MATLAB script for the SAR generation algorithm is available as well as the script for a fast estimation algorithm of the sample climacogram in very long timeseries and in many scales.

REFERENCES

Bickel, D. R. & Fruwirth, R. 2006 On a fast, robust estimator of the mode: comparisons to other robust estimators with applications. *Computational Statistics & Data Analysis* **50**, 3500–3530.

Cochran, W. G. 1934 The distribution of quadratic forms in a normal system, with applications to the analysis of covariance. *Mathematical Proceedings of the Cambridge Philosophical Society* **30** (2), 178–191. doi:10.1017/S0305004100016595.

Dimitriadis, P. 2017 *Hurst-Kolmogorov Dynamics in Hydrometeorological Processes and in the Microscale of Turbulence.* PhD Thesis, National Technical University of Athens, p. 167.

Dimitriadis, P. & Koutsoyiannis, D. 2015 Climacogram versus autocovariance and power spectrum in stochastic modelling for Markovian and Hurst–Kolmogorov processes. *Stochastic Environmental Research & Risk Assessment* **29** (6), 1649–1669.

Dimitriadis, P. & Koutsoyiannis, D. 2018 Stochastic synthesis approximating any process dependence and distribution. *Stochastic Environmental Research & Risk Assessment* **32** (6), 1493–1515. doi:10.1007/s00477-018-1540-2.

Dimitriadis, P., Koutsoyiannis, D. & Papanicolaou, P. 2016a Stochastic similarities between the microscale of turbulence and hydrometeorological processes. *Hydrological Sciences Journal* **61** (9), 1623–1640. doi:10.1080/02626667.2015.1085988.

Dimitriadis, P., Tegos, A., Oikonomou, A., Pagana, V., Koukouvinos, A., Mamassis, N., Koutsoyiannis, D. & Efstratiadis, A. 2016b Comparative evaluation of 1D and quasi-2D hydraulic models based on benchmark and real-world applications for uncertainty assessment in flood mapping. *Journal of Hydrology* **534**, 478–492.

Dimitriadis, P., Gournari, N. & Koutsoyiannis, D. 2016c Markov vs. Hurst-Kolmogorov behaviour identification in hydroclimatic processes. *European Geosciences Union General Assembly.* Geophysical Research Abstracts, Vol. 18, European Geosciences Union, Vienna, EGU2016-14577-4. doi:10.13140/RG.2.2.21019.05927.

Dimitriadis, P., Gournary, N., Petsiou, A. & Koutsoyiannis, D. 2018 How to adjust the fGn stochastic model for statistical bias when handling a single time series; application to annual flood inundation. In: *13th Hydroinformatics Conference*, 1–6 July 2018, Palermo, Italy.

Goulianou, T., Papoulakos, K., Iliopoulou, T., Dimitriadis, P. & Koutsoyiannis, D. 2019 Stochastic characteristics of flood impacts for agricultural insurance practices. *European Geosciences Union General Assembly* 2019, Geophysical Research Abstracts, Vol. 21, European Geosciences Union, Vienna, EGU2019-5891.

Gournary, N. 2017 *Probability Distribution of the Climacogram Using Monte Carlo Techniques.* Diploma Thesis, Department of Water Resources and Environmental Engineering, National Technical University of Athens, Athens (in Greek), p. 108.

Iliopoulou, T. & Koutsoyiannis, D. 2019 Revealing hidden persistence in maximum rainfall records. *Hydrological Sciences Journal.* doi.org/10.1080/02626667.2019.1657578.

Kendziorski, C. M., Bassingthwaighte, J. B. & Tonellato, P. J. 1999 Evaluating maximum likelihood estimation methods to determine the Hurst coefficient. *Physica A* **273** (3–4), 439–451.

Koutsoyiannis, D. 2002 The Hurst phenomenon and fractional Gaussian noise made easy. *Hydrological Sciences Journal* **47** (4), 573–595.

Koutsoyiannis, D. 2010 HESS opinions 'A random walk on water'. *Hydrology and Earth System Sciences* **14**, 585–601.

Koutsoyiannis, D. 2011 Hurst-Kolmogorov dynamics as a result of extremal entropy production. *Physica A: Statistical Mechanics and its Applications* **390** (8), 1424–1432.

Koutsoyiannis, D. 2016 Generic and parsimonious stochastic modelling for hydrology and beyond. *Hydrological Sciences Journal* **61** (2), 225–244.

Koutsoyiannis, D. 2018 *Climate Change Impacts on Hydrological Science: A Comment on the Relationship of the Climacogram with Allan Variance and Variogram.* ResearchGate.

Koutsoyiannis, D. 2019 Knowable moments for high-order stochastic characterization and modelling of hydrological processes. *Hydrological Sciences Journal* **64** (1), 19–33.

Koutsoyiannis, D. & Cohn, T. A. 2008 The Hurst phenomenon and climate (solicited), *European Geosciences Union General Assembly 2008, Geophysical Research Abstracts*, Vol. 10, Vienna, 11804, European Geosciences Union. doi:10.13140/RG.2.2.13303.01447.

Koutsoyiannis, D., Dimitriadis, P., Lombardo, F. & Stevens, S. 2018 *From fractals to stochastics: seeking theoretical consistency in analysis of geophysical data.* In: *Advances in Nonlinear Geosciences* (A. A. Tsonis, ed.). Springer, Cham, Switzerland, pp. 237–278.

Lombardo, F., Volpi, E., Koutsoyiannis, D. & Papalexiou, S. M. 2014 Just two moments! A cautionary note against use of high-order moments in multifractal models in hydrology. *Hydrology and Earth System Sciences* **18**, 243–255. doi:10.5194/hess-18-243-2014,.

Mandelbrot, B. B. 1963 The variation of certain speculative prices. *The Journal of Business* **36**, 394–419.

Mandelbrot, B. B. & Van Ness, J. W. 1968 Fractional Brownian motions, fractional noises and applications. *SIAM Review* **10**, 422–437.

Mandelbrot, B. B. & Wallis, J. R. 1968 Noah, Joseph and operational hydrology. *Water Resource Research* **4**, 909–918.

O'Connell, P. E., Koutsoyiannis, D., Lins, H. F., Markonis, Y., Montanari, A. & Cohn, T. A. 2016 The scientific legacy of Harold Edwin Hurst (1880–1978). *Hydrological Sciences Journal* **61** (9), 1571–1590. doi:10.1080/02626667.2015.1125998.

Papoulis, A. 1991 *Probability, Random Variables and Stochastic Processes*, 3rd edn. McGraw-Hill, New York.

Pearson, K. 1895 Contributions to the mathematical theory of evolution – II, Skew variation in homogeneous material. *Philosophical Transactions of the Royal Society of London* **186**, 343–414. Available from: https://royalsocietypublishing.org/doi/pdf/10.1098/rsta.1895.0010.

Serinaldi, F. & Kilsby, C. G. 2016 Understanding persistence to avoid underestimation of collective flood risk. *Water* **8**, 152.

Serinaldi, F. & Kilsby, C. G. 2018 Unsurprising surprises: the frequency of record-breaking and over-threshold hydrological extremes under spatial and temporal dependence. *Water Resources Research* **54** (9), 6460–6487.

Tyralis, H. & Koutsoyiannis, D. 2011 Simultaneous estimation of the parameters of the Hurst-Kolmogorov stochastic process. *Stochastic Environmental Research & Risk Assessment* **25** (1), 21–33.

Vanmarcke, E. 2010 *Random Fields: Analysis and Synthesis.* World Scientific, New Jersey, USA.

Volpi, E., Fiori, A., Grimaldi, S., Lombardo, F. & Koutsoyiannis, D. 2015 One hundred years of return period: strengths and limitations. *Water Resources Research* **51** (10), 8570–8585.

First received 11 February 2019; accepted in revised form 6 August 2019. Available online 16 September 2019

A drought monitoring framework for data-scarce regions

Roberto A. Real-Rangel IWA, Adrián Pedrozo-Acuña, J. Agustín Breña-Naranjo and Víctor H. Alcocer-Yamanaka

ABSTRACT

Drought monitoring is a critical activity for drought risk management; however, the lack of ground-based observations of climatological and hydrological variables in many regions of the world hinders an adequate follow-up and investigation of this phenomenon. This paper introduces a transparent framework for monitoring the spatio-temporal distribution of drought hazard based on uni- and multivariate standardized drought indices that use reanalysis datasets of hydrological variables available freely and globally. In the case study of the 2015–2017 East-Southwest drought in Mexico, the introduced framework successfully detected the spatial and temporal patterns of drought conditions, even in regions where a benchmark drought monitoring system failed to detect deficits. In addition, the ability of the introduced framework to detect drought impacts on the annual agricultural maize production in Mexico was evaluated using data of 1980–2018, yielding scores of the false alarm ratio =0.32, the probability of detection = 0.71, and the proportion correct = 0.68 for the analysis at the national scale. Currently, the framework provides a significant extension to the capabilities for national drought monitoring, and it is being used by the Mexican water authority in the decision-making process related to drought severity assessment.

Key words | drought monitoring, Mexico, Multivariate Standardized Drought Index (MSDI), Standardized Precipitation Index (SPI), Standardized Runoff Index (SRI), Standardized Soil Moisture Index (SSI)

Roberto A. Real-Rangel IWA (corresponding author)
Adrián Pedrozo-Acuña
J. Agustín Breña-Naranjo
Institute of Engineering,
National Autonomous University of Mexico,
Circuito Escolar s/n, Ciudad Universitaria,
Coyoacan, CDMX, Mexico 04510,
Mexico
E-mail: *rrealr@iingen.unam.mx*

Roberto A. Real-Rangel
Program of Master and Doctorate in Engineering,
National Autonomous University of Mexico,
Universidad 3000, Ciudad Universitaria, Coyoacan,
CDMX, Mexico 04510,
Mexico

Víctor H. Alcocer-Yamanaka
National Water Commission,
Insurgentes Sur 2416, Copilco El Bajo, Coyoacan,
CDMX, Mexico 04340,
Mexico

INTRODUCTION

Droughts are a temporary, recurrent event originating from the lack of precipitation that has potential effects on the water availability in a hydrologic system (Smakhtin & Schipper 2008). Although they are a typical feature of any climate, they might have negative consequences on a wide range of social, economic, and environmental sectors (Stahl *et al.* 2016). The severity of these impacts is directly related to the severity of the drought, but it is the lack of an adequate drought risk management that might turn an event into a disaster (Blauhut *et al.* 2016). Thus, efforts dedicated to reducing drought risk are generally oriented to protect and minimize the vulnerability of the human and environmental systems, including actions like monitoring natural indicators.

Drought monitoring consists of the compilation and analysis of data from natural variables associated with the onset, persistence, and termination of droughts (e.g., precipitation, runoff, and soil moisture) to quantify drought indices and communicate them to decision makers and the public (Wilhite 2000). Through drought indices, it is possible to characterize the severity of events, for which one of the most widely used approaches is the method of the reference level (Yevjevich 1967) based on the theory of statistical runs. This approach consists in selecting a reference value of a

doi: 10.2166/hydro.2019.020

given drought indicator or an index from which deviations will be analyzed: the greater the negative deviation, the more intense the deficit of water. A negative run, interpreted as a period of deficit, starts when the variable analyzed crosses the reference value downwards, and it lasts until a new value above the reference occurs. The time between the start and end of the run is the drought *duration*; the sum of the deviations during the deficit period is known as the *magnitude* of the drought, and the ratio of the magnitude to the duration is the *mean intensity* of the event (Figure 1). Duration, magnitude, and intensity are the basic temporal characteristics for drought definition.

While there is no such thing as 'the best drought index', for sake of operability, drought monitoring requires performing a selection of the more than 100 indices available in the literature, ideally based on criteria such as robustness, tractability, and transparency (Keyantash & Dracup 2002). Besides, this selection should also consider the drought aspect intended to be detected, such as the meteorological origins of the drought (meteorological drought), the agricultural impacts of droughts (agricultural drought), or the effects of dry spells on surface, subsurface, and groundwater hydrology (hydrological drought) (Wilhite & Glantz 1985). The signature of drought may be lagged during its passage through the catchment, provoking delays between the onsets of the different types of droughts, prolonging the duration, and attenuating the intensity of the anomalies in a process known as drought propagation (van Loon 2013). Following this rationale, a single drought indicator may not be sufficient to characterize the complex drought conditions (Hao & Singh 2015), which has motivated the use of indicators of different stages of the hydrological cycle in many drought monitoring systems.

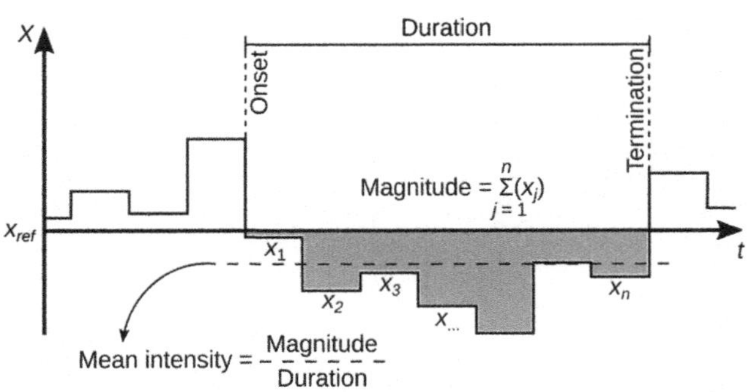

Figure 1 | Definition of the basic features of droughts.

There are several drought monitors at global, continental, and regional scales, differing in their data sources, spatial and temporal resolution, and the drought indices applied. For example, the Global Integrated Drought Monitoring and Prediction System (GIDMaPS; Hao *et al.* 2014; http://drought.eng.uci.edu), supported by the University of California Irvine, derived data of precipitation and soil moisture from NASA's reanalysis product MERRA-Land to compute three drought indices on a global scale: the Standardized Precipitation Index (SPI; McKee *et al.* 1993), the Standardized Soil Moisture Index (SSI; Hao & AghaKouchak 2013), and the Multivariate Standardized Drought Index (MSDI; Hao & AghaKouchak 2013). Authors demonstrated that the system successfully detected severe droughts like the Amazon, the Russian, and the Texas-Mexico droughts in 2010 and the East Africa drought of 2011; unfortunately, the production of GIDMaPS was interrupted on February 2016 when the MERRA-Land product was also discontinued. The European Drought Observatory (EDO; Sepulcre-Canto *et al.* 2012; http://edo.jrc.ec.europa.eu) of the European Commission reports a Combined Drought Indicator (CDI) that evaluates the degree of drought hazard for agriculture in Europe based on the SPI, anomalies in an indicator of soil moisture derived from a hydrological model, and the anomalies in a remote sensing-derived indicator of biomass and vegetation condition. These variables and indices are combined with different time-lags and aggregation periods to capture the temporal development of droughts, firstly detected in early precipitation deficits, then in the soil moisture and, finally, in their impacts on the vegetation conditions, to assess the severity level. Its capabilities for detecting events on a regional scale were qualitatively tested with the European drought of 2011, identifying the different stages of the agricultural drought in the most affected areas. The United States Drought Monitor (USDM; Svoboda *et al.* 2002; https://droughtmonitor.unl.edu/) reports a series of maps of drought indices and indicators, including the Palmer Drought Severity Index (PDSI; Palmer 1965), percentiles of soil moisture and daily streamflow, the percent of normal precipitation, and the SPI, among others. Likewise, it presents a map of the result of a process that involves the objectively weighted average of drought indices and the input of regional and local experts to define the affected regions.

In Mexico, since 2003, the National Meteorological Service of the National Water Commission (SMN-CONAGUA) has replicated the methodology of the USDM for Mexico: first, as a part of the North American Drought Monitor (NADM; Lawrimore *et al.* 2002; https://www.ncdc.noaa.gov/temp-and-precip/drought/nadm/) and, from 2014 on, also in the independent system Drought Monitor in Mexico (MSM; Lobato-Sánchez 2016; https://smn.conagua.gob.mx/es/climatologia/monitor-de-sequia/monitor-de-sequia-en-mexico). However, the procedure by which the USDM methodology is applied in Mexico differs from the original in two main aspects: the available input data for computing the drought indices and the indices blending procedure. The MSM derives precipitation and temperature data from the CONAGUA's Climatological Database, using only sites with at least 30 years of records and with less than 20% of missing data. The spatial density of the resultant subset (with 415 stations currently) does not meet the minimum recommendations for managing water resources (WMO 2009). Figure 2(a) shows a Voronoi diagram where each polygon encloses the region for which each station is the nearest available. It can be derived that 86% of the stations cover a region that exceeds 900 km^2, recommended for the least dense classification (coastal regions), excluding the polar/arid zone. This is a drawback for the robust methodology of the USDM, which may become weaker if only a few inputs are available (WMO & GWP 2016).

On the other hand, the MSM also incorporates the percent of storage available in the principal reservoirs of the country as a drought indicator. Figure 2(b) includes the basins of the 172 reservoirs monitored, which cover 36% (693,606 km^2) of the Mexican territory. Nevertheless, 62 of the reservoirs with the largest basins receive discharges influenced by the presence of hydropower- or irrigation-dams upstream, which might provoke a misrepresentation of the hydro-climatological conditions of the region. The main effect of these controls on drought characteristics is the attenuation of the intensity, and the lagging and prolongation of the drought period when propagating a deficit in precipitation in the basin to a deficit in streamflow (Wu *et al.* 2016), hindering the efficiency of the available storage in reservoirs as a drought indicator.

Besides the limitations of the inputs used, the methodology applied in the MSM is not transparent in that the weights used in the average of drought indicators are assigned subjectively by the climatology expert performing the analysis (Lobato-Sánchez 2016). This subjectivity inherited from the USDM methodology might be seen as an issue when using drought maps resulting from this procedure as the basis for decision-making for assigning resources to aid drought-affected regions (Kao & Govindaraju 2010). This is the case for Mexico, where CONAGUA, the highest authority in the water resources management, issues official opinions on drought conditions, based on the results of MSM, which are used by the Interior Ministry (SEGOB) to allocate financial resources (SHCP 2010).

This paper introduces a transparent framework for generating drought monitoring products in Mexico that applies a simple, reproducible methodology that derives

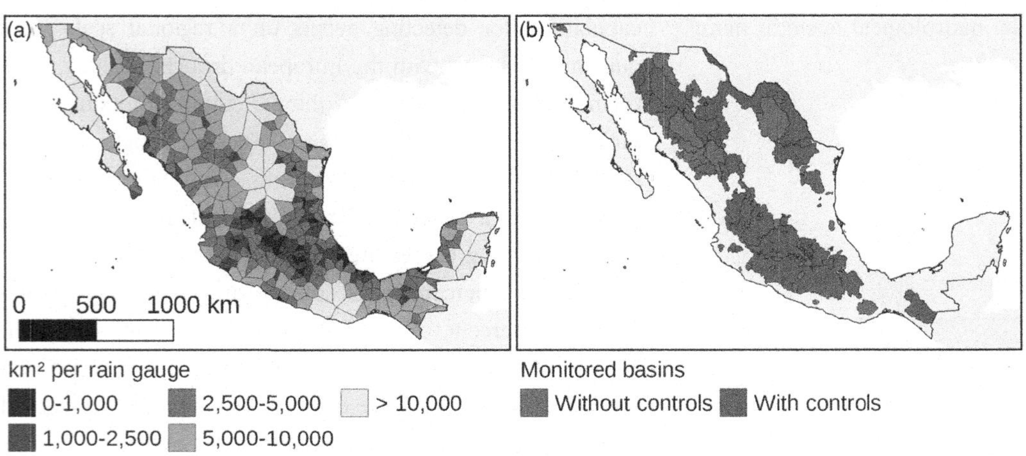

Figure 2 | Networks of direct observations used for drought monitoring in the MSM: (a) density of climatological stations and (b) basins of monitored reservoirs.

hydrological variables from a freely and globally available reanalysis product to compute a set of Standardized Drought Indices (SDI) at a regional scale based on the principle of the SPI.

DATA

The drought monitoring framework introduced in this paper uses as input monthly hydrological data from NASA's reanalysis product Modern-Era Retrospective Analysis for Research and Applications, version 2 (MERRA-2; Gelaro *et al.* 2017). MERRA-2 simulates the atmosphere dynamics with an atmospheric general circulation model (AGCM) and reproduces the exchange of water and energy fluxes between the atmosphere and the land surface with the land surface model (LSM) Catchment (Koster *et al.* 2000). Its datasets are available from January 1980 to the present, with a spatial resolution of $5/8° \times 1/2°$, and are published with a latency of 18–20 days in the website of the Goddard Earth Sciences Data and Information Services Center (GES DISC; https://disc.gsfc.nasa.gov). Although this latency period hinders using the product for real-time monitoring, given the slow-onset and creeping nature of this phenomenon, it is well suited for representing ongoing drought conditions.

The AGCM forces the LSM with precipitation, radiation, and near-surface air temperature and humidity, among other variables. In turn, the LSM delivers fluxes of moisture and energy to the AGCM, updating in the process its internal moisture and energy states (Reichle *et al.* 2017a). In the framework presented here, drought is monitored through the anomalies in precipitation (PRECTOTLAND), soil moisture in the root zone (RZMC), and the sum of overland flow, throughflow, and baseflow (RUNOFF + BASEFLOW), stored in the Land Surface Diagnostics MERRA-2 data collection (M2T1NXLND).

PRECTOTLAND is the precipitation flux generated with the AGCM and corrected with gauge- and satellite-based precipitation observations. The variability of RZMC is estimated by the LSM based on the transfers of moisture between three levels of the soil: the surface soil, the root zone, and the vadose zone. These transfers work to keep the system close to the assumed equilibrium soil moisture profile in the unsaturated zone. Evaporation and runoff processes are represented with different models depending on the distribution of the RZMC; for example, if a storm occurs in a saturated region, all the throughfall is converted to overland flow (RUNOFF); on the other hand, in a region where RZMC is subsaturated, but above the vegetation-specific wilting point, the precipitation contributes to RUNOFF only if there is an excess in its surface soil moisture. Finally, BASEFLOW is directly controlled by the mean water table depth in the catchment, defined through the moist transfers described above, and the mean of the topographic index in the catchment, computed at each point as $\ln(a/\tan\beta)$, with a being the upstream area that contributes flow to the given point and β being the local slope angle.

The performance of land surface variables of MERRA-2 was recently tested by Reichle *et al.* (2017b) against *in situ* measurements. In a comparison of the seasonal mean PRECTOTLAND against the Global Precipitation Climatology Project product, version 2.2 (GPCPv2.2; Adler *et al.* 2003; https://precip.gsfc.nasa.gov/gpcp_v2.2_data.html), it was found that the average magnitude of the bias for the period December–February is 0.7 mm day^{-1} and for June–August, 1.1 mm day^{-1}. Besides, an evaluation of the correlation coefficients (R) between monthly time series of both products in the period of 1979–2015, a global average R of 0.82 was found. Meanwhile, the root zone moisture field exhibits an unbiased root-mean-square error of 0.04 m^3 m^{-3} and an average R of 0.71 when compared to ground-based measurements in the USA, France, and Australia. Finally, reanalysis runoff estimates, compared to naturalized monthly streamflow gauge data in the USA, present skill values (anomaly R) ranging from 0.5 to 0.9.

METHODS

The framework presented here is synthesized as shown in Figure 3. The following subsections describe the components of the analysis pipeline.

Drought definition

Monitoring products generated with this framework use the SDI approach to quantify drought features based on the

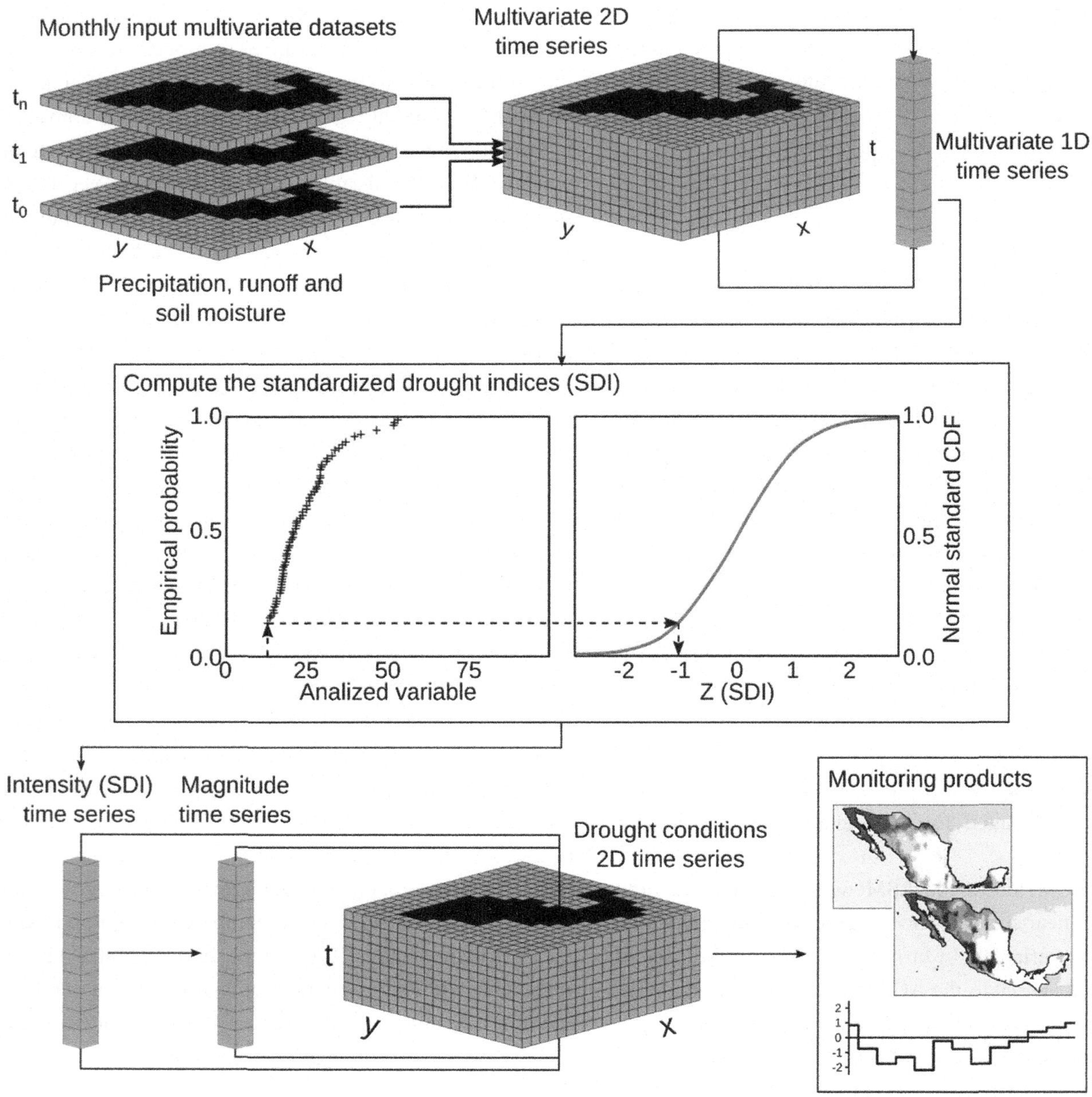

Figure 3 | Analysis pipeline of the monitoring framework proposed.

procedure originally proposed for the SPI. To calculate the SPI, a long-term precipitation record is fitted to a probability distribution, which is then normalized applying the inverse normal standard distribution. The resulting Z value for each observed precipitation is the SPI. Negative (positive) values of SPI are interpreted as a negative (positive) deviation from regular conditions; the higher the absolute value, the greater the intensity of the deviation. After its publication in 1993, the SPI methodology has been applied to other variables. For example, Hao & AghaKouchak (2013) found that the SSI captures the lagged onset of the deficit of soil moisture and the prolongation of the dry periods compared with the SPI. Likewise, Shukla & Wood (2008) observed similar results with the Standardized Runoff Index (SRI) that presented less variability than the SPI, resulting in longer periods of deficit of runoff than those of deficit of precipitation.

While powerful, the SPI has the drawback of being highly sensitive to the choice of the probability distribution function. Quiring (2009) found significant differences between SPI values calculated using the Pearson type III and the two-parameter gamma PDFs, mainly in the extreme values ($|SPI| \geq 1.5$), where mean absolute deviations ranging from 0.18 to 0.29 were found. This inconsistency hinders the comparison of drought conditions between regions where SPI uses different PDFs. In this regard, Farahmand & AghaKouchak (2015) proposed a workaround for this issue by using the empirical probability of each value in the long-term record, which is then transformed to the normal standard distribution, resulting in a nonparametric SDI. To this end, they used the following plotting position formula:

$$p(x_k) = \frac{m - 0.44}{n + 0.12} \quad (1)$$

and then

$$\mathrm{SDI} = \phi^{-1}(p) \quad (2)$$

where x_k is a given value in the long-term record of the variable X; n is the size of the sample; m is the number of occurrences of the value x_i for $x_i \leq x_k$; and ϕ is the standard normal distribution function.

The use of univariate drought indices along with multivariate indices offers a framework useful for risk assessment and management for a wide range of sectors, such as agriculture, hydropower generation, and public water supply, while keeping attention on the process of propagation of the deficit throughout the catchment. In this sense, Hao & AghaKouchak (2013, 2014) proposed the MSDI as an extension of the SDI approach to assess multiple variables using a single index, allowing the study of drought features from a broader view to detect different phases of a drought event, such as its onset and persistence. The non-parametric approach already described can also be applied in the MSDI, using the multivariate form of Equation (1), to compute $p(x_k, y_k, \ldots)$, where x and y would be two variables related to a phase of a drought; n is the number of sets $(x, y, \ldots)$ in the sample; and m is the number of occurrences of the set $(x_i, y_i, \ldots)$ for $x_i \leq x_k$ and $y_i \leq y_k$. An implicit effect of this methodology is to present more intense deficits when more than one of the considered variables detect, marginally, deficit conditions. This characteristic may be understood as the result of the effect of two (or more) threads acting simultaneously over an exposed system, which might be the desired behavior. However, it must be kept in mind when interpreting the results of MSDIs that negative values of this index are more probable (Erhardt & Czado 2018). This might be corrected by using lower intensity thresholds for the MSDIs than the other SDIs; however, that measure has not been explored in this analysis.

The monitoring framework presented here computes three univariate and three multivariate SDIs using the described approach: the SPI, for precipitation; the SRI, for runoff; the SSI, for soil moisture; the MSDI, for the combination of precipitation and runoff ($MSDI_{PR}$), precipitation and soil moisture ($MSDI_{PS}$), and precipitation, soil moisture, and runoff ($MSDI_{PRS}$). The different MSDIs allow us to detect and characterize drought events simultaneously through multiple phases of the hydrological cycle in the land surface, while the SPI, SSI, and SRI quantify deficits in each of its components. Furthermore, the SDIs are computed for different periods of temporal aggregation to provide drought monitoring products for analyses and decision-making support in different temporal scales.

The reference level that separates excess and deficit values in SDIs is 0.0; however, deviations from this reference lack a physical interpretation. Hence, to allow a clearer communication of the drought conditions, negative values of SDIs are usually classified as shown in Table 1.

This framework also computes the monthly drought magnitude, which represents the accumulated deficit since the onset of an event. McKee *et al.* (1993) reported the value of this indicator divided over an intensity threshold and interpreted its result as the drought duration if each

Table 1 | Drought intensity classification (modified from Svoboda *et al.* 2002)

Category	SDI range	Percentile chance
Exceptional drought (D4)	$\mathrm{SDI} \leq -2.0$	$P \leq 2$
Extreme drought (D3)	$-2.0 < \mathrm{SDI} \leq -1.6$	$2 < P \leq 5$
Severe drought (D2)	$-1.6 < \mathrm{SDI} \leq -1.3$	$5 < P \leq 10$
Moderate drought (D1)	$-1.3 < \mathrm{SDI} \leq -0.8$	$10 < P \leq 20$
Abnormally dry (D0)	$-0.8 < \mathrm{SDI} \leq -0.5$	$20 < P \leq 30$

month of the drought has the value of that threshold. Here, a threshold of 1.3 is adopted, which is the absolute value of intensity categorized as a 'severe drought' (see Table 1). Also, an arbitrary classification of its values is proposed to make maps of drought magnitude (M) clearer: M1 ($1 \leq M < 3$), M2 ($3 \leq M < 6$), M3 ($6 \leq M < 9$), M4 ($9 \leq M < 12$), and M5 ($M \geq 12$). Most of the regional and global drought monitoring systems do not include the drought magnitude as a monitoring product (see, for example, the GIDMaPS). This product would give insight into the progressive aggravation of drought conditions until the present, which might be helpful to define the priority of attention, especially for long-term impacts, like hydrological or ecological, even in the case when present intensity is not particularly high.

Data processing

Literature reports informatic tools to compute SDIs. For example, Farahmand & AghaKouchak (2015) introduced the open source package Standardized Drought Analysis Toolbox (SDAT) for MATLAB (available at http://amir.eng.uci.edu/sdat.php), which applies the nonparametric approach to compute uni- and multivariate SDIs on which the present framework is based. Their software allows only time series as inputs which leaves the alternative of implement SDAT in a loop structure throughout the region of interest; however, this approach becomes inefficient when working in regions formed by many grid cells. Besides, although the code is openly available, it is written in a proprietary programming language for the commercial software MATLAB, which requires purchasing a license. Tigkas *et al.* (2014) presented the Drought Indices Calculator (DrinC), programmed in Visual Basic 6. In contrast to the SDAT, DrinC has a full graphical user interface, which makes it more user-friendly. This tool computes the SPI, among other drought indices, based on time series data including the option of entering data from multiple points to compute the drought indices in a loop structure, using an approach similar to that of SDAT. However, given that the source code is not available, DrinC cannot be implemented as a part of a programming script, which would expedite setting the input data in the format required by the software.

Land Surface Diagnostics data collection of MERRA-2 contains 50 variables of flux and state of water and energy, for the almost 208,000 grid cells of the globe, stored in 1-month files in the Network Common Data Form (NetCDF) format of 196 MB each, summing more than 92 GB of data in all its temporal coverage (January 1980 to present). This amount of data might represent a limitation for its processing in a commercially available personal computer with constrained specifications due to the memory needed to load the data and perform calculations with it.

In this scenario, it was necessary to develop a set of scripts to process the input datasets and compute the SDIs optimizing the use of the computer resources. This framework processes MERRA-2 multivariate data taking advantage of recent developments of Python packages for geosciences, using the *xarray* package (Hoyer & Hamman 2017), which, in turn, uses the out-of-core analysis of datasets of the *Dask* library (Rocklin 2015). The *xarray* library implements data structures and an analytics toolkit for multidimensional labeled arrays based on the data model on which the NetCDF format is built. This package allows, among other key features, resampling and aggregating data along a specific dimension, which is used to compute the SDI for different temporal scales. On the other hand, the *Dask* package allows us to use the disk as an extension of the memory, which makes it possible to perform calculations with arrays of data bigger than the memory of the computer.

The software developed performs the following steps. First, the monthly datasets of MERRA-2 are combined into a single multivariate 2D time series. Then, the uni- and multivariate SDIs are computed in each cell in the analyzed area, using as input its associated time series. After this, the drought magnitude is derived by accumulating the values of the SDIs of each time step until a shift in the sign of the values of intensity is detected. Finally, the data arrays of intensity and magnitude are merged into one multivariate 2D time series, which is then exported as a spatial dataset in the NetCDF format that can be easily visualized or used to compose maps in the most widely used geographic information systems, or postprocessed for further analysis through a programming language, using one of the multiple libraries available for this end. The source code of the package developed is freely available online under the GNU GPL license at https://bitbucket.org/pysdi/pysdi.

The results of this procedure are used to derive maps and time series of drought intensity and magnitude, as well as time series of the fraction of the area affected by the different categories of the drought. These products are generated for regions defined by three different aspects: political division, water administration, and hydrography. Aimed at representing the spatial distribution of the values in each region, time series include the minimum and maximum value, and the 25th, 50th, and 75th percentiles of the data within the region.

RESULTS AND VALIDATION

Validating drought monitoring products is not an easy, nor straightforward task, given that directly measuring evidence of the severity of a drought event is virtually impossible. Alternatively, a common approach to assess the ability of an analysis to detect droughts is to qualitatively compare the timing and intensity of events detected with a given index (or set of indices) with documented impacts. For example, Palmer (1965) demonstrated the skills of the PDSI comparing the results of its methodology with the effects of historic droughts in Kansas and Iowa, in the USA. Also, Sepulcre-Canto *et al.* (2012) evaluated the skill of the CDI comparing its results with the timing of historic droughts and the variability of annual cereal yields in Europe.

In this work, two applications of the framework were implemented to validate its results. First, its drought monitoring products were compared against documented impacts and the conditions reported by the MSM during the 2015–2017 drought in East-Southwest Mexico. Second, the framework was applied to detect annual anomalies of maize yield in Mexico from 1980 to 2018. This latter analysis includes the selection of SDIs based on the strength of their correlation with the anomalies of crop yield and a quantitative validation of their drought monitoring performance using categorical indices.

Event monitoring: the 2015–2017 East-Southwest drought

From May 2015 to August 2017, a drought in the East (E) and Southwest (SW) regions of Mexico provoked costly impacts mainly in the agricultural sector, with Guerrero and Oaxaca (both in SW) being the most affected states. In the former case, impacts included the failure of 86,400 hectares of maize crops. In Oaxaca, the impacts included more than 80,200 hectares of crops, mainly of maize and sorghum, and 18,300 heads of cattle. During this long-lasting event, the storage volume of the Benito Juarez dam in Oaxaca – used for irrigation – dropped to 14% of its total capacity. Given the severity of the damages, the Federal Government issued Disaster Declarations for both states, which allowed the allocation of funds to implement response measures for the agricultural sector and urban water supply. At the end of the drought, Guerrero and Oaxaca, along with the rest of the states in the region (Hidalgo, Puebla, Tlaxcala and Veracruz in E; and Chiapas in SW) summed together 320,500 hectares of failed crop and the loss of 49,400 heads of cattle (SADER 2018a, 2018b).

Monthly maps of the SDIs were generated to compare the features of the drought events for each phase of the hydrological cycle along with the complete signal of the drought across the hydrological system. For this reason, the SDIs used include the SPI, SRI, SSI, and $MSDI_{PRS}$, while the other combinations of variables for the MSDI were omitted. To focus on capturing the monthly variation of the analyzed indicators, all the SDIs were computed for a temporal scale of 1 month. The results of the MSM, derived from the NADM maps (https://www.drought.gov/nadm/content/north-american-drought-monitor), are included along with the products of the framework introduced to visually compare the results of both systems.

In the E and SW regions, the rainy season normally starts in May; however, in June 2015, the amount of precipitation observed over a great area was significantly lower than expected. This meteorological drought was detected by the SPI (first row in Figure 4), which identified the highest intensities in the states of Chiapas and Oaxaca ($SPI_{min} = -2.16$ and -2.13, respectively; *exceptional droughts*; D4). A deficit of runoff was detected by the SRI (second row), starting simultaneously with the meteorological drought, covering a smaller area and with slightly lower, although still high, intensities ($SRI_{min} = -1.73$; *extreme drought*; D3). In July, the meteorological drought expanded to other regions, covering more than 44% of the country; however, its intensity in the studied area was lightly reduced.

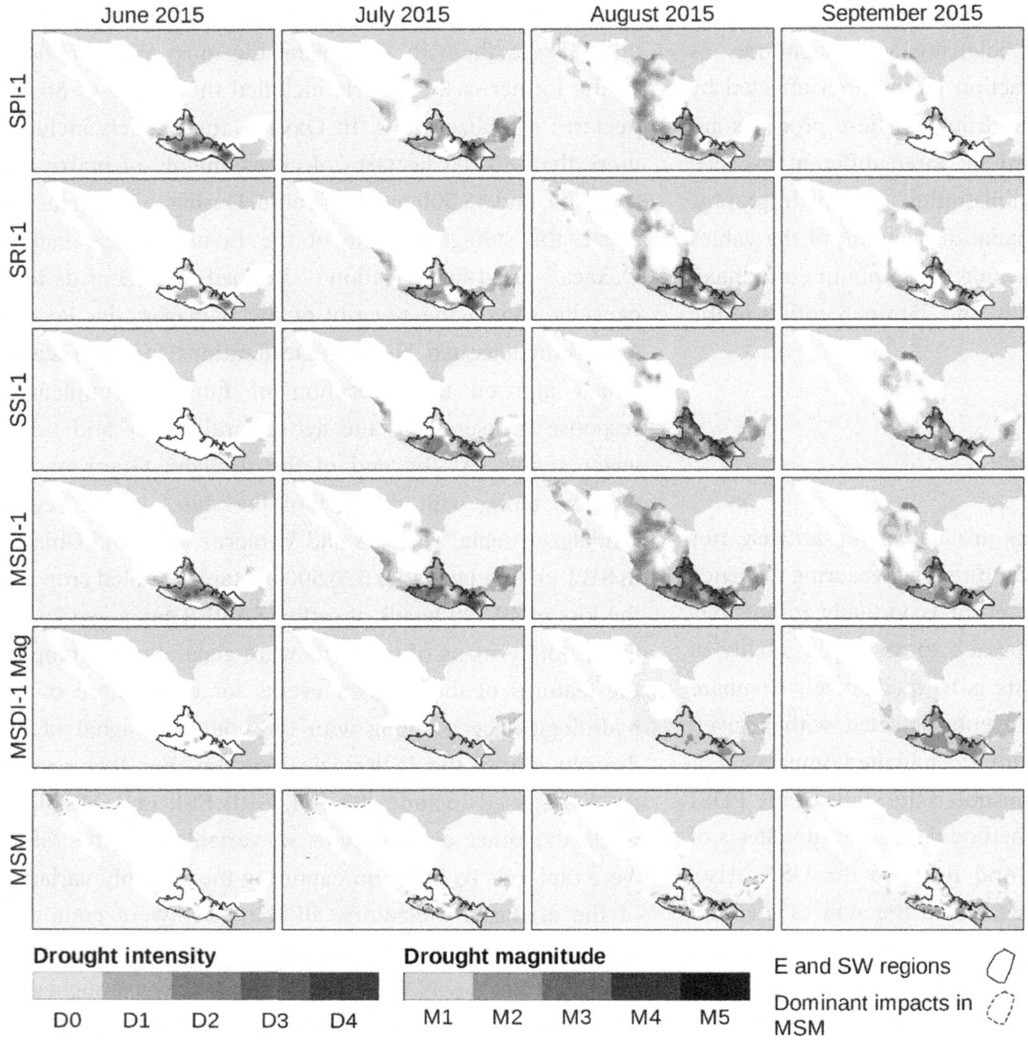

Figure 4 | Comparison of the drought maps of the framework introduced against the MSM during the onset of the Southwest 2015–2017 drought. MSM maps were composed with data from https://www1.ncdc.noaa.gov/pub/data/nidis/shapefiles/.

The runoff deficit expanded as well as the meteorological drought but, again, in a smaller area and with an attenuated intensity, comparatively. Moreover, in this month, the onset of an agricultural drought was detected with the SSI (third row) in the regions of interest, as well as in the neighboring regions, Chiapas being the state with the highest intensities, reaching values of $SSI_{min} = -2.17$ (*exceptional drought*; D4). From that moment on, runoff and soil moisture droughts closely followed the precipitation drought with a temporal lag of 1 month. The $MSDI_{PRS}$ (fourth row) shows the joint effects of the three previous variables, with deficits significantly more intense starting in July 2015, when the marginal deficits of precipitation, runoff and soil moisture are concurrently in extreme or exceptional categories.

Drought magnitude maps (fifth row in Figures 4 and 5) were derived with the drought intensity according to the $MSDI_{PRS}$ in order to capture the overall drought conditions. By means of this indicator, it is possible to detect that, although in September 2015, a great part of the country was under water deficit conditions, the drought in the E and SW regions lasted longer and had accumulated a larger deficit than the rest of the country. Moreover, it allows to identify areas where drought persisted until August 2017, even with SDIs reporting near-to-normal conditions. Here, accumulated deficit values were divided by an intensity

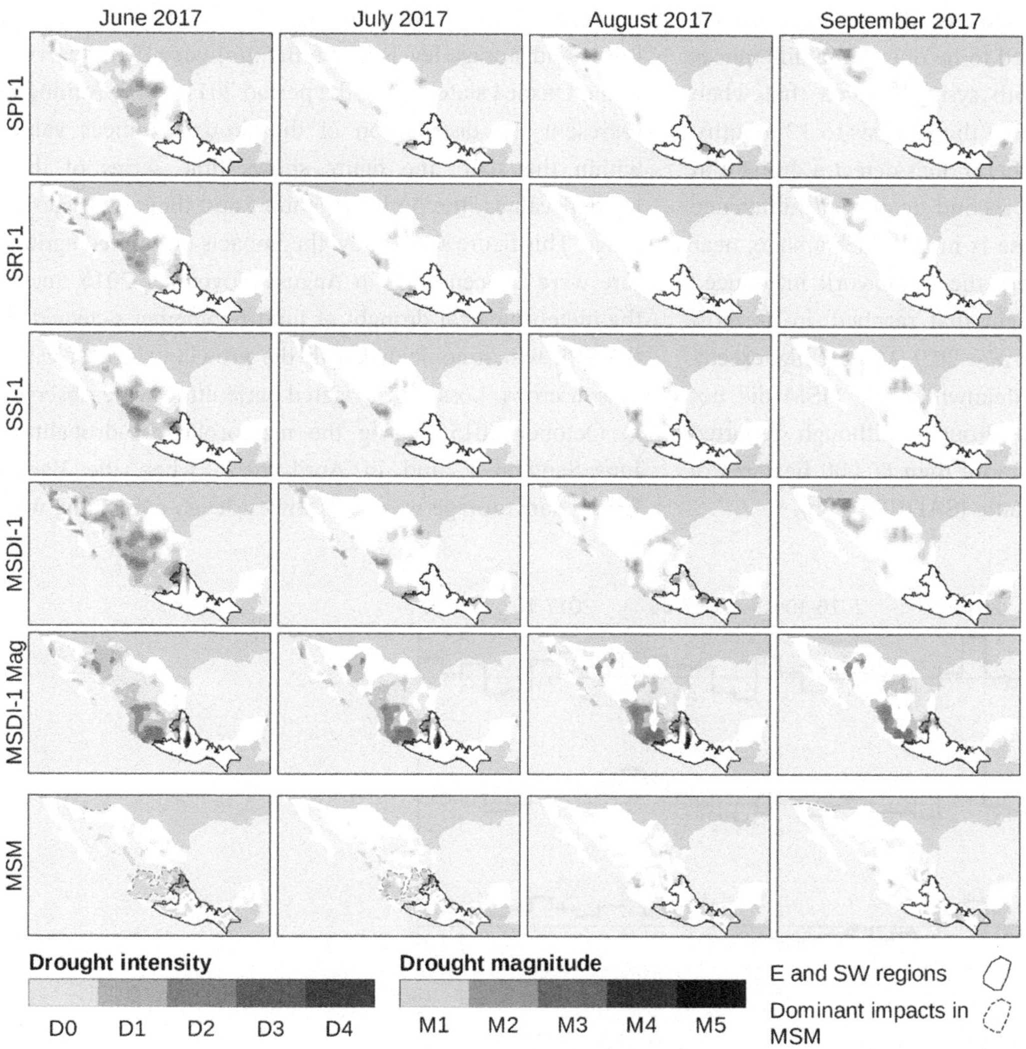

Figure 5 | Comparison of the drought maps of the framework introduced against the MSM during the termination of the Southwest 2015–2017 drought. MSM maps were composed with data from https://www1.ncdc.noaa.gov/pub/data/nidis/shapefiles/.

threshold of -1.3 to express the magnitude results in terms of the drought category 'severe' (see Table 1); however, this value might be modified to fit results to observed impacts.

The last row of Figures 4 and 5 shows MSM drought maps to contrast the results of both systems. However, the differences in the temporal scale of the inputs of each drought monitor make the comparison somewhat subjective. Drought intensities derived with the presented framework seem to systematically indicate more severe drought conditions given its temporal scale of 1 month, chosen here to track the variation of the monitored variables in the present without the influence of anomalies that occurred in the past, which may not be related to current water deficits (or excesses). Meanwhile, inputs of the MSM include the percent of the normal precipitation for temporal aggregation windows from 30 to 365 days into a single product.

The spatial and temporal distribution of the drought conditions were similarly represented by the two approaches mostly between the drought magnitude indicator and the MSM results. Nevertheless, it is worth highlighting three significant differences in the analyzed case. The first of these is outside the E and SW regions in the Northwest (NW) of Mexico. In the westernmost part of the boundary with the USA, the MSM reported severe drought conditions (Figure 4), while none of the SDIs of the system introduced here detected deficits in that area. This difference is also assumed to be due to the differences in the temporal scale of both systems, given that when increasing the temporal scale of the analysis to 12 months, the $MSDI_{PRS}$-12 (not shown here) does detect a drought in the mentioned location. A second important difference identified in the analyzed case is in the Puebla state, near to the Gulf of Mexico, where the framework introduced here detected a severe drought that reached, in its most affected point, a magnitude of $M = 20.9$ (M5, the highest category of its classification). Meanwhile, the MSM did not report the occurrence of a drought, although reported losses in this state included more than 60,000 hectares of maize and 10,800 heads of cattle (SADER 2018b).

Figure 6 shows the time series of the SDIs, agricultural losses and storage levels in the Benito Juarez dam observed in the Oaxaca state during the period 2015–2017. Aiming to represent the distribution of the drought indices values within the state, the figure shows time series of their median values (the 50th percentile) and their interquartile range. This figure shows how the impacts on rainfed agriculture were concentrated in August–November 2015 due to the meteorological drought of June–September detected by the SPI, which coincided with the growing stage of warm season crops. Losses in irrigated agriculture were observed in October 2015, during the meteorological drought of June–September, and in April 2017, when the Benito Juarez dam storage was at its lowest level, coinciding with

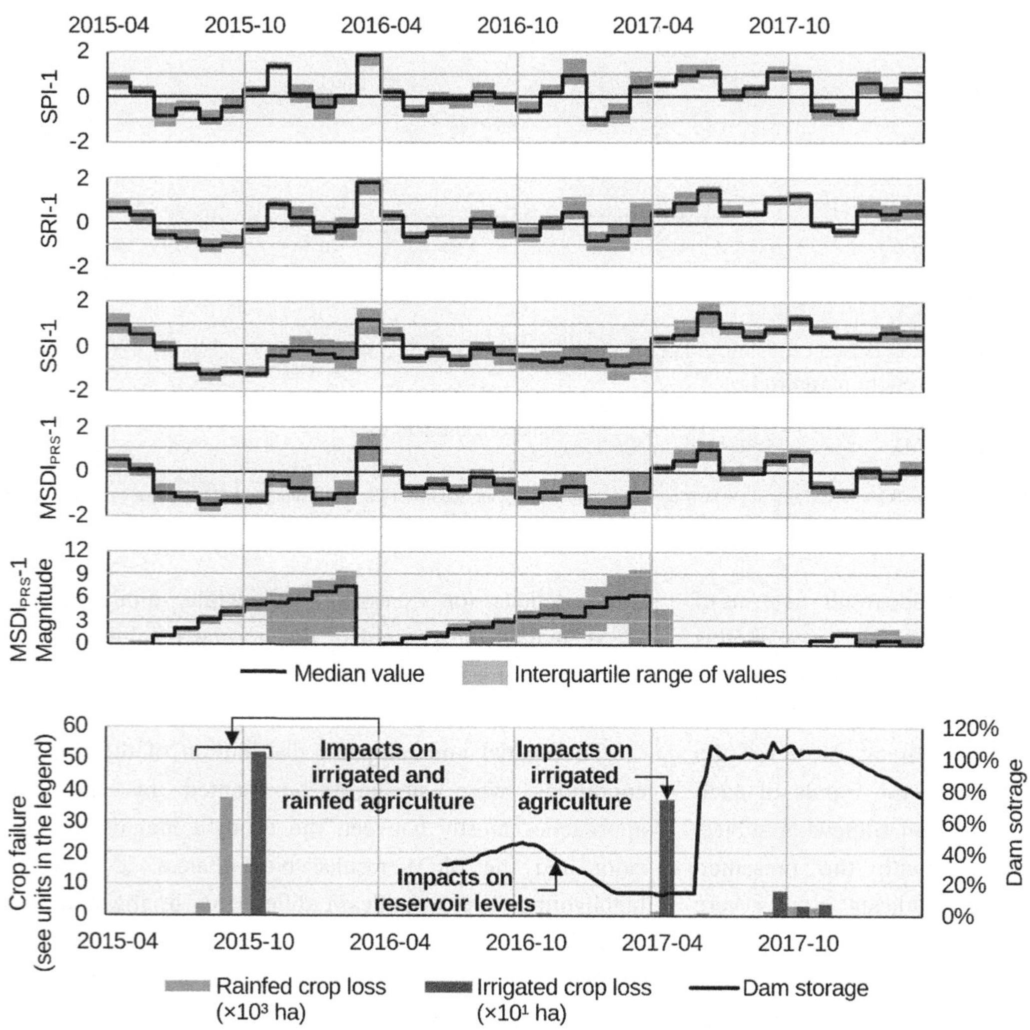

Figure 6 | SDIs with a temporal scale of 1 month and impacts reported during the 2015–2017 East-Southwest drought. Impacts data from SADER (2018a) and CONAGUA (2018).

the agricultural drought from May 2016 to March 2017 detected by the SSI.

The hydrological impacts observed in the drop of Benito Juarez dam storage were gradually aggravated from October 2016 to its critical value of 14% in February 2017. Due to lack of data before July 2016, it is difficult to distinguish how the storage correlates with the drought indices; however, a similarity with the results of the drought magnitude, given the cumulative nature of both variables, can be observed.

Evaluating annual agricultural impacts

The monitoring capabilities of the framework to detect impacts on agriculture was tested on the warm season rainfed maize annual yield in Mexico in the period 1980–2018 using data from the Agri-food and Fisheries Information Service (SIAP; http://infosiap.siap.gob.mx/gobmx/datosAbiertos_a.php) of the Ministry of Agriculture and Rural Development in Mexico (SADER). To this end, the combination of SDI, temporal scale and month of computation that better reflects the variation in the annual crop yield in each state was first determined. This included the six drought indices described in the subsection 'Drought definition', 12 different temporal scales (1, 2,..., 12 months) and the 12 months of the agricultural year of analysis. In this study, the correlation between the agricultural production of a given year and the drought conditions of previous years was not explicitly explored; however, there are cases in which the combination of temporal scale and month of computation implies including the conditions of the last months of the preceding year.

The crop yield was defined as the quantity of product harvested per unit of the planted area (in tons per hectare). Anomalies of crop yield were calculated using as a reference a locally weighted regression model, coupled with a multiplicative decomposition model (Lu Carbone & Gao 2017). The correlation between the anomalies of crop yield and SDIs was quantified with the Spearman's rank correlation coefficient (r_S), computed as follows (Zwillinger & Kokoska 2000):

$$r_S = 1 - \frac{6\sum_{i=1}^{n}(u_i - v_i)^2}{n(n^2 - 1)} \quad (3)$$

where u_i and v_i are the ranks of the ith value of the variables x and y, respectively; and n is the size of the sample. In order to estimate the confidence intervals of the correlations, a bivariate bootstrap approach was used. The correlation assigned to each drought index is the mean of the r_S computed for 1,000 bootstrapped samples, and the approximate lower and upper confidence limits were estimated as the $\alpha/2$ and $1-\alpha/2$ percentiles of the results, respectively (Howell 2013). The combination with the highest correlation was selected as the best indicator for drought monitoring for the rainfed maize production in each state.

To test the monitoring capabilities of the selected SDIs, each pair of values of SDI and crop yield anomaly was converted to dichotomous variables using the value of 0.0 as a threshold between negative and positive anomalies in order to apply the classifications of *hit*, *miss*, *false alarm*, and *correct negative* (Table 2).

For this evaluation, three indices were computed: the probability of detection (POD), the false alarm ratio (FAR), and the proportion correct (PC). POD computes the fraction of times when a reduction in observed crop yield was detected by the index as follows (Wilks 2006):

$$\mathrm{POD} = \frac{a}{a+c} \quad (4)$$

where a is the number of *hits* and c is the number of *misses*. The FAR quantifies the reliability of the drought indices as the fraction of times when a drought was detected by the SDI but no reduction in the crop yield was observed. The following expression is used to compute it:

$$\mathrm{FAR} = \frac{b}{a+b} \quad (5)$$

where b is the number of *false alarms*. Finally, the PC was computed to evaluate the accuracy of the SDIs. This

Table 2 | Contingency table for the detection of crop yield reduction due to droughts

	Yield reduction observed	No yield reduction observed
Drought indicated	Hit	False alarm
No drought indicated	Miss	Correct negative

metric computes the fraction of the times for which the SDI correctly represented the occurrence or not of a reduction in the crop yield as follows:

$$\mathrm{PC} = \frac{a + d}{a + b + c + d} \tag{6}$$

where d is the number of *correct negatives*. Possible values of these indices range from 0 to 1, being 1 the optimum value for PC and POD, while 0 is the optimum value for the FAR.

Table 3 shows the results of this evaluation, performed in 31 states of Mexico, with Baja California Sur being the only one excluded because it does not produce maize with rainfed agriculture. Temporal scales are denoted with a number attached to the SDI names. Results show that maize yield anomalies in 74% (23) of the states are better

Table 3 | SDI with the strongest correlation with the anomalies of the maize rainfed crop yield for each state

				r_s 95% confidence limits				
State	**SDI aggregation**	**Month of computation**	r_s	**Lower**	**Upper**	**POD**	**FAR**	**PC**
Aguascalientes	SPI-4	October	0.64	0.41	0.81	0.71	0.25	0.72
Baja California	SSI-5	May	0.69	0.44	0.86	0.65	0.08	0.72
Campeche	SRI-1	July	0.52	0.25	0.73	0.64	0.26	0.67
Chiapas	SPI-1	August	0.62	0.40	0.80	0.74	0.26	0.74
Chihuahua	SPI-5	September	0.44	0.18	0.70	0.69	0.39	0.69
Coahuila	SSI-1	August	0.40	0.12	0.65	0.70	0.26	0.72
Colima	SRI-5	April	0.55	0.30	0.75	0.67	0.20	0.69
Durango	$\mathrm{MSDI_{PS}}$-2	October	0.70	0.49	0.84	0.84	0.33	0.72
Guanajuato	SPI-6	November	0.79	0.64	0.90	0.78	0.18	0.82
Guerrero	$\mathrm{MSDI_{PS}}$-1	August	0.73	0.52	0.88	1.00	0.44	0.72
Hidalgo	SPI-1	August	0.32	0.01	0.58	0.57	0.40	0.56
Jalisco	SSI-12	November	0.67	0.46	0.83	0.67	0.33	0.69
Mexico City	$\mathrm{MSDI_{PRS}}$-6	May	0.55	0.28	0.77	0.80	0.29	0.67
Mexico State	$\mathrm{MSDI_{PS}}$-3	June	0.41	0.13	0.67	0.83	0.35	0.72
Michoacan	SPI-9	June	0.36	0.06	0.63	0.59	0.35	0.59
Morelos	$\mathrm{MSDI_{PS}}$-5	August	0.58	0.31	0.79	0.81	0.26	0.74
Nayarit	$\mathrm{MSDI_{PR}}$-1	June	0.41	0.13	0.65	0.67	0.26	0.69
Nuevo Leon	$\mathrm{MSDI_{PRS}}$-1	September	0.44	0.17	0.68	0.78	0.44	0.62
Oaxaca	SPI-2	August	0.62	0.38	0.79	0.68	0.32	0.69
Puebla	SPI-12	November	0.44	0.13	0.71	0.74	0.36	0.67
Queretaro	$\mathrm{MSDI_{PR}}$-2	September	0.50	0.21	0.71	0.76	0.30	0.69
Quintana Roo	SPI-2	February	0.53	0.25	0.76	0.67	0.26	0.69
San Luis Potosi	$\mathrm{MSDI_{PRS}}$-3	October	0.37	0.07	0.62	0.79	0.40	0.64
Sinaloa	$\mathrm{MSDI_{PS}}$-4	September	0.54	0.32	0.74	0.77	0.26	0.72
Sonora	SPI-1	August	0.47	0.15	0.71	0.50	0.35	0.55
Tabasco	SPI-3	March	0.41	0.12	0.63	0.57	0.40	0.56
Tamaulipas	SSI-2	September	0.69	0.46	0.84	0.68	0.35	0.67
Tlaxcala	SSI-1	June	0.47	0.16	0.75	0.69	0.39	0.69
Veracruz	SSI-1	June	0.47	0.17	0.71	0.71	0.43	0.64
Yucatan	SPI-2	August	0.45	0.15	0.68	0.63	0.33	0.67
Zacatecas	$\mathrm{MSDI_{PRS}}$-3	October	0.72	0.51	0.87	0.90	0.32	0.72

detected by SDIs that use precipitation as an input. This finding was unexpected; given the nature of the impacts evaluated, soil moisture was the variable expected to predominate among the other variables. However, the soil moisture is present as input in 48% (15) of the states, in the SSI and the $MSDI_{PS}$ and $MSDI_{PRS}$. In 52% (16) of the cases, the strongest correlation was found in SDIs with temporal scales of 1–2 months, which may be related to the duration of the critical period of water requirement of the corn plant in 30 days after the end of its tasseling stage, the period in which each day of water deficit may provoke a final yield reduction between 6% and 13% (Ruiz Corral *et al.* 2013).

The analysis of the drought monitoring performance of the SDIs on a national scale yielded values of FAR = 0.32, POD = 0.71, and PC = 0.68, which means that 32% of instances of impacts inferred through the SDIs were false alarms; 71% of the times that agricultural impacts were observed, they were successfully inferred through the SDIs; and 68% of inferences of the SDIs succeeded in depicting the observed conditions (impacts and no impacts observed). On the other hand, in the analysis on a state scale, 81% (25) of the states have values of FAR <0.4; 87% (27) yielded POD >0.6; and 87% (27) exhibited values of PC >0.6.

The results of this analysis are satisfactory in most of the states evaluated. However, four states are noted for their poor performance: Hidalgo, Michoacan, Sonora, and Tabasco, where applying the SDIs derived from this framework to detect drought impacts on annual maize production must be carried out considering its limitation or even avoided. The performance derived from this analysis might be constrained by the variations in the planting timing, which would hinder the possibility of using SDIs fixed in time to detect the agricultural impacts. Furthermore, the values of the SDIs used in this evaluation are computed as the median of the area in each state, while the actual rainfed agricultural surface fraction does not exceed 15% of the state surface in most cases. The analysis presented here can be refined using the drought conditions found exclusively in the agricultural surface from which the state annual production is derived. However, available data to perform such analysis in Mexico is limited to less than 16 years. Finally, the information reported in the records of the SIAP does not discriminate the impacts of droughts from other natural or biological hazards for crops, such as floods and plagues. This may affect the results of the evaluation in cases when evidence of impact is observed but no drought detected, which would be classified as a *miss* in the contingency table. Further analysis is still needed to assess the influence of these aspects in this exercise of validation.

CONCLUSIONS

In this paper, a transparent framework for drought monitoring based on global, freely available datasets is introduced. This framework characterizes drought severity based on non-parametric uni- and multivariate standardized indices, from which drought features, like duration, intensity, and magnitude, can be easily derived.

The data processing was carried out with a novel open source software package for Python that expedites the computation of the drought indices using multidimensional, multivariate datasets as input. The future development of the package (and the framework presented) includes exploring other data sources that might reduce the temporal latency of the publishing of new monitoring products, which currently is constrained to around 18–20 days due to the latency of the MERRA-2 products.

In a visual comparison of maps generated with the presented framework against drought maps of the MSM for the 2015–2017 East-Southwest drought in Mexico, the framework introduced here succeeded in reproducing the temporal and spatial distribution of the drought conditions of the MSM system, and, furthermore, it was capable of detecting regions severely affected by droughts that the MSM failed to detect. It is important to state here that, although reproducing the results of the MSM was not the target of this development, these results are encouraging, given that they demonstrate the usefulness of the framework following a transparent procedure completely based on freely available global reanalysis datasets.

In a quantitative evaluation of the framework's performance in detecting impacts on the annual maize yield through a contingency table approach, results obtained are satisfactory, with values of FAR, POD, and PC on the national scale of 0.32, 0.71, and 0.68, respectively. However, relatively poor metrics were obtained in the states of Hidalgo, Michoacan, Sonora, and Tabasco. These results might be improved in a further analysis that should include changes in the validation procedure that allow us to fix the computation of the SDI on the growing stage of the crop. Moreover, future work includes evaluating the performance of the framework to detect drought impacts on other crops and using these results to generate a set of warning thresholds for different applications.

Among the products generated in this system, drought magnitude emerges as a key indicator of drought severity due to its ability to represent drought persistency even under low drought intensity conditions. Moreover, it has the advantage of being derived from an easy, straightforward procedure (it is the accumulated deficit during a drought).

Given its ability to detect regional and country-wide droughts on a monthly temporal scale, the introduced drought monitoring framework has become operational and complementary to the drought monitoring efforts of the Mexican government since 2016 in the Multivariate Drought Monitor of Mexico (MOSEMM), supported by CONAGUA and the Institute of Engineering of the National Autonomous University of Mexico (II-UNAM). The simplicity and transparency of this framework, together with the use of globally, freely available data as input for computing the spatial and temporal distribution of drought conditions with open source informatic tools, make it an alternative drought monitoring tool for other regions where records of hydrological variables are scarce.

REFERENCES

Adler, R. F., Huffman, G. J., Chang, A., Ferraro, R., Xie, P. P., Janowiak, J., Rudolf, B., Schneider, U., Curtis, S., Bolvin, D., Gruber, A., Susskind, J., Arkin, P. & Nelkin, E. 2003 The Version-2 Global Precipitation Climatology Project (GPCP) Monthly Precipitation Analysis (1979–present). *Journal of Hydrometeorology* **4**, 21.

Blauhut, V., Stahl, K., Stagge, J. H., Tallaksen, L. M., De Stefano, L. & Vogt, J. 2016 Estimating drought risk across Europe from reported drought impacts, drought indices, and vulnerability factors. *Hydrology and Earth System Sciences* **20** (7), 2779–2800.

CONAGUA 2018 *Almacenamiento de las Principales Presas del País (Storage in the Main Dams of the Country)*. Available from: https://www.gob.mx/conagua/documentos/almacenamiento-de-las-principales-presas-del-pais (accessed 23 August 2019).

Erhardt, T. M. & Czado, C. 2018 Standardized drought indices: a novel univariate and multivariate approach. *Journal of the Royal Statistical Society: Series C (Applied Statistics)* **67** (3), 643–664.

Farahmand, A. & AghaKouchak, A. 2015 A generalized framework for deriving nonparametric standardized drought indicators. *Advances in Water Resources* **76**, 140–145.

Gelaro, R., McCarty, W., Suárez, M. J., Todling, R., Molod, A., Takacs, L., Randles, C. A., Darmenov, A., Bosilovich, M. G., Reichle, R., Wargan, K., Coy, L., Cullather, R., Draper, C., Akella, S., Buchard, V., Conaty, A., da Silva, A. M., Gu, W., Kim, G.-K., Koster, R., Lucchesi, R., Merkova, D., Nielsen, J. E., Partyka, G., Pawson, S., Putman, W., Rienecker, M., Schubert, S. D., Sienkiewicz, M. & Zhao, B. 2017 The Modern-Era Retrospective Analysis for Research and Applications, Version 2 (MERRA-2). *Journal of Climate* **30** (14), 5419–5454.

Hao, Z. & AghaKouchak, A. 2013 Multivariate standardized drought index: a parametric multi-index model. *Advances in Water Resources* **57**, 12–18.

Hao, Z. & AghaKouchak, A. 2014 A nonparametric multivariate multi-index drought monitoring framework. *Journal of Hydrometeorology* **15** (1), 89–101.

Hao, Z. & Singh, V. P. 2015 Drought characterization from a multivariate perspective: a review. *Journal of Hydrology* **527**, 668–678.

Hao, Z., AghaKouchak, A., Nakhjiri, N. & Farahmand, A. 2014 Global integrated drought monitoring and prediction system. *Scientific Data* **1**, 140001.

Howell, D. C. 2013 *Statistical Methods for Psychology*, 8th edn. Wadsworth, Belmont, CA, USA.

Hoyer, S. & Hamman, J. J. 2017 Xarray: N-D labeled arrays and datasets in python. *Journal of Open Research Software* **5**, 10.

Kao, S. C. & Govindaraju, R. S. 2010 A copula-based joint deficit index for droughts. *Journal of Hydrology* **380** (1–2), 121–134.

Keyantash, J. A. & Dracup, J. A. 2002 The quantification of drought: an evaluation of drought indices. *Bulletin of the American Meteorological Society* **83** (8), 1167–1180.

Koster, R. D., Suarez, M. J., Ducharne, A., Stieglitz, M. & Kumar, P. 2000 A catchment-based approach to modeling land surface processes in a general circulation model: 1. model structure. *Journal of Geophysical Research: Atmospheres* **105** (D20), 24809–24822.

Lawrimore, J., Heim Jr., R. R., Svoboda, M. D., Swail, V. & Englehart, P. J. 2002 Beginning a new era of drought monitoring across North America. *Bulletin of the American Meteorological Society* **83** (8), 1191–1192.

Lobato-Sánchez, R. 2016 El monitor de sequía en México (The drought monitor in Mexico). *Tecnología y Ciencias del Agua* **7** (5), 197–211.

Lu, J., Carbone, G. J. & Gao, P. 2017 Detrending crop yield data for spatial visualization of drought impacts in the United States,

1895–2014. *Agricultural and Forest Meteorology* **237–238**, 196–208.

McKee, T. B., Doesken, N. J. & Kleist, J. 1993 The relationship of drought frequency and duration to time scales. In: *Eighth Conference on Applied Climatology*. American Meteorological Society, Anaheim, pp. 179–184.

Palmer, W. C. 1965 *Meteorological Drought*. United States Weather Bureau, Washington, DC, p. 58.

Quiring, S. M. 2009 Developing objective operational definitions for monitoring drought. *Journal of Applied Meteorology and Climatology* **48** (6), 1217–1229.

Reichle, R. H., Liu, Q., Koster, R. D., Draper, C. S., Mahanama, S. P. P. & Partyka, G. 2017a Land surface precipitation in MERRA-2. *Journal of Climate* **30** (5), 1643–1664.

Reichle, R. H., Draper, C. S., Liu, Q., Girotto, M., Mahanama, S. P. P., Koster, R. D. & De Lannoy, G. J. M. 2017b Assessment of MERRA-2 land surface hydrology estimates. *Journal of Climate* **30** (8), 2937–2960.

Rocklin, M. 2015 Dask: parallel computation with blocked algorithms and task scheduling. In: *Python in Science Conference, Presented at the Python in Science Conference*, Austin, TX, pp. 126–132.

Ruiz Corral, J. A., Medina García, G., González Acuña, I. J., Flores López, H. E., Ramírez Ojeda, G., Ortiz Trejo, C., Byerly Murphy, K. F. & Martínez Parra, R. A. 2013 *Requerimientos agroecológicos de cultivos (Agroecological Crop Requirements) (Libro técnico núm. 3)*, 2nd edn. INIFAP-CIRPAC, Tepatitlán de Morelos, Jalisco, México.

SADER 2018a *Servicio de Información Agroalimentaria y Pesquera (Agrifood and Fisheries Information Service)*. Available from: https://www.gob.mx/siap (accessed 15 January 2019).

SADER 2018b *Información histórica de sequías (Historic Information on Droughts)*. Retrieved from: https://datos.gob.mx/busca/dataset/informacion-historica-de-sequia-de-sagarpa (accessed 22 May 2019).

Sepulcre-Canto, G., Horion, S., Singleton, A., Carrão, H. & Vogt, J. V. 2012 Development of a combined drought indicator to detect agricultural drought in Europe. *Natural Hazards and Earth System Science* **12** (11), 3519–3531.

SHCP 2010 *Acuerdo por el que se emiten las Reglas Generales del Fondo de Desastres Naturales (Agreement by Which the General Rules of the Natural Disaster Fund Are Issued)*. DOF: 03/12/2010.

Shukla, S. & Wood, A. W. 2008 Use of a standardized runoff index for characterizing hydrologic drought. *Geophysical Research Letters* **35** (2), 1–7.

Smakhtin, V. U. & Schipper, E. L. F. 2008 Droughts: the impact of semantics and perceptions. *Water Policy* **10** (2), 131.

Stahl, K., Kohn, I., Blauhut, V., Urquijo, J., de Stefano, L., Acácio, V., Dias, S., Stagge, J. H., Tallaksen, L. M., Kampragou, E., van Loon, A. F., Barker, L. J., Melsen, L. A., Bifulco, C., Musolino, D., de Carli, A., Massarutto, A., Assimacopoulos, D. & van Lanen, H. A. J. 2016 Impacts of European drought events: insights from an international database of text-based reports. *Natural Hazards and Earth System Sciences* **16** (3), 801–819.

Svoboda, M. D., LeComte, D., Hayes, M. J., Heim Jr., R. R., Gleason, K., Angel, J. R., Rippey, B., Tinker, R., Palecki, M., Stooksbury, D., Miskus, D. & Stephens, S. 2002 Drought monitor. *Bulletin of the American Meteorological Society* **83** (8), 1181–1190.

Tigkas, D., Vangelis, H. & Tsakiris, G. 2014 DrinC: a software for drought analysis based on drought indices. *Earth Science Informatics* **8** (3), 697–709.

van Loon, A. F. 2013 *On the Propagation of Drought: How Climate and Catchment Characteristics Influence Hydrological Drought Development and Recovery*. PhD Thesis, Wageningen University, Wageningen, NL.

Wilhite, D. A. 2000 Drought as a natural hazard: concepts and definitions. In: *Drought: a Global Assessment*, Vol. I (D. A. Wilhite, ed.). Routledge Publishers, London, pp. 3–18.

Wilhite, D. A. & Glantz, M. H. 1985 Understanding the drought phenomenon: the role of definitions. *Water International* **10** (3), 111–120.

Wilks, D. S. 2006 *Statistical Methods in the Atmospheric Sciences*. International Geophysics Series, 2nd edn. Academic Press, Amsterdam, Boston.

WMO (ed.) 2009 *Guide to Hydrological Practices – Volume I: Hydrology – From Measurements to Hydrological Information (WMO-No. 168)*, Vol. I, 6th edn. World Meteorological Organization, Geneva, Switzerland.

WMO & GWP 2016 In: *Handbook of Drought Indicators and Indices* (M. D. Svoboda & B. A. Fuchs, eds). Integrated Drought Management Tools and Guidelines Series Integrated Drought Management Programme (IDMP), Geneva.

Wu, J., Chen, X., Gao, L., Yao, H., Chen, Y. & Liu, M. 2016 Response of hydrological drought to meteorological drought under the influence of large reservoir. *Advances in Meteorology* **2016**, 2197142.

Yevjevich, V. 1967 An objective approach to definitions and investigations of continental hydrologic droughts. *Hydrology Papers* **23** (23), 25–25.

Zwillinger, D. & Kokoska, S. 2000 *CRC Standard Probability and Statistics Tables and Formulae*. CRC Press, Boca Raton.

First received 16 January 2019; accepted in revised form 27 October 2019. Available online 11 December 2019

Attribution of growing season vegetation activity to climate change and human activities in the Three-River Headwaters Region, China

Chen Chen IWA, Tiejian Li, Bellie Sivakumar, Jiaye Li and Guangqian Wang

ABSTRACT

Over the past century, vegetation change has been reported at global, national, and regional scales, accompanied by significant climate change and intensified human activities. Among the regions is the rangeland of the Three-River Headwaters Region (TRHR) in China. However, which factor dominates in causing vegetation change in this region is still under considerable debate, and how would the grasslands adapt to the changing environment is largely unknown. To address these issues, we attribute growing season vegetation activity to climate change and human activities, investigate the interactions among different driving variables, and explore the dynamic relationship between vegetation activity and the driving variables. We perform Mann–Kendall trend analysis, Pearson correlation analysis, and partial correlation analysis. The results indicate that the dominant factor for vegetation growth, during the period 1995–2014, was temperature for the southeastern and southern parts of the TRHR, precipitation for the western part, and solar radiation for the northeastern part. The regulation effects of temperature on precipitation and cloud cover contributed to vegetation growth, while grazing activity and population activity offset the positive contribution of climate change. The dynamic relationship between vegetation activity and the driving variables reflected the acclimatization and adaption processes of vegetation, which needs further investigation.

Key words | climate change, grazing, Normalized Difference Vegetation Index, partial correlation, Three-River Headwaters Region, vegetation dynamics

Chen Chen IWA
Tiejian Li (corresponding author)
Bellie Sivakumar
Jiaye Li
Guangqian Wang
State Key Laboratory of Hydroscience and Engineering,
Tsinghua University,
Beijing 100084,
China
E-mail: *litiejian@tsinghua.edu.cn*

Tiejian Li
Guangqian Wang
State Key Laboratory of Plateau Ecology and Agriculture,
Qinghai University,
Xining 810016,
China
and
Sanjiangyuan Collaborative Innovation Center,
Qinghai University,
Xining 810016,
China

Bellie Sivakumar
UNSW Water Research Centre, School of Civil and Environmental Engineering,
University of New South Wales,
Sydney, NSW 2052,
Australia
and
Department of Civil Engineering,
Indian Institute of Technology Bombay,
Powai, Mumbai 400076,
India

NOTATION

NDVI	Normalized Difference Vegetation Index
$NDVI_{GS}$	Growing season NDVI
GP	Growing season precipitation
GT	Growing season temperature
GC	Growing season cloud cover
LP	Annual livestock population
POP	Annual human population
$R_{NDVI\text{-}GP}$	Partial correlation coefficient between $NDVI_{GS}$ and GP
$R_{NDVI\text{-}GT}$	Partial correlation coefficient between $NDVI_{GS}$ and GT
$R_{NDVI\text{-}GC}$	Partial correlation coefficient between $NDVI_{GS}$ and GC
$R_{NDVI\text{-}LP}$	Partial correlation coefficient between $NDVI_{GS}$ and LP
$R_{NDVI\text{-}POP}$	Partial correlation coefficient between $NDVI_{GS}$ and POP

doi: 10.2166/hydro.2019.003

INTRODUCTION

In the terrestrial biosphere, vegetation plays a vital role in providing food and habitats for humankind and animals (de Jong *et al.* 2013). Vegetation is one of the most important components of the ecosystem, as it bridges water, energy, and carbon transfer between the land and the atmosphere (Chen *et al.* 2013; Mao *et al.* 2016; Gao *et al.* 2017). Among the indicators of vegetation activity, the Normalized Difference Vegetation Index (NDVI) is a remote-sensed proxy for vegetation cover and photosynthetic capacity, and has been commonly used in vegetation-change studies (Peng *et al.* 2013; Wu *et al.* 2015). Using this index, vegetation change, including greening trend and browning trend, has been reported in global, national, and regional scales, generally based on simple linear models (Liu *et al.* 2014; Mao *et al.* 2016) or seasonal models (de Jong *et al.* 2011; Forkel *et al.* 2013). The study of vegetation dynamics under climate change has the potential to provide a theoretical basis for ecosystem-based adaptation and thus has become a major topic of global change research (Chen *et al.* 2013; Mao *et al.* 2016; Gao *et al.* 2017).

Vegetation activity is driven by both climatic factors and anthropogenic factors. In general, climatic factors include, among others, precipitation, temperature, and solar radiation, which entangle and interact with each other, imposing complex and varying limitations on vegetation activity (Nemani *et al.* 2003). For instance, Karnieli *et al.* (2010) reported that the relationship between NDVI and land surface temperature is regulated by water and solar radiation. Apart from climatic factors, vegetation change may be induced by anthropogenic drivers, such as land use change, fertilization, irrigation, urbanization, and grazing, among others (Diaz *et al.* 2007; Neigh *et al.* 2008; Zhu *et al.* 2016). Recently, attempts have been made to disentangle the contributions of multiple climatic and anthropogenic drivers from each other. For instance, Jiang *et al.* (2017) applied a residual analysis trend method to distinguish between the effects of precipitation, temperature, and the excessive exploitation of water resource, oil, and natural gas extraction on vegetation dynamics in Central Asia. Liu *et al.* (2015) analyzed the correlations of global NDVI trends with climatic and human-induced factors using the Human Influence Index. For the rangeland ecosystem, grazing effect is one of the most important anthropogenic drivers and has been assessed by Diaz *et al.* (2007) and Walker *et al.* (2009). However, which factor dominates in causing vegetation change in rangeland ecosystems is still under considerable debate (Archer 2004). Some believe that overgrazing and other anthropogenic impacts are the primary causes of grassland change, while others believe that climate change is the main driver (Dong *et al.* 2010). Debate on this is particularly relevant and important for large semi-arid rangeland regions, such as the Qinghai-Tibet Plateau in China (Wang *et al.* 2015; Li *et al.* 2018). Li *et al.* (2018) reported that four views exist for alpine grassland changes: climatic driver, non-climatic driver, the combination of climatic and anthropogenic drivers, or the alternation of climatic and anthropogenic drivers. They concluded that spatiotemporal heterogeneity of human activity intensity should be taken into account in the models for attribution. We address this debate and discuss the spatiotemporal heterogeneity of anthropogenic drivers in the present study.

In recent decades, the Qinghai-Tibet Plateau has undergone significant climate warming, with the magnitude of warming reported for the region to be larger than that for the Northern Hemisphere and for the entire globe (Kuang & Jiao 2016). The increase in precipitation has not been as pronounced as the increase in temperature (Xie *et al.* 2010). Nevertheless, it has shown more complicated spatially heterogeneous pattern (Li *et al.* 2014), which has resulted in different impacts on vegetation growth. Generally, grassland activity is extremely sensitive to temperature and precipitation changes, and also to non-climatic factors, including grazing, fires, nitrogen deposition, and rising CO_2 levels (Zelikova *et al.* 2014; Li *et al.* 2018). With aggravating global climate change and increasing human activities, quantifying the effects of different driving variables on vegetation dynamics has become an important issue for undertaking countermeasures and management policies (Zhang *et al.* 2016). The gap in knowledge lies in how to detect and assess the impacts of climate change and human activities on the vegetation dynamics (Gao *et al.* 2016). To our knowledge, only a few studies have assessed the impacts of climate change on vegetation activities and the influencing mechanisms (Reyer *et al.* 2013; Gao *et al.* 2017). The present study attempts to bridge the gap in the current studies.

Over the past century, numerous methods have been developed to detect and assess the influence of variables/ components in dynamic systems. In terms of vegetation ecology, the application of the partial correlation analysis to study the effects of climatic factors on vegetation activity has been attempted by a few studies, with some promising outcomes. Piao *et al.* (2014) argued that the weakening relationship between inter-annual temperature variability and northern sphere vegetation activity is induced by the effect of drought trends and increasing occurrence of extreme hot days. Cong *et al.* (2017) found that the impact of warming on vegetation activity in the Tibetan Plateau is more positive during periods with more precipitation. Peng *et al.* (2013) and Yang *et al.* (2017) have tried to explain the asymmetric responses of vegetation to daytime and night-time warming. In general, these studies have mainly focused on the effects of only climatic variables on vegetation dynamics. Only a few studies have discussed the interactions among different variables on vegetation dynamics (Zhang *et al.* 2013b; Cong *et al.* 2017), especially the interactions between climatic variables and anthropogenic variables. This provides the motivation for the present study.

In this study, we attempt to properly attribute the growing season vegetation activity to both climatic factors (precipitation, temperature, and cloud cover) and human activities (grazing effect and population pressure) in the Three-River Headwaters Region (TRHR), the hinderland of the Qinghai-Tibet Plateau. For this purpose, we perform Pearson correlation analysis and partial correlation analysis. First, we conduct trend analysis for the growing season vegetation, growing season climatic variables, and anthropogenic variables. To do this, we use NDVI data from the GIMMS3 g NDVI dataset, precipitation data from the Climate Hazards Group Infrared Precipitation with Stations (CHIRPS), temperature data from ERA-Interim, cloud cover data from the Climate Research Unit (CRU, version TS 4.01), and livestock production and human population data from the Qinghai Statistical Almanacs. Then, we perform Pearson correlation analysis and partial correlation analysis to detect the dominant factor(s) governing the vegetation dynamics. Here, we endeavour to figure out the interactions among the climatic variables as well as those between climatic variables and anthropogenic variables. Finally, we investigate the dynamic relationship between vegetation activity and the driving factors, and describe the mechanism that might be responsible for this dynamic relationship.

MATERIALS AND METHODS

Study area

The TRHR (Figure 1) is the source region for three major rivers in China, namely the Yangtze River, the Yellow River, and the Lancang River. It covers an area of 35,000 km^2 and features a fluctuating terrain, dense river networks, extensive snowy mountains, and criss-crossing glaciers (Zhang *et al.* 2016). The region has a plateau continental climate and belongs to climate zones varying from semi-arid to semi-humid to humid areas from the northwest to the southeast. Along with Tibet, the region is regarded as the Asian Water Tower (Immerzeel *et al.* 2010), with altitude ranging from 2,000 m to nearly 6,800 m. Mainly covered with alpine meadow and alpine steppe, the region plays a key role in the ecological security of China due to its sensitive and fragile ecological environment (Zhang *et al.* 2016).

Over the past half a century, the TRHR has experienced significant climate change (Li *et al.* 2010; Cuo *et al.* 2013; Wang *et al.* 2018). Overall, climate change in the TRHR has supposedly altered the biogeochemical cycles, hydrological regime, and vegetation activity profoundly (Chen *et al.* 2013; Mao *et al.* 2016; Shi *et al.* 2016). Apart from climate change, human activities in the TRHR have intensified during the past few decades. On the one hand, urbanization, land use change, overgrazing, and population growth have resulted in rangeland degradation in the TRHR (Harris 2010; Pan *et al.* 2017). On the other hand, ecological protection and restoration programmes have significantly improved the grassland quality and reduced grazing pressure, as revealed by Pan *et al.* (2017) and Zhang *et al.* (2017). The complexity associated with climate change and human activities and, hence, on their effects on vegetation has led to a key question: what are the major causes of vegetation change in the TRHR (Wang *et al.* 2015)?

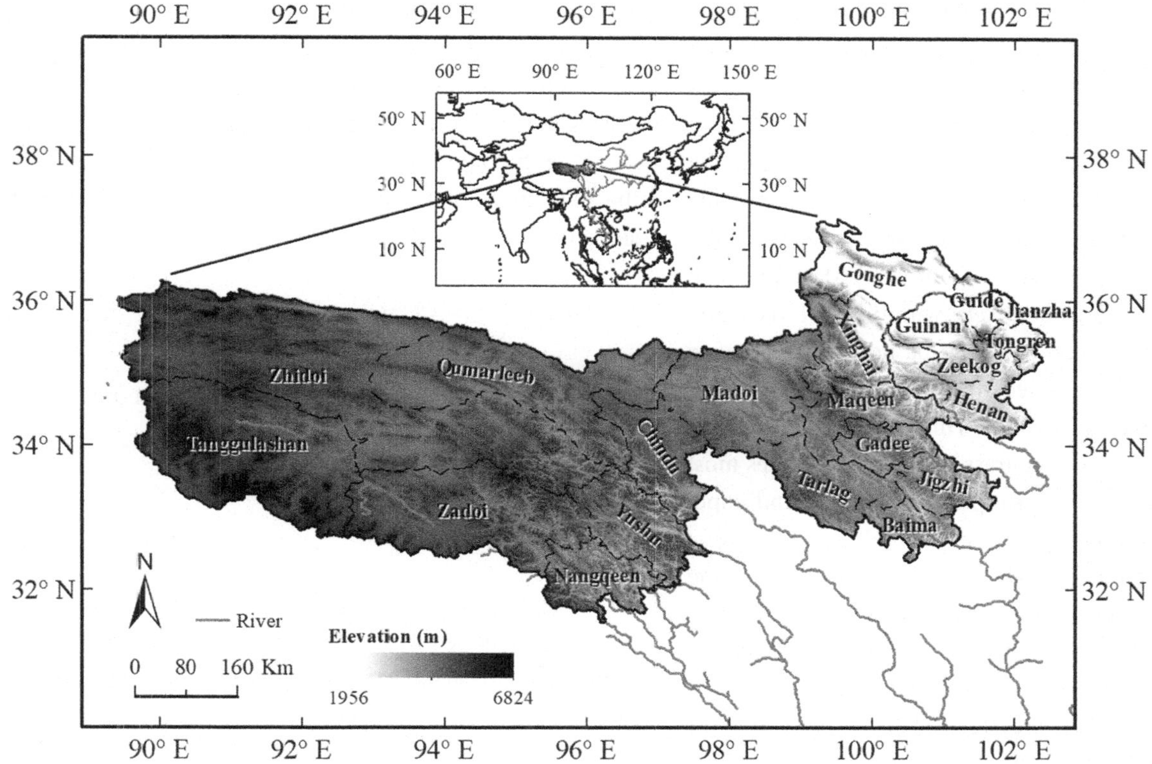

Figure 1 | The TRHR in China. Please refer to the online version of this paper to see this figure in colour: http://dx.doi.org/10.2166/hydro.2019.003.

Data

Normalized Difference Vegetation Index

The NDVI data used in this study is the NDVI3 g version from the Global Inventory Monitoring and Modelling Studies (GIMMS) group, derived from the NOAA/AVHRR land dataset. The dataset spans from July 1981 to December 2015 at a spatial resolution of 0.083° and at a temporal resolution of 15 days (Tucker *et al.* 2005). Monthly NDVI is composited from the fortnightly data using the Maximum Value Composite (MVC) method to remove noises from clouds and aerosols (Holben 1986). Areas with sparse or no vegetation cover (NDVI < 0) are masked out.

Climate data

The climate data used in this study are precipitation, temperature, and cloud cover. They are obtained from CHIRPS, ERA-Interim, and CRU TS 4.01, respectively. Monthly CHIRPS product, developed by the Climate Hazards Group, has a spatial resolution of 0.05° and available from 1981 to real time. The ERA-Interim dataset is a global atmosphere reanalysis dataset available from 1979 to real time. We make use of monthly-mean surface temperature at 2 m with a resolution of 0.125°. The CRU TS 4.01 dataset is interpolated from over 4,000 meteorological stations with spatial autocorrelation functions and has a spatial resolution of 0.5° (Harris *et al.* 2014). Besides, we use the gridded precipitation and temperature datasets from the China Meteorological Administration (CMA) to test the robustness of our analyses. The CMA gridded datasets have a spatial resolution of 0.5° and are interpolated from data obtained at the ground-level gauging stations.

Livestock and population

Livestock data (beef and mutton production) and human population data from 1995 to 2014 are acquired from the records of the Qinghai Statistical Almanacs. The datasets are grouped by the county scale.

In this study, growing season NDVI ($NDVI_{GS}$), growing season precipitation (GP), growing season temperature (GT), growing season cloud cover (GC), annual livestock production (LP), and annual human population (POP) are used in the analysis. The growing season is defined as the 5-month period from May to September, and the growing season value is taken as the average of the five monthly values. $NDVI_{GS}$, GP, GT, and GC for the TRHR and each county are averaged from pixels within the TRHR and each county, respectively. It should be noted that growing season vegetation is assumed to be related to annual LP and POP, due to data availability. This assumption is reasonable, for the growing season grass biomass constitutes most of the total annual biomass supporting livestock and population in the study area.

We select different periods for different analysis. In the trend analysis, $NDVI_{GS}$, GP, GT, and GC are selected from 1982 to 2014 based on the temporal coverage of $NDVI_{GS}$, to reflect the temporal variations as long as possible, while LP and POP are selected from 1995 to 2014 due to data availability. In the Pearson correlation analysis and partial correlation analysis, all the variables are selected from 1995 to 2014, since the anthropogenic variables are available only over this period.

A panel presentation of the overall datasets is presented in Figure 2. Spatially, the distribution of $NDVI_{GS}$, GP, and GC roughly follows an increasing pattern from the northwestern part to the southeastern part, and the distribution of GT, LP, and POP exhibits an increasing pattern from the western part to the eastern part. Temporally, all the six variables show an upward trend from 1982 to 2014.

Methods

Trend analysis

The non-parametric Mann–Kendall test, which is widely used in meteorology, hydrology and ecology studies (Liuzzo *et al.* 2015; Mao *et al.* 2016; Shi *et al.* 2017), is adopted for trend detection in vegetation and the driving variables. The assumption of the Mann–Kendall test is that the data are serially independent (Liuzzo *et al.* 2015). Therefore, the trend-free pre-whitening method is applied to the data to eliminate the influence of autocorrelation before the application of Mann–Kendall test (Yue & Wang 2002). Considering that the coefficients obtained for lag-2 and lag-3 autocorrelation are small, only the lag-1 autocorrelation signal is removed from the original signal.

After the application of the Mann–Kendall trend test, the magnitude of trend is estimated using the Thiel–Sen slope method (Hirsch *et al.* 1982):

$$\beta = \text{Median}\left(\frac{x_i - x_j}{i - j}\right), \ \forall j < i \tag{1}$$

where β is the estimate of the slope for change rate, and x_i refers to the observation for the i-th year.

Pearson correlation analysis

To check the relationship between the vegetation and the driving variables, the Pearson correlation analysis is performed. The Pearson correlation coefficient is defined as follows:

$$\rho_{X,Y} = \frac{\text{cov}(X, Y)}{\sigma_X \sigma_Y} \tag{2}$$

where $\rho_{X,Y}$ is the Pearson correlation coefficient beween X and Y, cov(X,Y) is the covariance, and σ_X and σ_Y are the standard deiation of X and Y, respectively.

Partial correlation analysis

To investigate the net effect of one independent variable on the dependent variable, there are two ways to control the effects of other variables: experimental and statistical (Garrett & Woodworth 1971). The experimental approach may reduce drastically the size of samples by selecting the appropriate observations. The statistical approach can eliminate the effects of other variables statistically without losing any samples. In the present study, the statistical approach is adopted. In particular, the partial correlation analysis is performed, as such an analysis is widely used in vegetation studies (Peng *et al.* 2013; Shen *et al.* 2015a; Wu *et al.* 2015; Cong *et al.* 2017). The function of the partial correlation analysis is to find the correlation between two variables by holding the other variables constant.

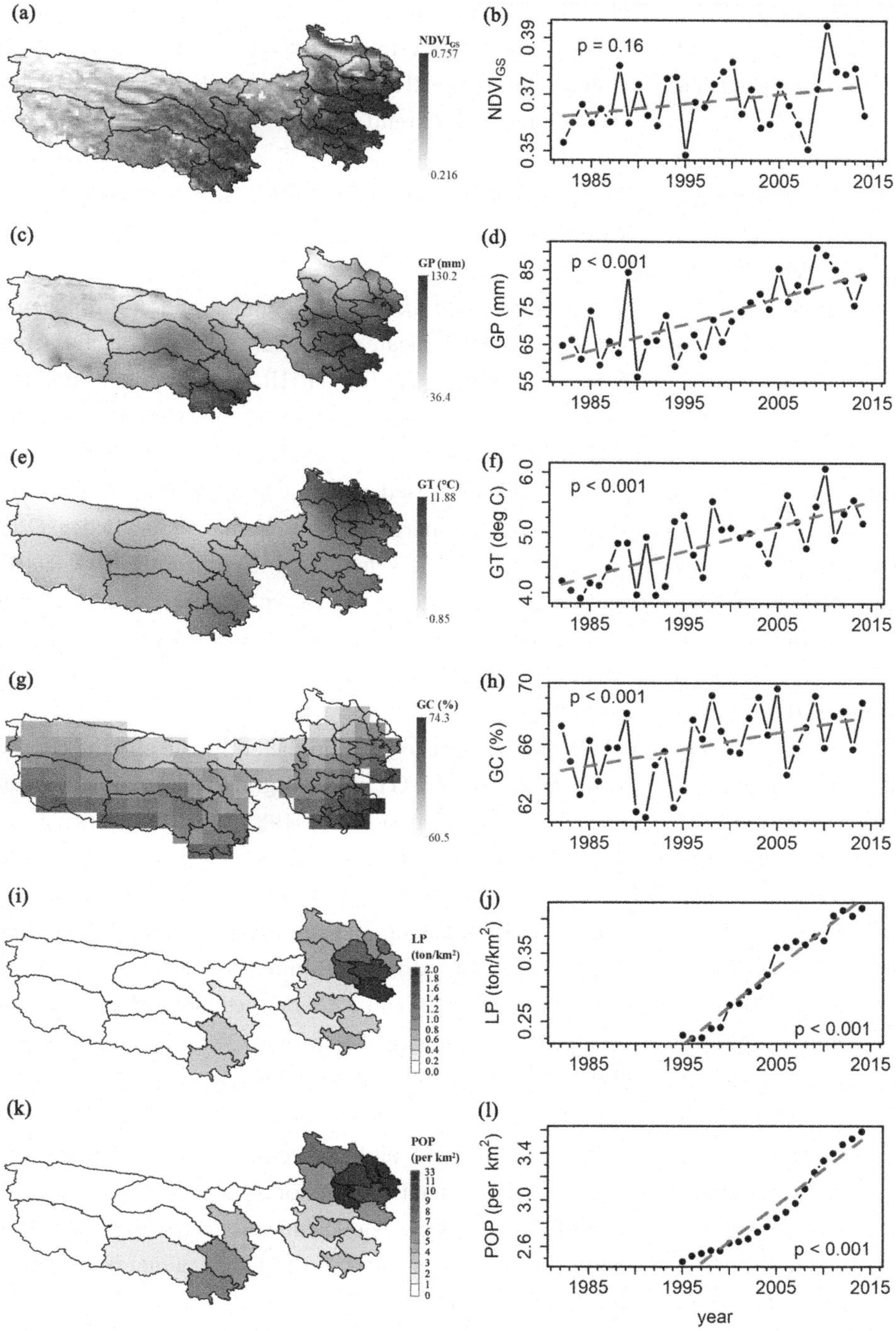

Figure 2 | Spatial and temporal distribution of vegetation and the driving variables. (1) Spatial distribution of (a) $NDVI_{GS}$, (c) GP, (e) GT, (g) GC, (i) LP, and (k) POP. $NDVI_{GS}$, GP, GT, and GC are averaged from 1982 to 2014, and LP and POP are averaged from 1995 to 2014. (2) Temporal variation of (b) $NDVI_{GS}$, (d) GP, (f) GT, (h) GC, (j) LP, and (l) POP in the TRHR.

In general, the partial correlation coefficient can be computed as:

$$R_{12,34\ldots n} = \frac{R_{12,34\ldots(n-1)} - R_{1n,34\ldots(n-1)}R_{2n,34\ldots(n-1)}}{\sqrt{1 - R^2_{1n,34\ldots(n-1)}}\sqrt{1 - R^2_{2n,34\ldots(n-1)}}} \tag{3}$$

where $R_{12,34\ldots n}$ refers to partial correlation coefficient between variables 1 and 2 while removing the effect of variables 3, 4, ..., n. For example, if we denote 1 for NDVI, 2 for precipitation, 3 for temperature, 4 for cloud cover, 5 for livestock production, and 6 for population, then $R_{12,3456}$

represents the partial correlation coefficient between NDVI and precipitation when the effects of temperature, cloud cover, livestock production, and population are partialed out.

In this study, we conduct the partial correlation analysis between $NDVI_{GS}$ and only any one of the six driving variables at a time, controlling the other four variables. We try to disentangle the coordination effects between different variables in performing the partial correlation analysis.

RESULTS

Spatiotemporal change in vegetation

First, we apply the trend analysis to investigate the regional vegetation change in the TRHR. Overall, $NDVI_{GS}$ in the TRHR experienced, during 1982–2014, a slightly greening trend (0.36%/decade, $p = 0.16$) (Figure 2(b)). This result is in agreement with that reported by some earlier studies (Liu *et al.* 2014; Pan *et al.* 2017).

Next, we analyze the trend in $NDVI_{GS}$ data at grid and county scales in the TRHR for the same period (from 1982 to 2014). The distribution of greening and browning trend of vegetation is shown in Figure 3. As seen, the grids with greening trend outnumbered those with browning trend, and about 42.2% of grids in total experienced a significant vegetation change from 1982 to 2014 ($p < 0.05$). The areas with a browning trend, accounting for 24.1% of the total number of grids, were mainly located in the southeastern and southern parts of the TRHR, but only 19.3% of the browning grids show significant negative trend ($p < 0.05$, Figure 3(a)). The areas with a greening trend, accounting for 75.9% of the total number of grids, were distributed in the western and northeastern parts of the TRHR. Here, 49.4% of the greening grids exhibited significant positive trend ($p < 0.05$).

The trend of $NDVI_{GS}$ at the county scale is consistent with that at the grid scale (Figure 3(b)), with browning trend located in the southeastern and southern parts and greening trend located in the northeastern and western parts. The browning trend ranged from −0.08%/decade in the Tongren County to −0.22%/decade in the Gadee County, while the greening trend ranged from 0.09%/decade in the Baima County to 0.81%/decade in the Jianzha County. Overall, the trend of $NDVI_{GS}$ in three out of 22 counties in the TRHR was significant, and all of them showed a greening trend.

Spatiotemporal change in climatic and anthropogenic variables

We perform trend analysis for climatic and anthropogenic variables, and investigate the spatial and temporal characteristics of the trend in the TRHR. During the past three decades, climatic and anthropogenic variables in the TRHR have experienced a significant change. From 1982 to 2014, GP increased with a change rate of 0.78 mm/yr ($p < 0.001$), GT increased with a rate of 0.043 deg C/yr ($p < 0.001$), and GC increased significantly ($p < 0.001$) with a change rate of 0.11%/yr (Figure 2(d), 2(f), and 2(h)). Moreover, a significant increasing trend ($p < 0.001$) was found in LP and POP from 1995 to 2014 (Figure 2(j) and 2(l)).

With reference to trend in the grid scale and the county scale, both GP and GT in most grids and counties experienced an upward trend from 1982 to 2014 (Figure 3(c)–3(f)). The areas with a significant increase in GP were located in the northeastern and western parts of the TRHR, and the maximum rate was as large as 1.45 mm/yr (Figure 3(c) and 3(d)). It was found that 89.4% grids of the whole region, or 19 out of 22 counties, showed a significant change in GP. The rate of change in GT was found to be uniformly positive in all grids and counties ($p < 0.05$), ranging from 0.019 to 0.067 °C/yr (Figure 3(e) and 3(f)). The trend in the northwestern and southeastern parts is among the strongest. The spatial distribution of change in GC showed remarkable patterns. The significant increasing trend was mainly in the western part of the TRHR, while the decreasing trend was distributed in the southern part (Figure 3(g) and 3(h)).

From 1995 to 2014, LP and POP went up rapidly in most counties (Figure 3(i) and 3(j)). The rate of change in LP ranged from −0.01 to 0.08 ton/km^2/yr, with the maximum rate observed in the northeastern part of the TRHR. The rate of change in LP was relatively smaller in the western part. The distribution of change in POP was similar to that of LP.

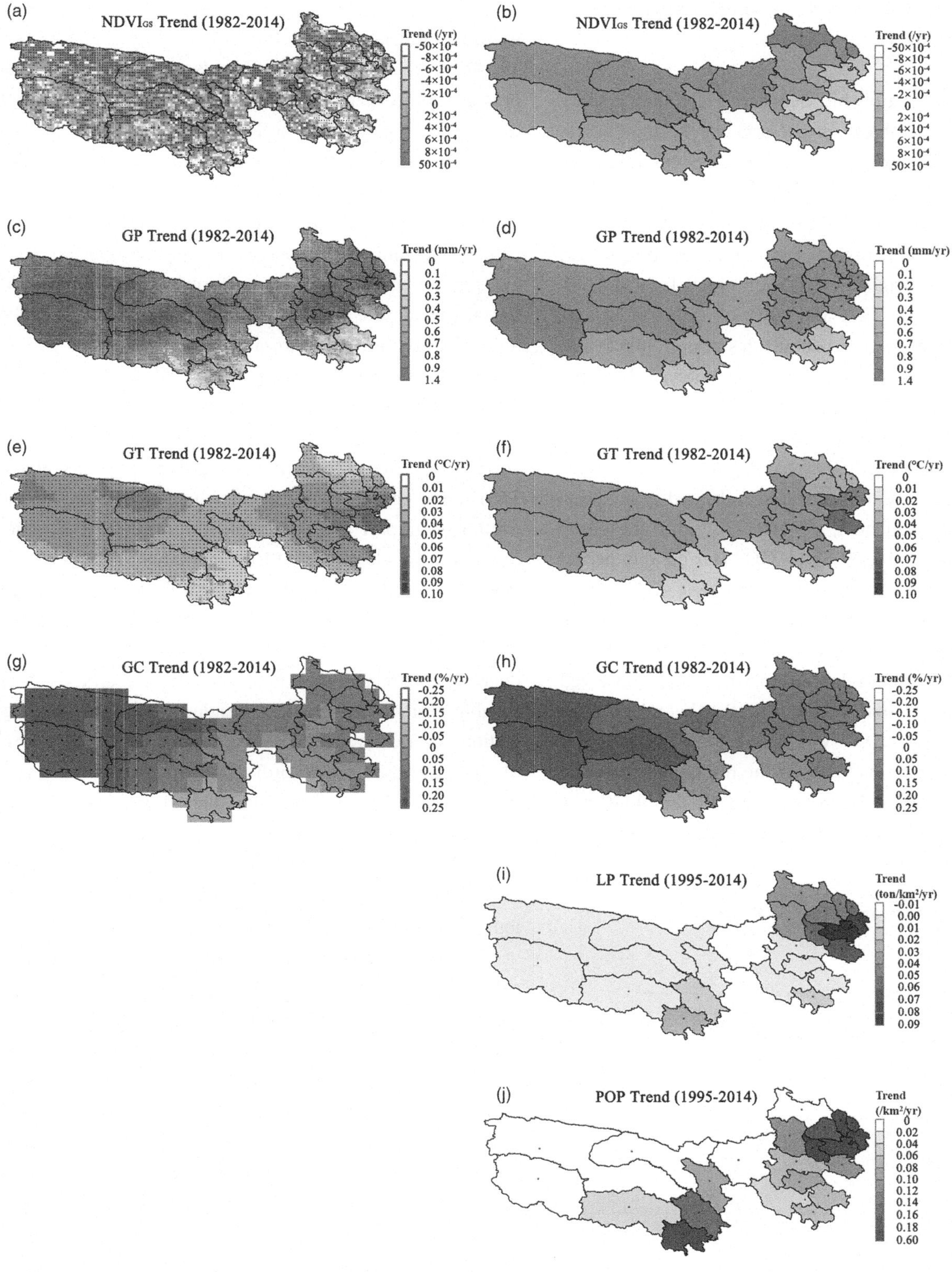

Figure 3 | Distribution of trend in vegetation and the driving variables in the TRHR. (a) $NDVI_{GS}$ trend at the county scale; (b) $NDVI_{GS}$ trend at the grid scale; (c) GP trend at the county scale; (d) GP trend at the grid scale; (e) GT trend at the county scale; (f) GT trend at the grid scale; (g) GC trend at the county scale; (h) GC trend at the grid scale; (i) LP trend at the county scale; (j) POP trend at the county scale. The dots indicate significance at $p < 0.05$.

Pearson correlation between vegetation activity and the driving variables

To examine the relationship between vegetation activity and the driving variables, we perform the Pearson correlation analysis. Since the anthropogenic variables are available only at the county scale, we first check the relationship between vegetation activity and climatic variables based on the grid scale. For the period 1995–2014, we aggregate the original vegetation ($NDVI_{GS}$) and climatic data (i.e. GP, GT, and GC) of different spatial resolutions into a single constant resolution (0.5°) and then conduct the Pearson correlation analysis between them, i.e. $NDVI_{GS}$ and GP, $NDVI_{GS}$ and GT, $NDVI_{GS}$ and GC. We use the Pearson correlation coefficient as an indicator of the relationship.

Figure 4 presents the relationship, indicated by the Pearson correlation coefficient, of $NDVI_{GS}$ with the three climatic variables GP, GT, and GC. During the period 1995–2014, the correlation between $NDVI_{GS}$ and GP was positive in over 75% of the total number of grids, but the positive relationship was significant in only about 25% of the grids. The negative relationship was mainly observed in the southeastern part of the TRHR. The relationship between $NDVI_{GS}$ and GT was also mainly positive, but negative relationship was found in the northeastern part of the TRHR. The relationship between $NDVI_{GS}$ and GC was relatively weak, with the median of the Pearson correlation

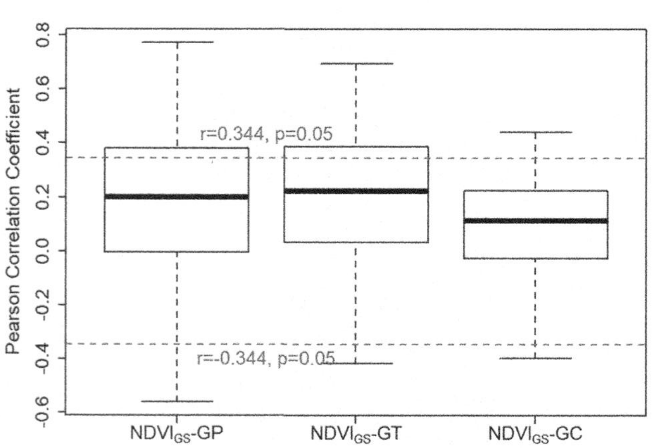

Figure 4 | Relationship of $NDVI_{GS}$ with climatic variables at the grid scale. The boxplots represent 0%, 25%, 50%, 75% and 100% quantiles of Pearson correlation coefficients for all grids, and the dashed line stands for significant level at $p = 0.05$.

coefficients only slightly above 0. The positive relationship was mainly distributed in the western part of the TRHR. The Pearson correlation coefficients seem to suggest that precipitation and temperature were the main constraints for vegetation growth for the TRHR. The impacts of precipitation on vegetation growth was complicated, as the range was huge for the coefficients.

With respect to the relationship between vegetation activity and climatic variables as well as anthropogenic variables at the county scale, we perform a similar analysis as the above for the period of 1995–2014, as data for the anthropogenic variables are available only for this period. The results are generally in agreement with those obtained for the grid scale. Overall, $NDVI_{GS}$ was positively correlated with GP and GT (Figure 5(a) and 5(b)), and was negatively correlated with GT only in the northeastern part of the TRHR. The response of vegetation to temperature for the northeastern part was different from that of the other parts, as two clusters could be identified. This might be due to the situation that the temperature is relatively high in this region (Figure 2(e)) while water availability is limited (Figure 2(c)) (see the Discussion section below for further details). The relationship between $NDVI_{GS}$ and GC was roughly positive (Figure 5(c)). Besides, growing-season vegetation coverage was higher with more LP and POP across the whole region (Figure 5(d) and 5(e)). As discussed previously, the counties with the highest livestock density (Figure 2(i)) and population density (Figure 2(k)) were located in the eastern parts of the TRHR, and this spatial variability might lead to slightly different vegetation behaviour to LP and POP between the eastern and western parts.

In terms of spatial heterogeneity, the results also indicated different responses of vegetation to the driving variables. For the period 1995–2014, $NDVI_{GS}$ was negatively correlated with GP in the southeastern part of the TRHR, positively correlated with GC and POP, respectively, in the western and northeastern parts. Except for the northeastern part, the relationship between $NDVI_{GS}$ and GT was positive across the entire region, with a particularly significant relationship in the southeastern and southern parts of the TRHR ($p < 0.05$). The correlation between $NDVI_{GS}$ and LP was positive in 16 out of 22 counties, indicating that better vegetation coverage could have supported more livestock, as is normally the case.

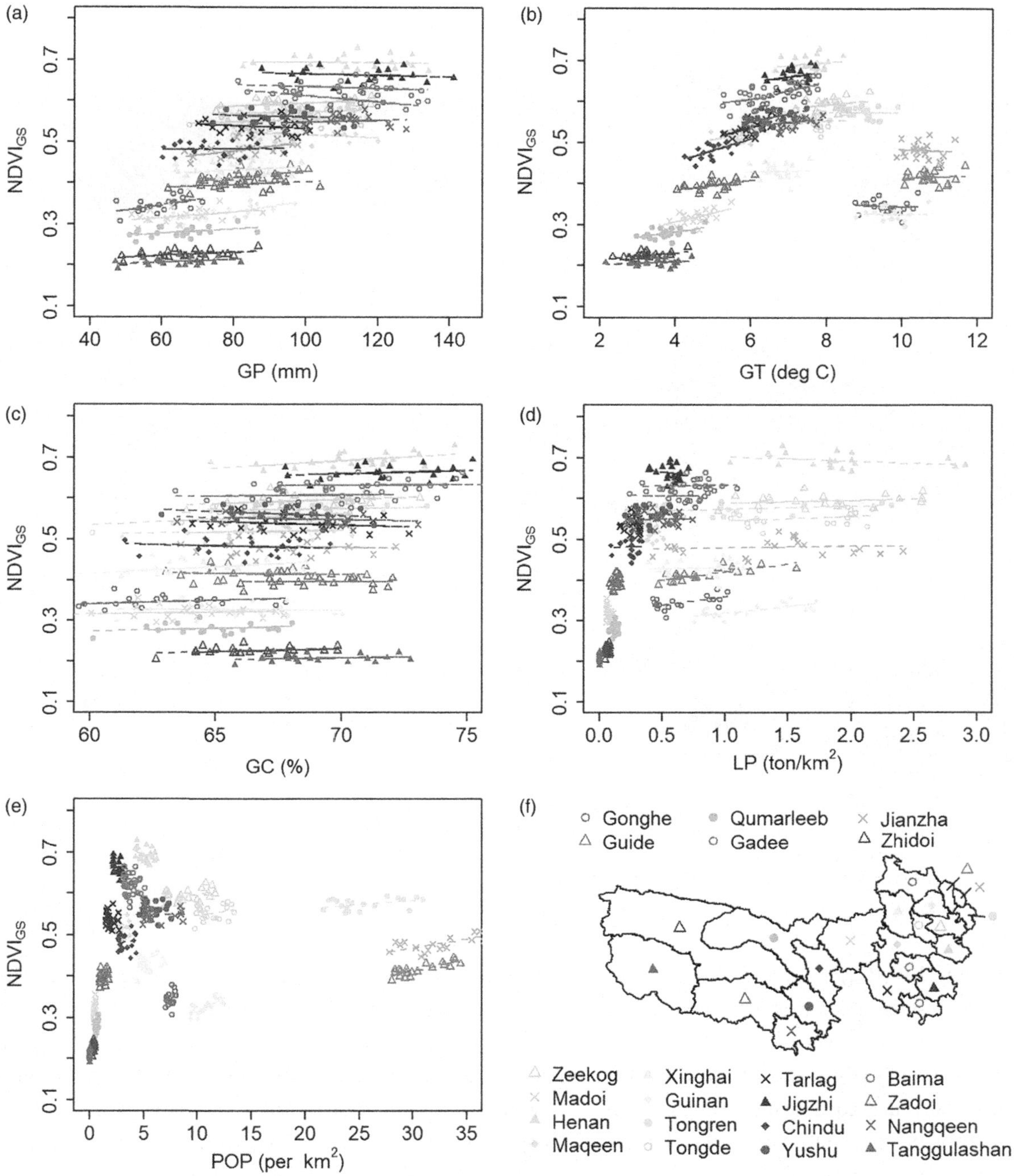

Figure 5 | Relationship of $NDVI_{GS}$ with the driving variables at the county scale. (a) $NDVI_{GS}$ and GP; (b) $NDVI_{GS}$ and GT; (c) $NDVI_{GS}$ and GC; (d) $NDVI_{GS}$ and LP; (e) $NDVI_{GS}$ and POP; and (f) locations of counties.

Discriminating climatic and anthropogenic drivers with partial correlation analysis

To further distinguish between climatic drivers and anthropogenic drivers of vegetation change, we first detrend all the variables to focus on correlations in interannual variability over the period 1995–2014. We then calculate the partial correlation coefficient between the vegetation and climatic/anthropogenic variables, i.e. between $NDVI_{GS}$ and GP ($R_{NDVI\text{-}GP}$), $NDVI_{GS}$ and GT ($R_{NDVI\text{-}GT}$), $NDVI_{GS}$ and GC

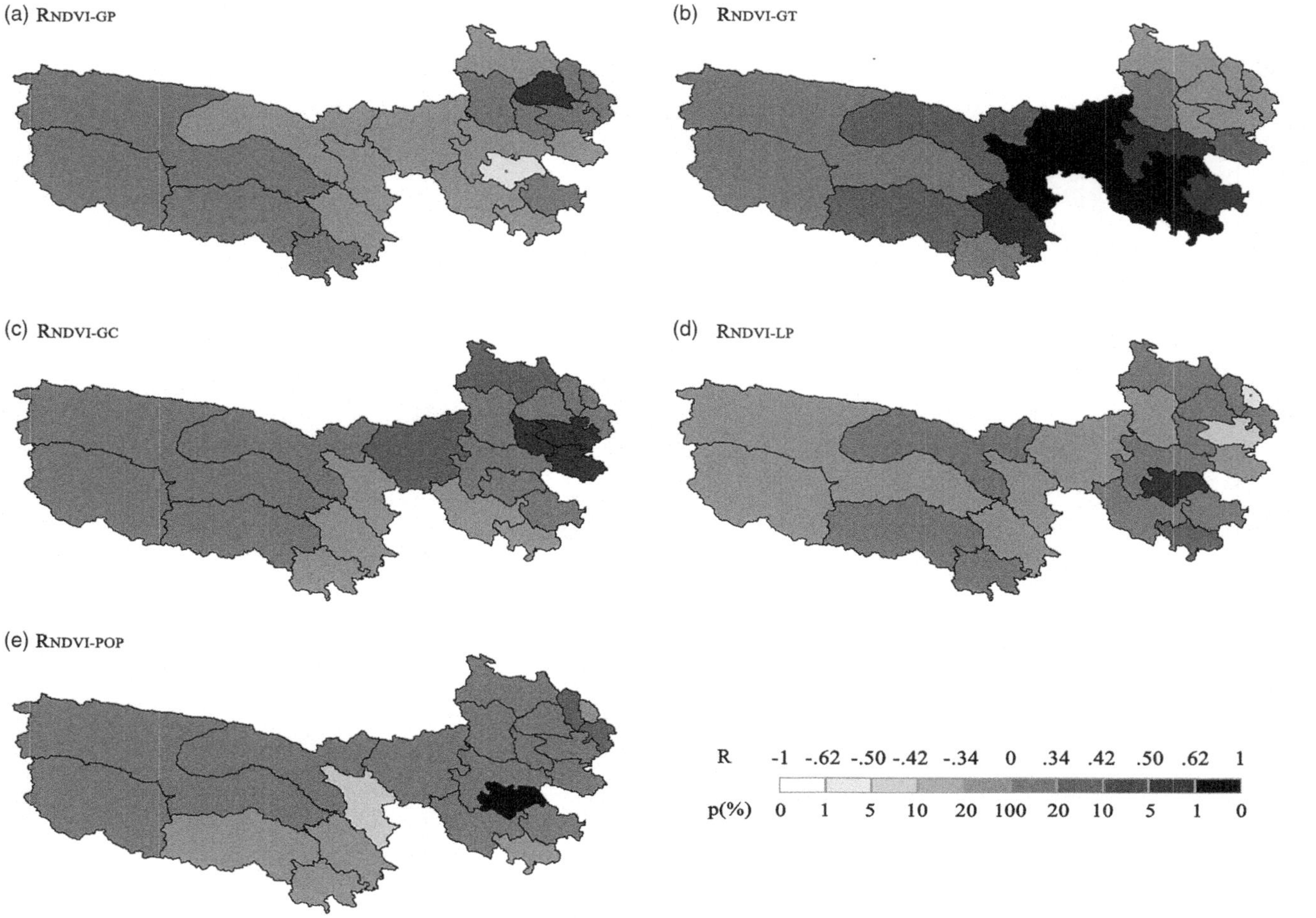

Figure 6 | Partial correlation coefficients of $NDVI_{GS}$ to the driving factors. (a) $R_{NDVI\text{-}GP}$; (b) $R_{NDVI\text{-}GT}$; (c) $R_{NDVI\text{-}GC}$; (d) $R_{NDVI\text{-}LP}$; and (e) $R_{NDVI\text{-}POP}$. The labels on the colour bar, $R=\pm0.62$, $R=\pm0.50$, $R=\pm0.42$, and $R=\pm0.34$, correspond to the 1%, 5%, 10%, and 20% significance levels.

($R_{NDVI\text{-}GC}$), $NDVI_{GS}$ and LP ($R_{NDVI\text{-}LP}$), and $NDVI_{GS}$ and POP ($R_{NDVI\text{-}POP}$). The results show (not presented here) that, over the period 1995–2014, vegetation activity was enhanced by climate warming ($R_{NDVI\text{-}GT}=0.43$, $p=0.10$) in the TRHR, whereas it was degraded by increasing grazing pressure ($R_{NDVI\text{-}LP}=-0.28$, $p=0.29$), with the former having a more dominant effect. Precipitation ($R_{NDVI\text{-}GP}=0.05$, $p=0.84$), cloud cover ($R_{NDVI\text{-}GC}=0.18$, $p=0.51$), and population ($R_{NDVI\text{-}POP}=0.09$, $p=0.75$) had no significant impacts on the greening trend of vegetation for the TRHR during 1995–2014. These observations are also in good agreement with those reported by earlier studies (Piao *et al.* 2014; Cong *et al.* 2017; Chen *et al.* 2018).

We conduct a similar analysis at the county scale for the period 1995–2014. Figure 6 presents the partial correlation coefficients between $NDVI_{GS}$ and the driving factors. The partial correlation coefficients between $NDVI_{GS}$ and GP, controlling the effects of GT, GC, LP, and POP, showed a remarkable spatial pattern (Figure 6(a)). The results suggest that $NDVI_{GS}$ in the northeastern and western parts of the TRHR was positively correlated with GP over the period 1995–2014, but the link was weak ($0.1< p < 0.2$). However, $NDVI_{GS}$ in the southeastern and southern parts of the TRHR was negatively correlated with GP, but again the correlation was not significant ($p > 0.2$). The relationship between $NDVI_{GS}$ and GT also showed a spatial pattern (Figure 6(b)). A significant positive partial correlation between $NDVI_{GS}$ and GT existed in the southeastern and southern parts of the TRHR, while a weak negative correlation between $NDVI_{GS}$ and GT existed in the northeastern part. The partial correlation coefficient between $NDVI_{GS}$ and GC was negative in the southeastern part of the TRHR, while it was positive in the other parts (Figure 6(c)). These results are consistent with those reported in a previous study (Liu *et al.* 2014).

The effects of anthropogenic factors on vegetation variation during 1995–2014 in the TRHR were complicated. $NDVI_{GS}$ in the western part of the TRHR was negatively correlated with LP (Figure 6(d)), indicating that the counties therein faced an increasing grazing pressure. The eastern part of the TRHR exhibited positive $R_{NDVI\text{-}LP}$ values. The link between vegetation activity and grazing effects was weak ($p > 0.1$) for most counties. The relationship between $NDVI_{GS}$ and POP was positive in most areas, except for the southern part (Figure 6(e)).

To test the robustness of our analysis, we substitute CHIRPS and ERA-Interim datasets with CMA-gridded datasets and perform the same analysis. We find that the partial correlation coefficients between $NDVI_{GS}$ and the driving factors show spatial patterns similar to those observed with our earlier analysis (Supplementary Figures S1 and S2, available with the online Supplementary Material). Since the definition of 'growing season' (May–September) might also have an impact on the results, we perform the analysis also considering a slightly extended period of growing season, i.e. from April to October. The results indicate some differences in the magnitude of partial correlation coefficients, but nevertheless similar distributions (Supplementary Figure S3).

Since it is reasonable to consider that the climatic factor with the biggest absolute value of partial correlation coefficient is the primary climatic driving factor of vegetation change (Peng *et al.* 2015), we now extend the definition of the primary climatic driving factor to primary driving factor (since the anthropogenic factors have only less influence) and compare the absolute values of $R_{NDVI\text{-}GP}$, $R_{NDVI\text{-}GT}$, $R_{NDVI\text{-}GC}$, $R_{NDVI\text{-}LP}$, and $R_{NDVI\text{-}POP}$ for each county. The results indicate that the primary driving factor during 1995–2014 was temperature in most counties of the TRHR, including in the southeastern and southern parts. For counties where GP was relatively low in the western part, the primary driving factor was precipitation. Besides, the results also reveal that vegetation activity in northeastern part was mainly driven by cloud cover (solar radiation).

DISCUSSION

Trend analysis of different periods

Due to limitations in data availability for the different variables, we considered the period 1982–2014 for analyzing the trends in $NDVI_{GS}$, GP, GT, and GC, but only the period 1995–2014 for detecting the trend in LP and POP. Therefore, as one would normally expect for complex dynamic systems, such as the one under consideration in this study, the trend obtained for the shorter period (1995–2014) can be a little different from the one for the longer period (1982–2014). The results obtained for the two periods reflect this to a certain extent. Although the distribution of the trend of $NDVI_{GS}$ for the shorter period was similar to that of the longer period, the magnitude of the trend was found to be larger. This difference could be attributed to the fact that the enhancement in vegetation activity became stronger after 1995 in the TRHR. The GP across the region was relatively stable from 1982 to 1995, but the trend accelerated after that. This observation is consistent with the one reported in a previous study that the precipitable water in the atmosphere exhibited an increase since the 1990s (Zhang *et al.* 2013a). Past studies have reported that the TRHR has been exhibiting a generally uniform warming trend since 1979 (Kuang & Jiao 2016). The results from our study certainly reflect this as well. The biggest difference in the results obtained based on the analysis of different periods is that GT in the northeastern part of the TRHR experienced a slightly decreasing trend over the period 1995–2014, while the area went through an overall increasing trend over the period 1982–2015. Besides, GC over the period 1982–1995 showed a significant increasing trend, while it experienced a slightly decreasing trend during 1995–2014. Therefore, the overall GC trend was positive from 1982 to 2014.

Interactions of climatic and anthropogenic variables on vegetation activity

Climatic variables and anthropogenic factors might interact with each other to impact vegetation activity, as has been revealed by several previous studies (Piao *et al.* 2014; Cong *et al.* 2017; Pan *et al.* 2017). In this study, we examined the effects of interactions among the climatic variables as well as those between climatic variables and anthropogenic variables. The results reveal possible reasons as to why $NDVI_{GS}$ responded to the driving factors in a heterogeneous manner across the TRHR.

Vegetation in the western and northeastern parts responded to GP more sensitively, while it responded negatively in the southeastern part (Figure 6(a)). This heterogeneity could have been mainly due to temperature difference. As temperature is suggested as the main constraint on vegetation activity (Kato *et al.* 2006), warm weather could contribute to the effect of precipitation on vegetation. This is supported by the positive relationship between $R_{NDVI\text{-}GP}$ and GT obtained in our analysis (Figure 7(b)). The heterogeneity could also be explained by the difference in water availability. In arid zones, water could be the main factor that put a limitation on vegetation growth, as a higher metabolic rate requires more water (Cong *et al.* 2017). The positive relationship might be explained by the limitation of water in the western and northeastern parts of the TRHR, as these parts belong to semi-arid climate zones (Figure 2(c)). In humid zones, where precipitation exceeds the amount the vegetation needs, more water means higher relative humidity and less solar radiation (Ukkola *et al.* 2016). This puts a negative impact on vegetation growth and explains why the relationship between $NDVI_{GS}$ to GP was negative in the southeastern part of the TRHR. This mechanism is supported by the negative relationship between $R_{NDVI\text{-}GP}$ and GP (Figure 7(a)).

$R_{NDVI\text{-}GT}$ was significantly and negatively correlated with GT ($r = -0.60$, $p = 0.003$, Figure 7(d)), while weakly correlated with GP ($r = 0.21$, $p = 0.355$, Supplementary Figure S4) and GC ($r = 0.25$, $p = 0.265$, Supplementary Figure S4). Past studies have reported that climate warming contributed to the greening trend in the Tibetan Plateau and in the TRHR (Piao *et al.* 2015; Zhu *et al.* 2016). Such findings support the positive $R_{NDVI\text{-}GT}$ value in most areas of the TRHR. When the temperature reaches above the range of vegetation acclimation conditions, the relationship between $NDVI_{GS}$ and GT might become negative because of extreme hot days (Piao *et al.* 2014; Seneviratne *et al.* 2014) or drought (Zhang *et al.* 2015). It could be further found that $R_{NDVI\text{-}GT}$ significantly correlated negatively with both LP ($r = -0.57$, $p = 0.006$, Figure 7(e)) and POP ($r = -0.65$, $p = 0.001$, Figure 7(f)). These results suggest that both grazing pressure and human activity offset the contribution of climate warming on vegetation growth in the TRHR, and in general.

The results also indicate that the effect of GC on vegetation was regulated by GT, as a positive relationship between $R_{NDVI\text{-}GC}$ and GT was found ($r = 0.24$, $p = 0.274$, Figure 7(g)). Grassland in the TRHR normally favours cloudy and warm weather during the growing season. The coordination of climatic variables and anthropogenic variables might drive vegetation towards a favourable direction. The wet and cloudy weather facilitates the grassland to flourish, thus making it possible for the grassland to raise more livestock. This mechanism is supported by the significant positive correlation observed between $R_{NDVI\text{-}GC}$ and LP ($r = 0.63$, $p = 0.002$, Figure 7(h)) and the positive correlation between $R_{NDVI\text{-}LP}$ and GP ($r = 0.34$, $p = 0.125$, Figure 7(i)). Moreover, vegetation was enhanced by warm weather in most areas of the TRHR, and population pressure on vegetation was relieved by warming climate and increasing precipitation. This is confirmed by the positive correlation between $R_{NDVI\text{-}POP}$ and GT ($r = 0.23$, $p = 0.305$, Supplementary Figure S4) and the positive correlation between $R_{NDVI\text{-}GP}$ and POP ($r = 0.49$, $p = 0.021$, Figure 7(c)).

Both the Pearson correlation analysis and the partial correlation analysis support that temperature was the dominant variable for vegetation growth in most areas of the TRHR in recent decades and that vegetation has become better over the past three decades with warming climate and increasing precipitation. However, vegetation might also be driven in an opposite direction with variations in climatic and anthropogenic variables. Increasing precipitation and cloud cover (Figure 3) could have made a positive contribution to vegetation growth in the TRHR (Figure 7(c), 7(h), and 7(i)), but the relationship between vegetation activity and temperature was weakening with an increase in temperature (Figure 7(d)). This means that the role of climate warming has become less dominant, or even negative. Moreover, the continuously increasing grazing and population have posed pressure on vegetation activity, which might offset the contribution of climate warming further.

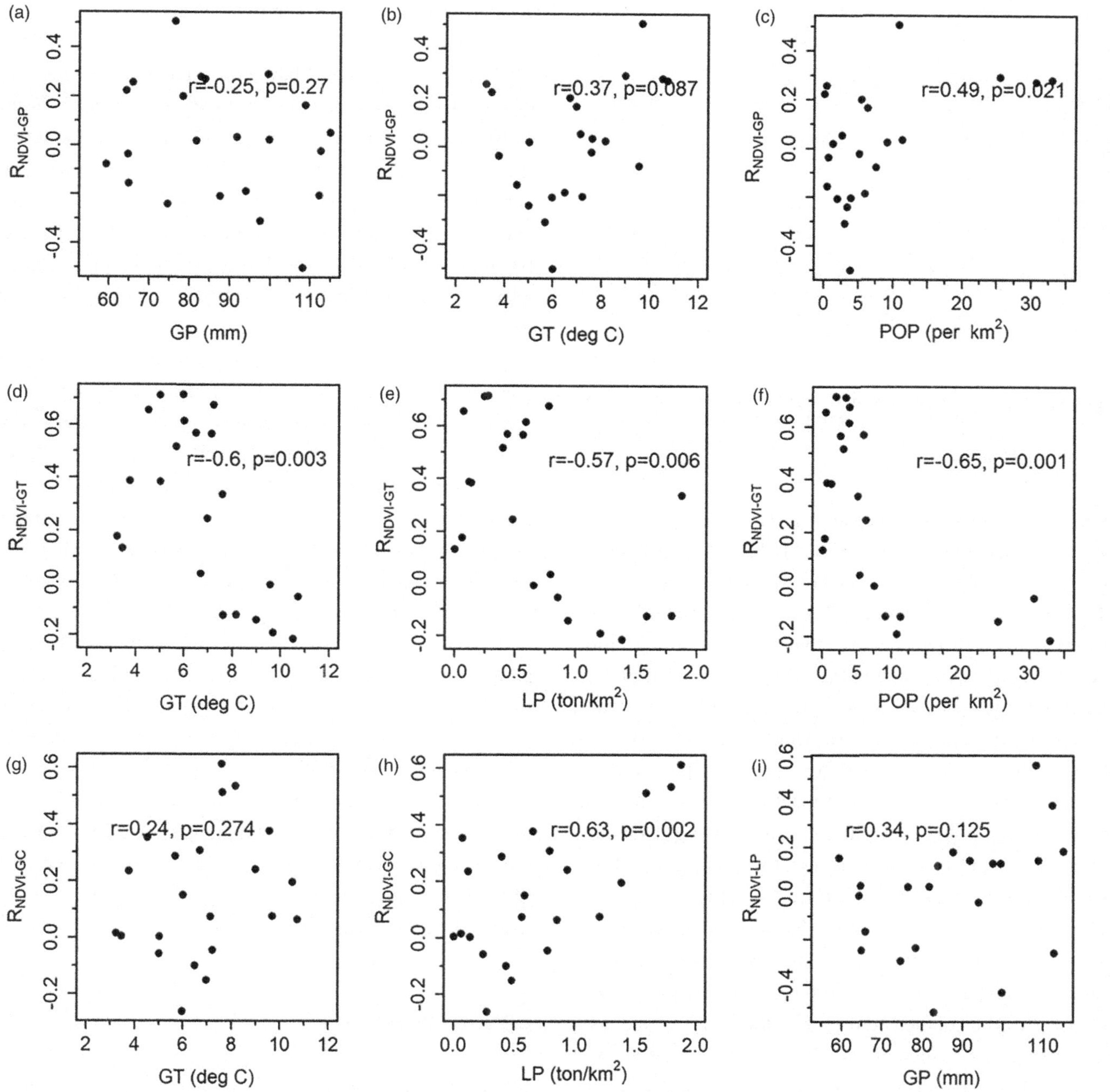

Figure 7 | Partial correlation coefficients of $NDVI_{GS}$ with GP, GT, GC, LP, and POP for the different counties in the TRHR. GP, GT, GC, LP, and POP are calculated as the mean of 20 years' (1995–2014) growing season temperature, growing season precipitation, growing season cloud cover, livestock production, and population within a county. Each point represents a county. (a) $R_{NDVI\text{-}GP}$ to GP; (b) $R_{NDVI\text{-}GP}$ to GT; (c) $R_{NDVI\text{-}GP}$ to POP; (d) $R_{NDVI\text{-}GT}$ to GT; (e) $R_{NDVI\text{-}GT}$ to LP; (f) $R_{NDVI\text{-}GT}$ to POP; (g) $R_{NDVI\text{-}GC}$ to GT; (h) $R_{NDVI\text{-}GC}$ to LP; and (i) $R_{NDVI\text{-}LP}$ to GP. The sub-figures are selected from Supplementary Figure S4 based on the significant level.

Dynamic relationship between vegetation activity and the driving factors

We assess the dynamic relationships between vegetation activity and the driving variables by performing the partial correlation analysis with an 11-year moving window. Figure 8 presents the results from this analysis. The results reveal that $R_{NDVI\text{-}GP}$, $R_{NDVI\text{-}GT}$, $R_{NDVI\text{-}GC}$, $R_{NDVI\text{-}LP}$, and $R_{NDVI\text{-}POP}$ changed with time over the past two decades in the TRHR. As seen, $R_{NDVI\text{-}GT}$ exhibited a slightly downward

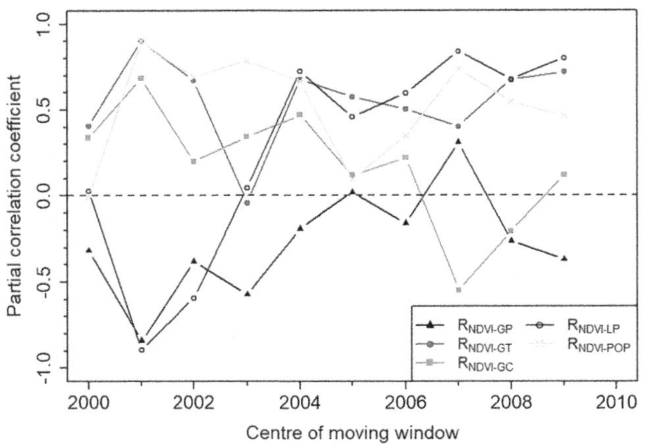

Figure 8 | Changes in the partial correlation coefficients between $NDVI_{GS}$ and GP, $NDVI_{GS}$ and GT, $NDVI_{GS}$ and GC, $NDVI_{GS}$ and LP, and $NDVI_{GS}$ and POP with an 11-year moving window. A value on the *x*-axis corresponds to the centre of the 11-year moving window.

trend, indicating the weakening relationship between vegetation dynamics and GT in the TRHR. This observation has also been made by a previous study (Piao *et al.* 2014). However, $R_{NDVI\text{-}GP}$ exhibited an upward trend, which means that GP became more dominant in the vegetation growth. With the warming of the TRHR, evapotranspiration increased and soil moisture decreased, and precipitation became more important in providing water for vegetation growth. $R_{NDVI\text{-}LP}$ experienced a significant upward trend, indicating that vegetation became better under warming climate and increasing precipitation, thus making it possible for the grassland to raise more livestock, as discussed earlier. There is no clear trend in $R_{NDVI\text{-}GC}$ and $R_{NDVI\text{-}POP}$.

At the county scale, the relationships between vegetation and its driving variables also went through significant changes. An increasing trend in $R_{NDVI\text{-}GP}$ was observed in the southeastern part of the TRHR, whereas a decreasing trend was found in the northeastern part (Figure 9). An increasing trend in $R_{NDVI\text{-}GT}$ and $R_{NDVI\text{-}LP}$ was mainly observed in the northeastern part and western part, respectively. These temporal changes in the response of vegetation activity to climatic variables have also been reported in other studies, based on different periods and study areas (Zhang *et al.* 2013b; Piao *et al.* 2014; Cong *et al.* 2017). Although these results are only based on 20 years of data, which is too short to capture the long-term trend, they still suggest the need to exercise caution in using the results from one period to constrain the response of vegetation to its driving variables in another period (Piao *et al.* 2014).

Multiple reasons may have contributed to the variability of the response of vegetation to the driving variables in the TRHR, including climatic and non-climatic considerations. It is now clear that both climate change and anthropogenic change were responsible for this variability, and still other effects, such as frost damage during spring, fire disturbances (Huxman *et al.* 2004), land use and land management changes (Epstein *et al.* 2013), may have also played important roles.

Our results also seem to suggest that the direction of temporal changes in the relationship (between vegetation and the driving variables) was to offset the restrictions that climate and human activity put on vegetation. More specifically, the change in $R_{NDVI\text{-}GP}$ was negatively correlated with $R_{NDVI\text{-}GP}$ ($r = -0.37$, $p = 0.093$, Supplementary Figure S5a), the change in $R_{NDVI\text{-}GT}$ was negatively correlated with $R_{NDVI\text{-}GT}$ ($r = -0.66$, $p = 0.001$, Supplementary Figure S5b), and the change in $R_{NDVI\text{-}GC}$ was negatively correlated with $R_{NDVI\text{-}GC}$ ($r = -0.19$, $p = 0.399$, Supplementary Figure S5c). That is to say, the magnitude of the response of vegetation to changes was to alleviate the pressure that these changes put on vegetation. This function might be fulfilled by adjusting respiration, photosynthesis, or evaporation (Shen *et al.* 2015b). Under intensified climate change and human activities, species must acclimate, adapt, move, or die (Corlett & Westcott 2013). The mechanism of this acclimatization and adaption needs further investigation.

CONCLUSIONS

In this study, we analyzed the changes in the growing season vegetation, growing season climate change, and annual human activities in the TRHR. We attributed the growing season vegetation activity to climate change and human activities by applying the Pearson correlation analysis and the partial correlation analysis. We also investigated the interactions among the climatic variables as well as between climatic variables and anthropogenic variables, and explored the dynamic relationship between vegetation activity and the driving factors. The results indicated that

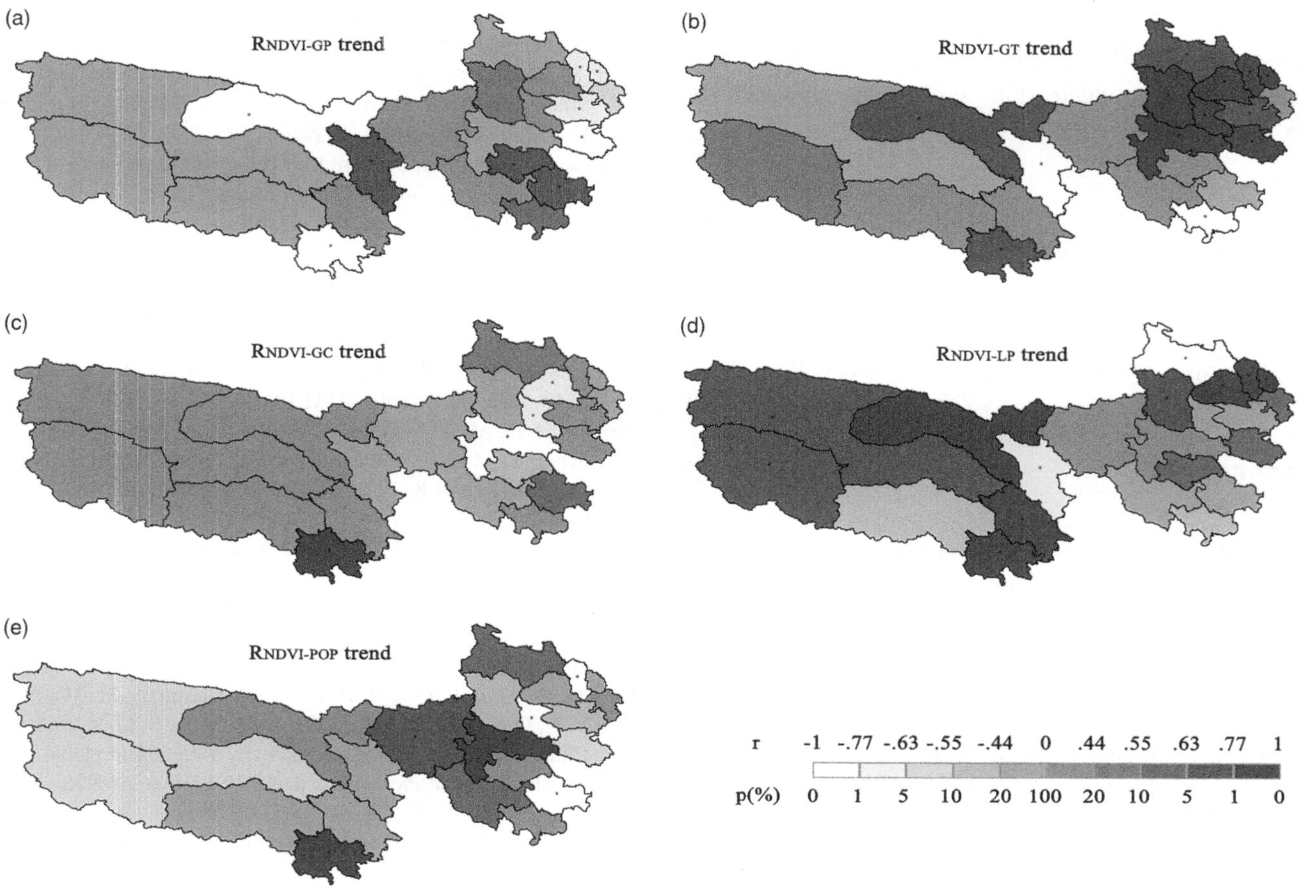

Figure 9 | Changes in the partial correlation coefficients by applying 11-year moving windows at the county scale. (a) Change in $R_{NDVI\text{-}GP}$; (b) change in $R_{NDVI\text{-}GT}$; (c) change in $R_{NCVI\text{-}GC}$; (d) change in $R_{NDVI\text{-}LP}$; and (e) change in $R_{NDVI\text{-}POP}$. The change means the Pearson correlation coefficient between partial correlation coefficient and time. The labels on the colour bar, $r = \pm 0.77$, $r = \pm 0.63$, $r = \pm 0.55$, and $r = \pm 0.44$, correspond to the 1%, 5%, 10%, and 20% significance levels. The dots indicate significance at $p < 0.05$.

vegetation in the TRHR experienced a slightly greening trend during 1982–2014, accompanied by significant changes in the climatic and anthropogenic variables. The southeastern and southern parts of TRHR exhibited a browning trend, while the western and northeastern parts exhibited a greening trend.

During the period 1995–2014, the dominant factor for vegetation dynamics was temperature for the southeastern and southern parts of the TRHR, precipitation for the western part, and solar radiation for the northeastern part. The regulation effects of temperature on precipitation and cloud cover contributed to vegetation growth, while grazing activity and population activity offset the contribution of warming climate. Although vegetation has become better over the past three decades with warming climate and increasing precipitation, it might be driven in the opposite direction under climate change and intensified human activities. The dynamic relationship between vegetation activity and the driving variables reflected the acclimatization and adaption processes of vegetation to climate change and human activities. This mechanism, however, needs further investigation, which we will look into in our future studies.

ACKNOWLEDGEMENTS

This paper was financially supported by the Research and Development Special Fund for Public Welfare Industry of the Ministry of Water Research in China (No. 201501028) and the National Key Research and Development Program of China (2017YFC0403600 and 2016YFE0201900). The first author gratefully acknowledges the Tsinghua Scholarship (2018015) for the financial support for a 6-month overseas graduate study visit at the University of New South Wales.

SUPPLEMENTARY MATERIAL

The Supplementary Material for this paper is available online at https://dx.doi.org/10.2166/hydro.2019.003.

REFERENCES

Archer, E. R. M. 2004 Beyond the 'climate versus grazing' impasse: using remote sensing to investigate the effects of grazing system choice on vegetation cover in the Eastern Karoo. *Journal of Arid Environments* **57** (3), 381–408. doi:10.1016/s0140-1963(03)00107-1.

Chen, H., Zhu, Q. A., Peng, C. H., Wu, N., Wang, Y. F., Fang, X. Q., Gao, Y. H., Zhu, D., Yang, G., Tian, J. Q., Kang, X. M., Piao, S. L., Ouyang, H., Xiang, W. H., Luo, Z. B., Jiang, H., Song, X. Z., Zhang, Y., Yu, G. R., Zhao, X. Q., Gong, P., Yao, T. D. & Wu, J. H. 2013 The impacts of climate change and human activities on biogeochemical cycles on the Qinghai-Tibetan Plateau. *Global Change Biology* **19** (10), 2940–2955. doi:10.1111/gcb.12277.

Chen, C., Li, T. J., Bellie, S., Li, J. Y. & Wang, G. Q. 2018 Vegetation change analyses considering climate variables and anthropogenic variables in the Three-River Headwaters Region. In: *13th International Conference on Hydroinformatics*. EPic Series in Engineering, pp. 419–427.

Cong, N., Shen, M. G., Yang, W., Yang, Z. Y., Zhang, G. X. & Piao, S. L. 2017 Varying responses of vegetation activity to climate changes on the Tibetan Plateau grassland. *International Journal of Biometeorology* **61** (8), 1433–1444. doi:10.1007/s00484-017-1321-5.

Corlett, R. T. & Westcott, D. A. 2013 Will plant movements keep up with climate change? *Trends in Ecology & Evolution* **28** (8), 482–488. doi:10.1016/j.tree.2013.04.003.

Cuo, L., Zhang, Y. X., Wang, Q. C., Zhang, L. L., Zhou, B. R., Hao, Z. C. & Su, F. G. 2013 Climate change on the Northern Tibetan Plateau during 1957–2009: spatial patterns and possible mechanisms. *Journal of Climate* **26** (1), 85–109. doi:10.1175/jcli-d-11-00738.1.

de Jong, R., de Bruin, S., de Wit, A., Schaepman, M. E. & Dent, D. L. 2011 Analysis of monotonic greening and browning trends from global NDVI time-series. *Remote Sensing of Environment* **115** (2), 692–702. doi:10.1016/j.rse.2010.10.011.

de Jong, R., Schaepman, M. E., Furrer, R., De Bruin, S. & Verburg, P. H. 2013 Spatial relationship between climatologies and changes in global vegetation activity. *Global Change Biology* **19** (6), 1953–1964. doi:10.1111/gcb.12193.

Diaz, S., Lavorel, S., McIntyre, S., Falczuk, V., Casanoves, F., Milchunas, D. G., Skarpe, C., Rusch, G., Sternberg, M., Noy-Meir, I., Landsberg, J., Zhang, W., Clark, H. & Campbell, B. D. 2007 Plant trait responses to grazing – a global synthesis. *Global Change Biology* **13** (2), 313–341. doi:10.1111/j.1365-2486.2006.01288.x.

Dong, S. K., Li, J. P., Li, X. Y., Wen, L., Zhu, L., Li, Y. Y., Ma, Y. S., Shi, J. J., Dong, Q. M. & Wang, Y. L. 2010 Application of design theory for restoring the 'black beach' degraded rangeland at the headwater areas of the Qinghai-Tibetan Plateau. *African Journal of Agricultural Research* **5** (25), 3542–3552. doi:10.5897/AJAR10.005.

Epstein, H. E., Myers-Smith, I. & Walker, D. A. 2013 Recent dynamics of arctic and sub-arctic vegetation. *Environmental Research Letters* **8** (1), 015040. doi:10.1088/1748-9326/8/1/015040.

Forkel, M., Carvalhais, N., Verbesselt, J., Mahecha, M. D., Neigh, C. S. R. & Reichstein, M. 2013 Trend change detection in NDVI time series: effects of inter-annual variability and methodology. *Remote Sensing* **5** (5), 2113–2144. doi:10.3390/rs5052113.

Gao, Q. Z., Zhu, W. Q., Schwartz, M. W., Ganjurjav, H., Wan, Y. F., Qin, X. B., Ma, X., Williamson, M. A. & Li, Y. 2016 Climatic change controls productivity variation in global grasslands. *Scientific Reports* **6**. doi:10.1038/srep26958.

Gao, J. B., Jiao, K. W., Wu, S. H., Ma, D. Y., Zhao, D. S., Yin, Y. H. & Dai, E. F. 2017 Past and future effects of climate change on spatially heterogeneous vegetation activity in China. *Earths Future* **5** (7), 679–692. doi:10.1002/2017ef000573.

Garrett, H. E. & Woodworth, R. S. 1971 *Statistics in Psychology and Education*. David McKay Company, Inc., New York, p. 491.

Harris, R. B. 2010 Rangeland degradation on the Qinghai-Tibetan plateau: a review of the evidence of its magnitude and causes. *Journal of Arid Environments* **74** (1), 1–12. doi:10.1016/j.jaridenv.2009.06.014.

Harris, I., Jones, P. D., Osborn, T. J. & Lister, D. H. 2014 Updated high-resolution grids of monthly climatic observations – the CRU TS3.10 dataset. *International Journal of Climatology* **34** (3), 623–642. doi:10.1002/joc.3711.

Hirsch, R. M., Slack, J. R. & Smith, R. A. 1982 Techniques of trend analysis for monthly water-quality data. *Water Resources Research* **18** (1), 107–121. doi:10.1029/WR018i001p00107.

Holben, B. N. 1986 Characteristics of maximum-value composite images from temporal AVHRR data. *International Journal of Remote Sensing* **7** (11), 1417–1434. doi:10.1080/01431168608948945.

Huxman, T. E., Smith, M. D., Fay, P. A., Knapp, A. K., Shaw, M. R., Loik, M. E., Smith, S. D., Tissue, D. T., Zak, J. C., Weltzin, J. F., Pockman, W. T., Sala, O. E., Haddad, B. M., Harte, J., Koch, G. W., Schwinning, S., Small, E. E. & Williams, D. G. 2004 Convergence across biomes to a common rain-use efficiency. *Nature* **429** (6992), 651–654. doi:10.1038/nature02561.

Immerzeel, W. W., van Beek, L. P. H. & Bierkens, M. F. P. 2010 Climate change will affect the Asian water towers. *Science* **328** (5984), 1382–1385. doi:10.1126/science.1183188.

Jiang, L. L., Jiapaer, G., Bao, A. M., Guo, H. & Ndayisaba, F. 2017 Vegetation dynamics and responses to climate change and human activities in Central Asia. *Science of the Total Environment* **599**, 967–980. doi:10.1016/j.scitotenv.2017.05.012.

Karnieli, A., Agam, N., Pinker, R. T., Anderson, M., Imhoff, M. L., Gutman, G. G., Panov, N. & Goldberg, A. 2010 Use of NDVI and land surface temperature for drought assessment: merits and limitations. *Journal of Climate* **23** (3), 618–633. doi:10.1175/2009jcli2900.1.

Kato, T., Tang, Y. H., Gu, S., Hirota, M., Du, M. Y., Li, Y. N. & Zhao, X. Q. 2006 Temperature and biomass influences on interannual changes in CO_2 exchange in an alpine meadow on the Qinghai-Tibetan Plateau. *Global Change Biology* **12** (7), 1285–1298. doi:10.1111/j.1365-2486.2006.01153.x.

Kuang, X. X. & Jiao, J. J. 2016 Review on climate change on the Tibetan Plateau during the last half century. *Journal of Geophysical Research-Atmospheres* **121** (8), 3979–4007. doi:10.1002/2015jd024728.

Li, L., Yang, S., Wang, Z. Y., Zhu, X. D. & Tang, H. Y. 2010 Evidence of warming and wetting climate over the Qinghai-Tibet Plateau. *Arctic Antarctic and Alpine Research* **42** (4), 449–457. doi:10.1657/1938-4246-42.4.449.

Li, Y. K., Liao, J. J., Guo, H. D., Liu, Z. W. & Shen, G. Z. 2014 Patterns and potential drivers of dramatic changes in Tibetan Lakes, 1972–2010. *PLos ONE* **9**(11). doi:10.1371/journal.pone.0111890.

Li, L. H., Zhang, Y. L., Liu, L. S., Wu, J. S., Li, S. C., Zhang, H. Y., Zhang, B. H., Ding, M. J., Wang, Z. F. & Paudel, B. 2018 Current challenges in distinguishing climatic and anthropogenic contributions to alpine grassland variation on the Tibetan Plateau. *Ecology and Evolution* **8** (11), 5949–5963. doi:10.1002/ece3.4099.

Liu, X. F., Zhang, J. S., Zhu, X. F., Pan, Y. Z., Liu, Y. X., Zhang, D. H. & Lin, Z. H. 2014 Spatiotemporal changes in vegetation coverage and its driving factors in the Three-River Headwaters Region during 2000–2011. *Journal of Geographical Sciences* **24** (2), 288–302. doi:10.1007/s11442-014-1088-0.

Liu, Y., Li, Y., Li, S. C. & Motesharrei, S. 2015 Spatial and temporal patterns of global NDVI trends: correlations with climate and human factors. *Remote Sensing* **7** (10), 13233–13250. doi:10.3390/rs71013233.

Liuzzo, L., Bono, E., Sammartano, V. & Freni, G. 2015 Analysis of spatial and temporal rainfall trends in Sicily during the 1921–2012 period. *Theoretical and Applied Climatology* **126** (1–2), 113–129. doi:10.1007/s00704-015-1561-4.

Mao, T. X., Wang, G. X. & Zhang, T. 2016 Impacts of climatic change on hydrological regime in the Three-River Headwaters Region, China, 1960–2009. *Water Resources Management* **30** (1), 115–131. doi:10.1007/s11269-015-1149-x.

Neigh, C. S. R., Tucker, C. J. & Townshend, J. R. G. 2008 North American vegetation dynamics observed with multi-resolution satellite data. *Remote Sensing of Environment* **112** (4), 1749–1772. doi:10.1016/j.rse.2007.08.018.

Nemani, R. R., Keeling, C. D., Hashimoto, H., Jolly, W. M., Piper, S. C., Tucker, C. J., Myneni, R. B. & Running, S. W. 2003 Climate-driven increases in global terrestrial net primary production from 1982 to 1999. *Science* **300** (5625), 1560–1563. doi:10.1126/science.1082750.

Pan, T., Zou, X. T., Liu, Y. J., Wu, S. H. & He, G. M. 2017 Contributions of climatic and non-climatic drivers to grassland variations on the Tibetan Plateau. *Ecological Engineering* **108**, 307–317. doi:10.1016/j.ecoleng.2017.07.039.

Peng, S. S., Piao, S. L., Ciais, P., Myneni, R. B., Chen, A. P., Chevallier, F., Dolman, A. J., Janssens, I. A., Penuelas, J., Zhang, G. X., Vicca, S., Wan, S. Q., Wang, S. P. & Zeng, H. 2013 Asymmetric effects of daytime and night-time warming on Northern Hemisphere vegetation. *Nature* **501** (7465), 88–92. doi:10.1038/nature12434.

Peng, J., Li, Y., Tian, L., Liu, Y. X. & Wang, Y. L. 2015 Vegetation dynamics and associated driving forces in Eastern China during 1999–2008. *Remote Sensing* **7** (10), 13641–13663. doi:10.3390/rs71013641.

Piao, S. L., Nan, H. J., Huntingford, C., Ciais, P., Friedlingstein, P., Sitch, S., Peng, S. S., Ahlstrom, A., Canadell, J. G., Cong, N., Levis, S., Levy, P. E., Liu, L. L., Lomas, M. R., Mao, J. F., Myneni, R. B., Peylin, P., Poulter, B., Shi, X. Y., Yin, G. D., Viovy, N., Wang, T., Wang, X. H., Zaehle, S., Zeng, N., Zeng, Z. Z. & Chen, A. P. 2014 Evidence for a weakening relationship between interannual temperature variability and northern vegetation activity. *Nature Communications* **5**, 5018. doi:10.1038/ncomms6018.

Piao, S. L., Yin, G. D., Tan, J. G., Cheng, L., Huang, M. T., Li, Y., Liu, R. G., Mao, J. F., Myneni, R. B., Peng, S. S., Poulter, B., Shi, X. Y., Xiao, Z. Q., Zeng, N., Zeng, Z. Z. & Wang, Y. P. 2015 Detection and attribution of vegetation greening trend in China over the last 30 years. *Global Change Biology* **21** (4), 1601–1609. doi:10.1111/gcb.12795.

Reyer, C. P. O., Leuzinger, S., Rammig, A., Wolf, A., Bartholomeus, R. P., Bonfante, A., de Lorenzi, F., Dury, M., Gloning, P., Abou Jaoude, R., Klein, T., Kuster, T. M., Martins, M., Niedrist, G., Riccardi, M., Wohlfahrt, G., de Angelis, P., de Dato, G., Francois, L., Menzel, A. & Pereira, M. 2013 A plant's perspective of extremes: terrestrial plant responses to changing climatic variability. *Global Change Biology* **19** (1), 75–89. doi:10.1111/gcb.12023.

Seneviratne, S. I., Donat, M. G., Mueller, B. & Alexander, L. V. 2014 No pause in the increase of hot temperature extremes. *Nature Climate Change* **4** (3), 161–163. doi:10.1038/nclimate2145.

Shen, M. G., Piao, S. L., Cong, N., Zhang, G. X. & Janssens, I. A. 2015a Precipitation impacts on vegetation spring phenology on the Tibetan Plateau. *Global Change Biology* **21** (10), 3647–3656. doi:10.1111/gcb.12961.

Shen, M. G., Piao, S. L., Jeong, S. J., Zhou, L. M., Zeng, Z. Z., Ciais, P., Chen, D. L., Huang, M. T., Jin, C. S., Li, L. Z. X., Li, Y., Myneni, R. B., Yang, K., Zhang, G. X., Zhang, Y. J. & Yao, T. D. 2015b Evaporative cooling over the Tibetan Plateau induced by vegetation growth. *Proceedings of the National Academy of Sciences of the United States of America* **112** (30), 9299–9304. doi:10.1073/pnas.1504418112.

Shi, H. Y., Li, T. J., Wang, K., Zhang, A., Wang, G. Q. & Fu, X. D. 2016 Physically based simulation of the streamflow decrease caused by sediment-trapping dams in the middle Yellow River. *Hydrological Processes* **30** (5), 783–794. doi:10.1002/hyp.10649.

Shi, H., Li, T. & Wei, J. 2017 Evaluation of the gridded CRU TS precipitation dataset with the point raingauge records over the Three-River Headwaters Region. *Journal of Hydrology* **548**, 322–332. doi:10.1016/j.jhydrol.2017.03.017.

Tucker, C. J., Pinzon, J. E., Brown, M. E., Slayback, D. A., Pak, E. W., Mahoney, R., Vermote, E. F. & El Saleous, N. 2005 An extended AVHRR 8-km NDVI dataset compatible with MODIS and SPOT vegetation NDVI data. *International Journal of Remote Sensing* **26** (20), 4485–4498. doi:10.1080/01431160500168686.

Ukkola, A. M., Prentice, I. C., Keenan, T. F., van Dijk, A., Viney, N. R., Myneni, R. B. & Bi, J. 2016 Reduced streamflow in water-stressed climates consistent with CO_2 effects on vegetation. *Nature Climate Change* **6** (1), 75. doi:10.1038/nclimate2831.

Walker, D. A., Leibman, M. O., Epstein, H. E., Forbes, B. C., Bhatt, U. S., Raynolds, M. K., Comiso, J. C., Gubarkov, A. A., Khomutov, A. V., Jia, G. J., Kaarlejarvi, E., Kaplan, J. O., Kumpula, T., Kuss, P., Matyshak, G., Moskalenko, N. G., Orekhov, P., Romanovsky, V. E., Ukraientseva, N. G. & Yu, Q. 2009 Spatial and temporal patterns of greenness on the Yamal Peninsula, Russia: interactions of ecological and social factors affecting the Arctic Normalized Difference Vegetation Index. *Environmental Research Letters* **4** (4), 045004. doi:10.1088/1748-9326/4/4/045004.

Wang, P., Lassoie, J. P., Morreale, S. J. & Dong, S. K. 2015 A critical review of socioeconomic and natural factors in ecological degradation on the Qinghai-Tibetan Plateau, China. *Rangeland Journal* **37** (1), 1–9. doi:10.1071/rj14094.

Wang, X. J., Pang, G. J. & Yang, M. X. 2018 Precipitation over the Tibetan Plateau during recent decades: a review based on observations and simulations. *International Journal of Climatology* **38** (3), 1116–1131. doi:10.1002/joc.5246.

Wu, D. H., Zhao, X., Liang, S. L., Zhou, T., Huang, K. C., Tang, B. J. & Zhao, W. Q. 2015 Time-lag effects of global vegetation responses to climate change. *Global Change Biology* **21** (9), 3520–3531. doi:10.1111/gcb.12945.

Xie, H., Ye, J. S., Liu, X. M. & E, C. Y. 2010 Warming and drying trends on the Tibetan Plateau (1971–2005). *Theoretical and Applied Climatology* **101** (3–4), 241–253. doi:10.1007/s00704-009-0215-9.

Yang, Z. Y., Shen, M. G., Jia, S. G., Guo, L., Yang, W., Wang, C., Chen, X. H. & Chen, J. 2017 Asymmetric responses of the end of growing season to daily maximum and minimum temperatures on the Tibetan Plateau. *Journal of Geophysical Research: Atmospheres* **122** (24), 13278–13287. doi:10.1002/2017jd027318.

Yue, S. & Wang, C. Y. 2002 Applicability of prewhitening to eliminate the influence of serial correlation on the Mann-Kendall test. *Water Resources Research* **38** (6), 1068. doi:10.1029/2001wr000861.

Zelikova, T. J., Blumenthal, D. M., Williams, D. G., Souza, L., LeCain, D. R., Morgan, J. & Pendall, E. 2014 Long-term exposure to elevated CO_2 enhances plant community stability by suppressing dominant plant species in a mixed-grass prairie. *Proceedings of the National Academy of Sciences of the United States of America* **111** (43), 15456–15461. doi:10.1073/pnas.1414659111.

Zhang, D. L., Huang, J. P., Guan, X. D., Chen, B. & Zhang, L. 2013a Long-term trends of precipitable water and precipitation over the Tibetan Plateau derived from satellite and surface measurements. *Journal of Quantitative Spectroscopy & Radiative Transfer* **122**, 64–71. doi:10.1016/j.jqsrt.2012.11.028.

Zhang, L., Guo, H. D., Ji, L., Lei, L. P., Wang, C. Z., Yan, D. M., Li, B. & Li, J. 2013b Vegetation greenness trend (2000 to 2009) and the climate controls in the Qinghai-Tibetan Plateau. *Journal of Applied Remote Sensing* **7**, 073572. doi:10.1117/1.jrs.7.073572.

Zhang, K., Kimball, J. S., Nemani, R. R., Running, S. W., Hong, Y., Gourley, J. J. & Yu, Z. B. 2015 Vegetation greening and climate change promote multidecadal rises of global land evapotranspiration. *Scientific Reports* **5**. doi:10.1038/srep15956.

Zhang, Y., Zhang, C. B., Wang, Z. Q., Chen, Y. Z., Gang, C. C., An, R. & Li, J. L. 2016 Vegetation dynamics and its driving forces from climate change and human activities in the Three-River Source Region, China from 1982 to 2012. *Science of the Total Environment* **563**, 210–220. doi:10.1016/j.scitotenv.2016.03.223.

Zhang, L. X., Fan, J. W., Zhou, D. C. & Zhang, H. Y. 2017 Ecological protection and restoration program reduced grazing pressure in the Three-River Headwaters Region, China. *Rangeland Ecology & Management* **70** (5), 540–548. doi:10.1016/j.rama.2017.05.001.

Zhu, Z. C., Piao, S. L., Myneni, R. B., Huang, M. T., Zeng, Z. Z., Canadell, J. G., Ciais, P., Sitch, S., Friedlingstein, P., Arneth, A., Cao, C. X., Cheng, L., Kato, E., Koven, C., Li, Y., Lian, X., Liu, Y. W., Liu, R. G., Mao, J. F., Pan, Y. Z., Peng, S. S., Penuelas, J., Poulter, B., Pugh, T. A. M., Stocker, B. D., Viovy, N., Wang, X. H., Wang, Y. P., Xiao, Z. Q., Yang, H., Zaehle, S. & Zeng, N. 2016 Greening of the Earth and its drivers. *Nature Climate Change* **6** (8), 791–795. doi:10.1038/nclimate3004.

First received 31 December 2018; accepted in revised form 10 July 2019. Available online 23 August 2019

Hydraulic head uncertainty estimations of a complex artificial intelligence model using multiple methodologies

E. Tapoglou, E. A. Varouchakis, I. C. Trichakis and G. P. Karatzas

ABSTRACT

The purpose of this study is to examine the uncertainty of various aspects of a combined artificial neural network (ANN), kriging and fuzzy logic methodology, which can be used for the spatial and temporal simulation of hydraulic head in an aquifer. This simulation algorithm was applied in a study area in Miami – Dade County, USA. The percentile methodology was applied as a first approach in order to define the ANN uncertainty, resulting in wide prediction intervals (PIs) due to the coarse nature of the methodology. As a second approach, the uncertainty of the ANN training is tested through a Monte Carlo procedure. The model was executed 300 times using different training set and initial random values, and the training results constituted a sensitivity analysis of the ANN training to the kriging part of the algorithm. The training and testing error intervals for the ANNs and the kriging PIs calculated through this procedure can be considered narrow, taking into consideration the complexity of the study area. For the third and final approach used in this work, the uncertainty of kriging parameter was calculated through the Bayesian kriging methodology. The derived results prove that the simulation algorithm provides consistent and accurate results.

Key words | artificial neural networks, Bayesian uncertainty, fuzzy logic, kriging, uncertainty analysis

E. Tapoglou (corresponding author)
E. A. Varouchakis
I. C. Trichakis
G. P. Karatzas
School of Environmental Engineering,
Technical University of Crete,
Chania 73100,
Greece
E-mail: *etapoglou@isc.tuc.gr*

INTRODUCTION

Data-driven models and statistical models have both been extensively used in groundwater modeling, especially in the last decades (Delbari *et al.* 2014; Maiti & Tiwari 2014). A new concept, presented in Tapoglou *et al.* (2014), allowed for the combination of two separate methodologies, artificial neural networks (ANNs) and kriging for the spatial and temporal simulation of hydraulic head using a fuzzy logic system in a 7,800 m^2 area covering Bavaria region, Germany. In this study, to examine the uncertainty of various aspects of the combined methodology and to prove the effectiveness of this approach providing better simulation results, the same methodology was applied in a smaller study area in Miami, Dade County, USA (1,460 m^2). Preliminary results of this study are presented in Tapoglou *et al.* (2018).

To provide a measure of the uncertainty, an analysis comprising of three different methodologies was performed. The first methodology corresponded to the uncertainty of kriging parameters, while the remaining two estimated the uncertainty derived from the ANNs. The remainder of this paper is organized as follows: the last part of the introduction presents briefly the simulation algorithm while in the next section the methodology is described, followed by some general information about the study area and the numerical results of the methodology. The last part includes the most important findings and the conclusion of this study.

Simulation algorithm

The simulation algorithm used in this work is an evolution of the one presented in Tapoglou *et al.* (2014) and involves the use of ANNs, kriging and a fuzzy logic methodology in order to simulate the water table changes in a temporal and spatial scale. The main idea of the conceptual

doi: 10.2166/hydro.2019.137

framework is that the ANN simulates the temporal changes in the hydraulic head, while kriging interpolates the results in a spatial manner to cover the whole area, creating a two-dimensional map from a set of point measurements.

The temporal simulation took place in multiple locations where data were available, diversifying the inputs as necessary in order to acquire the optimal ANN results. Owing to these different needs for input parameters and in order to get the best possible results, one ANN was developed for every location with available hydraulic head data. Many studies have followed a similar approach in the past, since it provides a higher flexibility in input parameter selection, combined with reduced computational effort for training compared to creating a single ANN for all the outputs (Sahoo *et al.* 2017).

Data-driven methods, like ANNs in this study, require the existence of a large dataset of input and target values. These prerequisites include information on meteorological and hydrological parameters measured on a daily basis (same as the time step in this study) as input data and hydraulic head measurements in the same time step as the target values of the output parameter during training. The selection of input parameters is focused on those who are linked directly or indirectly to the water budget. The ANNs used available data both for training and simulation of the current state, while the trained ANNs are able to predict future status of the aquifer, when evaluating different scenarios.

The results of all ANNs were subsequently interpolated using the kriging methodology. Ordinary kriging had been used, and the linear, exponential and power-law variograms were tested. The interpolation was conducted in a grid inside the convex polygon of the wells. This grid is a rectilinear grid, where all observation points fall exactly at a node. The nodes of the grid where observed data are not available are the prediction points. To exclude from the simulation grid nodes that fall far outside the study area and far from any observation point, Delaunay triangles were created from the known data locations. Nodes that fell outside the area covered by at least one triangle were not considered as prediction points.

To improve the initial results, a fuzzy logic system was used as a means to combine the two methodologies. More specifically, the neighborhood used in kriging interpolation was defined through fuzzy logic, combining the distance between the prediction point and data points with the ANN training and testing error in every location. For every prediction point involved in the training and testing process and for every time step, a different variogram is calculated and the hydraulic head is simulated. A detailed description of the functionality of the original methodology, followed also in this study, is found in Tapoglou *et al.* (2014). The main enhancement of the methodology in this study consists of the improved selection of ANN input parameters to reduce ANN error and the subsequently applied metamodel for uncertainty quantification of the different components of this study's methodology.

METHODOLOGIES

The methodologies followed in the present paper will cover the uncertainties involved in the simulation algorithm described in Tapoglou *et al.* (2014). Model uncertainty can be attributed to many reasons. In this work, we have identified and grouped the sources of uncertainty in two main categories, uncertainty derived from the ANN training process and uncertainty due to the kriging interpolation methodology.

In the past, various methodologies have been used in different applications for the calculation of model uncertainty. Monte Carlo simulation and bootstrap methods are very common and can give a perspective of the uncertainty involved. The objective of an uncertainty analysis is to show the effects of model inputs on the model simulation results (Eslamian 2014). In this study, different methodologies were used for the determination of uncertainty derived from different parts of the developed algorithm (Figure 1). As far as the ANN part of the simulation algorithm is concerned, two methodologies were followed: the percentile methodology and a Monte Carlo simulation for the uncertainty of the ANN's training. The percentile methodology is simple and easy to apply during ANN training, but studies have shown that it produces larger confidence intervals (Jiang *et al.* 2013). The second methodology produces narrower intervals which take into account the value and individual characteristics of each data point, producing more meaningful results. These results later became the

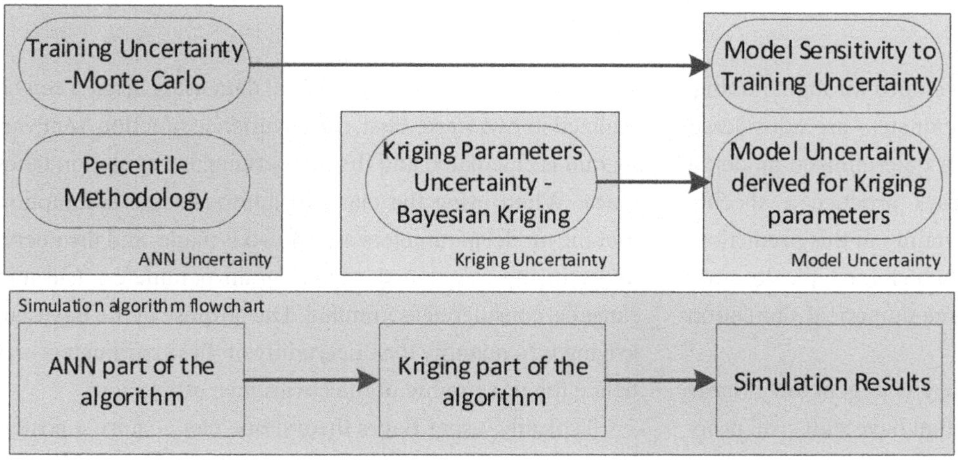

Figure 1 | Methods tested in each simulation step.

input for the Kriging part of the algorithm, constituting a sensitivity analysis of the kriging interpolation method to the ANN training. ANN training uncertainty can be attributed to a number of factors ranging from initialization of the weights and other processes involving random number generators. Uncertainty due to the Kriging parameters is also studied using the Bayesian kriging methodology.

Uncertainty in ANN results

Uncertainty due to the ANN model has been extensively studied in the past, using both bootstrap-based methods (Trichakis *et al.* 2011; Kasiviswanathan & Sudheer 2013) and Monte Carlo simulation methods (Jiang *et al.* 2013; Dehghani *et al.* 2014).

In ANNs, the source of uncertainty may be attributed to two reasons, model structure and training dataset (Dybowski & Roberts 2001). In this study, the model structure is fixed and only the uncertainty derived from the training dataset is considered. This uncertainty can be attributed to not only the imperfection in data collection tools or techniques, but also to the available amount of training data and how representative of the reality they are. Moreover, not all the possible realizations of the variables involved are available in the training set, as well as the training set itself is only one of a large number of possibilities. For example, the training dataset can be short or might not contain extreme values that are nevertheless possible to occur in the area. This will increase the uncertainty of the model when it is extrapolating.

By using these two methodologies, the prediction intervals (PIs) can be calculated.

Percentile prediction intervals

In this approach, the PIs can be calculated through the statistical characteristics of the model errors that occurred when reproducing those observed in the field data (Shrestha & Solomatine 2006).

This approach takes a large number of simulated variables of interest and orders them from smallest to largest. A 90% PI is then computed by identifying the lower PI as the 5th percentile and the upper PI as the 95th percentile. This leaves 90% of the simulated estimates within this range, while dividing the remaining 10% of the simulated values equally into the upper and lower tails. The advantage of this method is that it does not make any distributional assumptions and it does not require the distribution to be symmetric. However, evaluating the PI using this methodology yields larger confidence intervals compared to other methodologies.

In order to use this methodology to evaluate every ANN's PIs, for every ANN involved separately, the difference between the real and the simulated values is calculated for every time step studied. These differences are then sorted, and the 5th and 95th percentile values are identified. The values are then added to all simulated outputs, concluding in two new time series which represent the 5% upper and lower limit of the prediction.

Monte Carlo method

Conventionally, a single ANN is developed through training, and once completed, the ANN parameters are considered fixed. This way, an ANN becomes a deterministic model; a given input data vector will always produce a specific output value. To calculate the uncertainty in this prediction, Monte Carlo methods can be used, which are typically computer-based techniques for brute force numerical simulation of probabilistic processes.

The first step in this methodology is to generate a large number of sets of random numbers that have statistical properties similar to those of the real variables. If there are actual measurements describing the variables, it is also possible to select randomly among the measured values, as an approximation to the distribution underlying the particular values measured in the field. Using each set of values as input variables, the output variable is calculated. This process is then repeated several times to test whether the conclusions drawn are sensitive to the particular random values that have been generated. The number of repetitions necessary to achieve a particular level of precision is problem-dependent. The more iterations are conducted, the more accurate the prediction intervals will be. However, the computational costs may be enormous for a large number of iterations.

The uncertainty of the ANN results in this case can be assessed mainly by the random initialization of values (i.e. the ANN synaptic weights), as well as in the random selection of training and testing datasets. To investigate the uncertainty in the predictions of the ANN, multiple different executions of the algorithm, with random initial values for the neural weights and different training and testing sets, were performed. All the random number generators were set to choose a random integer number with uniform distribution. The examination of the uncertainty attributed to the architecture and structure of the ANNs would require large computational cost; hence, it is not examined in this study.

Uncertainty in kriging results

The uncertainty of the kriging algorithm is evaluated through the Bayesian kriging methodology, which can be used to estimate the uncertainty attributed to the kriging parameters.

Bayesian kriging

The process of interpolating spatial data using kriging can be realized in two steps. First, the covariance function/semivariogram is calculated and then the kriging interpolation takes place. When using the classical approach, the assumption that the model parameters are known is made, and the uncertainty in the calculation of variogram parameters (sill and range) is considered as minimal. The purpose of the Bayesian kriging is to quantify the uncertainty of these parameters and hence the uncertainty of the covariance estimation.

Typically, using Bayes theory, one can acquire a posteriori distribution of parameters using the prior distribution and the initial values of the parameters. Using a posteriori distribution, a prediction of the parameters can be done. In this way, a model 'averaging' over the unknown model parameters is constructed.

However, when using spatial data, the prior cannot be calculated easily. In Bayesian kriging (Pilz & Spöck 2008), the posterior distribution of the parameters involved (semivariogram and/or transformation parameters) is specified by means of simulations, which correspond to the uncertainty of covariance parameters.

To implement Bayesian kriging methods, first the appropriate variogram model must be fitted to the experimental data, using the standard procedure followed when using the variogram estimation (first step in kriging interpolation methodology). Using the parameters derived and the covariance function of the model, the covariance matrix is constructed.

A common method for testing a statistical model is the use of artificial data. In order to do so, the statistical properties have to be embedded in the dataset and then the model must be examined for the presence of these effects and how they behave under different experimental conditions.

Uniform random numbers generation is the first step in this procedure. A generally approved method of correlating random numbers with a known covariance matrix (C) is by finding a matrix U, according to the following equation:

$$U^T \cdot U = C \tag{1}$$

Matrix U is a triangular matrix and can be derived using decomposition methods. Matrix decomposition is commonly used in the Monte Carlo method for simulating systems with multiple correlated variables.

Using this matrix, correlated random numbers R_c can be generated from uncorrelated numbers R, by multiplying them with this matrix in the following equation:

$$R_c = R \cdot U \quad (2)$$

These correlated random values have the same statistical characteristics with the initial, observed in the field, data. These random values are then fitted into a variogram and the value in an unknown location is estimated using kriging.

Cholesky decomposition is one way of providing these random correlated values. The covariance matrix of a vector Y can be given as follows: $C = E(YY^T)$. If X is a random vector, consisting of uncorrelated random values uniform in [0,1], then $E(XX^T) = I$.

The Cholesky decomposition of the correlation matrix is given as $C = L \cdot L^T$. Note that it is possible to obtain a Cholesky decomposition of C since by definition the covariance matrix C is symmetric and positive definite.

When the random vector X is multiplied by L ($Z = L \cdot X$), statement (3) can be formed:

$$\begin{aligned} E(Z \cdot Z^T) &= E((L \cdot X) \cdot (L \cdot X^T)) = E(L \cdot X \cdot X^T \cdot L^T) \\ &= L \cdot E(X \cdot X^T) \cdot L^T = L \cdot I \cdot L^T = L \cdot L^T = C \end{aligned} \quad (3)$$

In this way, it can be proved that the random vector Z has the desired covariance matrix C.

The distribution of the simulated values using the kriging of different correlated random values can be used in order to derive the confidence intervals of the process. In this way, the distribution of the parameters involved (sill, range etc.) in the process is used as a prior knowledge, while the posterior distribution reflects the uncertainty of the covariance estimation (Pilz & Spöck 2008).

Uncertainty in simulation algorithm results

The overall simulation algorithm uncertainty was evaluated following two methodologies. The first one included the use of the Monte Carlo results of the ANN uncertainty analysis and can also be characterized as a sensitivity analysis of the kriging to the ANN training. The results of the analysis described in the section 'Bayesian kriging' can also serve as uncertainty analysis of the simulation algorithm to the kriging parameter estimation process (see Figure 1).

Monte Carlo method

In this case, the results of the Monte Carlo simulation in training the ANNs were used. More specifically, the multiple different training results (different neural weights) were used for the evaluation of the hydraulic head, resulting in multiple different values for every time step and for every prediction point. Using the 5th and 95th percentile, it is possible to determine the 90% prediction interval. These results were then passed forward to the kriging part of the algorithm, generating the range the simulation results are within, for every prediction point and time step separately.

STUDY AREA

The area in which the proposed methodology is applied is located in Miami, Dade County, Florida, USA. A total number of 30 wells within Biscayne aquifer inside or around the urban area of Miami were used for the water table simulation. The available data covered the time period from 25 April 2010 to 22 February 2014, with daily time step, and included the hydraulic head, the discharge and water level in four nearby rivers and meteorological data from two stations. The data used in this study regarding water levels were acquired through United States Geological Survey (USGS), while the meteorological data from the National Climatic Data center of the National Oceanic and Atmospheric Administration (NOAA). All the data used were directly or indirectly linked to the aquatic equilibrium. The locations of (a) 30 hydraulic head measuring stations, (b) four surface water stations and (c) two meteorological stations are depicted in Figure 2.

Biscayne aquifer is one of the most productive aquifers in the world; it is unconfined and shallow with the average hydraulic head often being less than 1 m. It is wedge-shaped, extending from a few meters thickness at the west coast to as much as 45 m thick near the east coast. As analyzed by USGS (2014), geologically, the study area is dominated by Miami Limestone and Kay Largo Limestone. It consists of

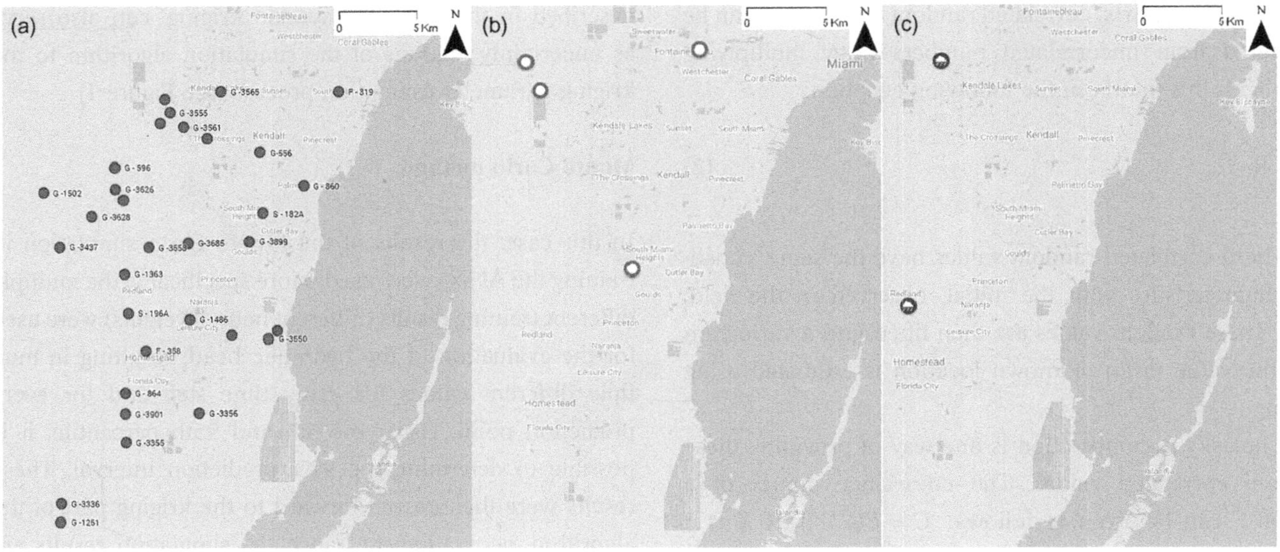

Figure 2 | (a) Hydraulic head measuring stations, (b) surface water measuring stations and (c) meteorological stations.

very porous limestone, created by secondary dissolution (Fish & Stewart 1991).

Water supply and water management systems, like regulated channels, can affect the spatial correlation of water level between monitoring wells. Natural factors can affect the spatial correlation of the data for the groundwater level. One such natural factor, rainfall, can be intense and extremely localized. Therefore, distant wells are less likely to show influence from the same rainfall events. Land use is another factor that could affect correlation. Rainfall-runoff in urbanized areas is different than in natural settings. In urban settings, pavement and storm sewers direct water into streams and channels; in natural settings, rainfall can readily infiltrate into the soil. Evapotranspiration in urban settings is also likely to be different than the one in a rural area. Because of the combined effect of all of these factors, correlation of the water level data in distant wells is generally expected to be small.

RESULTS

Preprocessing of available data

An important factor that can easily be neglected when choosing ANN parameters is that of the time lag, i.e. the time between an event happening (for example, a rainfall occurence) and the appearance of its effect in the hydraulic head dataset. When rainfall and change of groundwater level are the parameters under consideration, this lag represents the time that it takes for the water to penetrate the soil and reach the aquifer. The correlation coefficient between time series A for a parameter and time series B for the hydraulic head change can be used to determine the optimal lag between the parameter and the hydaulic head change in the following equation:

$$\mathrm{Correl}(A, B) = \frac{\sum (a - \bar{a}) \cdot (b - \bar{b})}{\sqrt{\sum (a - \bar{a})^2 \cdot (b - \bar{b})^2}} \tag{4}$$

where a denotes the value of the parameter, b denotes the value of the hydraulic head change and $\bar{a}$, $\bar{b}$ are the average values of time series A and B, respectively.

The selection of the lag of each parameter was determined by the highest correlation between input and output of the ANN. Parameters considered to have time lag include precipitation (rainfall, snowfall, etc.) and parameters related to surface water recharge or discharge (water level in a lake or river, flow rate, etc.). More than one time lag can have approximately equal correlation coefficient between an input parameter and the hydraulic head change.

The selection of input parameters relies on the outcome of the correlation coefficient analysis. In some cases, the time lag between the input and the output parameter is apparent, and the input parameter with the appropriate lag (the one which has the highest correlation) can enter the model. In other cases, the effect of the input parameter is apparent in more than 1 day. This is particularly true for rainfall, since the effect of a rainfall event, due to the relatively small groundwater velocities, is visible for more than 1 day, creating a hydrograph, similar to the one describing surface runoff. This effect is reflected in the values of the correlation coefficients between the input and the output variables (in this case, rainfall and hydraulic head change). These coefficients get their highest values in a range of consecutive days rather than in a single day with a single time lag. Therefore, for these parameters, all the values of the input parameter with a different time lag that had an effect on the output constituted part of the input dataset. The input parameters to the ANNs were a combination of the parameters with the highest correlation coefficient. The ANN output parameter is always the hydraulic head.

For each one of the ANNs, the following input parameters were used:

- Precipitation for two meteorological stations (3 days)
- Soil temperature
- Air temperature
- Humidity
- Surface water levels from two stations (2 days)
- Discharge from one station (best day)

In most occasions, the best time lag, in terms of correlation coefficient, was between 0 and 1 day for the surface water level and discharge, and for the rainfall, this time lag varied from 0 to 3 days, with the most common being 0 and 1 day. These results are to be expected in this study area, since the aquifer is highly productive, with water flowing through easily. Therefore, a small time lag of 0–1 days and at most 3 days also seems to have a physical meaning. The correlation coefficient values, of the time lagged parameters in respect to the output parameter that were finally used, varied from 0.4 to 0.8.

There is not a universal rule about the optimal architecture of an ANN. However, according to the rule of Fine (1999), using three times as many training patterns as network parameters (weights) is adequate to achieve good generalization. Taking into account that nine input parameters, one output and 1,100 training sets were used, two architectures were examined; the first one with one hidden layer with 32 hidden nodes (Architecture 1) and the second one with two hidden layers with 17 and 12 hidden nodes, respectively (Architecture 2).

Three variogram models were also tested for the best architecture, the linear, the exponential and the power-law, in order to choose the most appropriate for the data at hand. These variograms were chosen in order to have one variogram model per type: the simplest models, the linear, one differentiable, the exponential, and one non-differentiable, the power-law.

The root-mean-squared error (RMSE) and the Nash-Sutcliffe efficiency (NSE) coefficient (Equation (5)) were used as a measure of error.

$$\mathrm{NSE} = 1 - \frac{\sum_{t=1}^{T} (Q_{\mathrm{m}}^{t} - Q_{\mathrm{o}}^{t})^2}{\sum_{t=1}^{T} (Q_{\mathrm{o}}^{t} - \overline{Q_{\mathrm{o}}})^2} \tag{5}$$

where Q_{o}^{t} is the observed value at time step t, and Q_{m}^{t} is the simulated value at the same time step.

Simulation results

The ANN performances for each one of the two possible architectures (Architecture 1 – one hidden layer with 32 hidden nodes and Architecture 2 – two hidden layers with 17 and 12 hidden nodes) are presented in terms of RMSE in Table 1.

In terms of average minimum and maximum values, the second architecture has better results, both for the training and testing error. To evaluate the performance of each one of the 30 ANNs developed, they were divided into three categories, depending on their training and testing RMSE performance (low, medium and high RMSE values). The

Table 1 | Training and testing RMSE for two architectures

RMSE (m)	Architecture 1	Architecture 2
Average training error	1.16×10^{-3}	6.63×10^{-4}
Average testing error	1.20×10^{-3}	7.92×10^{-4}
Maximum training error	2.19×10^{-3}	1.78×10^{-3}
Maximum testing error	3.42×10^{-3}	2.98×10^{-3}
Minimum training error	2.46×10^{-4}	2.23×10^{-4}
Minimum testing error	2.63×10^{-4}	1.26×10^{-4}

results will be presented for one representative well for each category. The training RMSE, testing RMSE and the overall NSE coefficient for the representative well of each category are summarized in Table 2.

The first category includes the ANNs with the best performance, represented by well 11. The second category has average results, represented by well 3, while the third category has the worst RMSE results amongst the ANNs studied. It should be noted that the NSE values are not very close to the optimal value, which equals one ($NSE_{optimal} = 1$). This can be attributed to the low hydraulic head values that should be simulated by the model. As indicated by Schaefli & Gupta (2007), it is not possible to achieve very high NSE values when simulating small absolute values; however, it is an indication of the comparative performance of the model. Nevertheless, the NSE values are well above zero, which means that our model describes much better the output than a simple average output value could have.

The simulation results as hydraulic head for the three representative wells are depicted in Figure 3.

Next, the algorithm, with the use of fuzzy logic, defines the appropriate neighbors, which will be used for the kriging interpolation. The number of neighbors for each prediction point is set to 20. This number should be chosen carefully so as to allow for the 30 pairs per distance class, which, according to Journel & Huijbregts (1978), are necessary in order to acquire a good fitting of the experimental variogram to the theoretical model.

To evaluate the results, cross-validation was performed for all three variograms studied. For every data point, for 10% of the time steps, the observed values at that data point were ignored, and the algorithm follows as if the data point does not exist. In this way, three values are available: the observed in the field, the ANN simulated and the kriging simulated. In the end of the process, the kriging simulated values are compared to the real and ANN simulated, in order to derive an independent measure of performance. In terms of three error indicators, RMSE, mean absolute error (MAE) and bias, the results are 1.14, 0.89 and −0.0485 for the linear variogram, 0.962, 0.7047 and 0.0499 for the exponential and 0.98, 0.7422 and 0.0446 for the power correspondingly.

For all three error indicators used, the exponential variogram yielded better results and will be used henceforth. The cross-validation results can be summarized in Figure 4, where observed values are plotted against the simulated hydraulic head. The closer the simulation results are in the $x = y$ line, the better the simulation is.

As it is apparent in Figure 4, the kriging interpolation tends to correct any misjudgments of the hydraulic head performed by the ANNs. The improvement in the ANN results is also reflected in the error indicators. More specifically, the RMSE, NSE and MAE values of the data used for the cross-validation when only using ANNs are 0.58, 0.59 and 0.77, while when using ANNs together with kriging are 0.8, 0.41 and 0.49, respectively.

Uncertainty results

ANN prediction confidence intervals

Using the initial simulation results, it is possible to calculate the maximum difference between observed and simulated values. By applying the 95% and 5% percentile error to all simulated values, the 95% and 5% percentile of the simulation results can be calculated. The prediction interval width for the three representative wells is 1.18 (−0.43 to 0.75), 1.15 (−0.59 to 0.56) and 1.29 (−0.58 to 0.71) for wells 11, 3 and 18 correspondingly. A noteworthy fact is that with this methodology, each well gets a different width of prediction interval, and the fact that while the results for one well may be better in terms of RMSE, there is no guarantee that the interval width will also be narrower. This width

Table 2 | Performance of the three representative wells

Category number	Representative well	Training error of the representative well (m)	Testing error of the representative well (m)	NSE of the representative well	Number of wells within the category
1	11	2.77×10^{-4}	3.13×10^{-4}	0.69	9
2	3	5.68×10^{-4}	4.19×10^{-4}	0.65	13
3	18	1.11×10^{-3}	6.23×10^{-4}	0.58	8

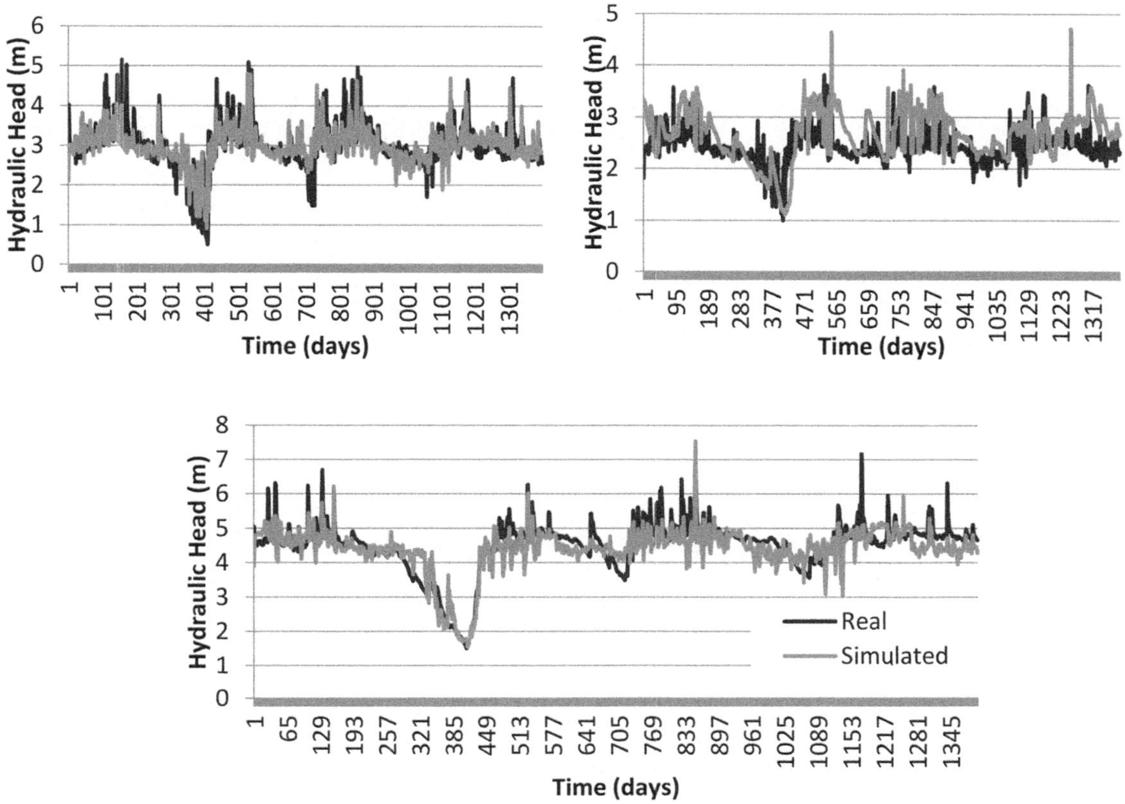

Figure 3 | Real and simulated hydraulic head for wells 11, 3 and 18.

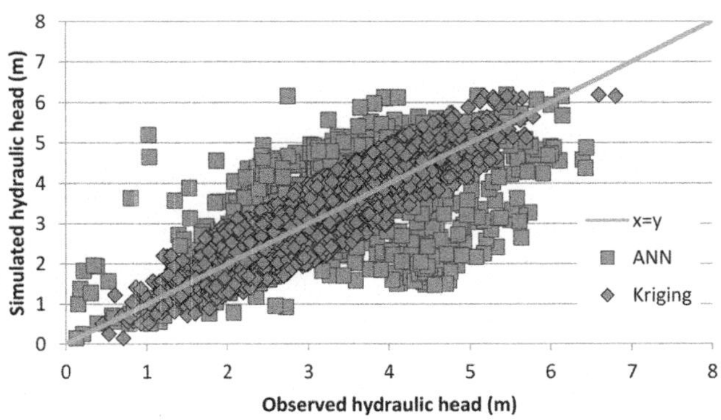

Figure 4 | Cross-validation kriging and ANN results versus observed data.

depends on the sensitivity of the output to the ANN training. In this case, well 11, albeit having an RMSE lower than well 3, ends up with slightly wider prediction intervals.

The results for the three representative wells are presented in Figure 5.

While the 90% prediction intervals were constructed using the training dataset, this does not necessarily mean that the testing data will always fall within these intervals as well. For the vast majority of available data points, especially in wells 3 and 18, the observed values of the testing dataset fall within the 90% prediction interval of the ANN, yet the total coverage is lower than the nominal value of 90%. In well 11, which has the best fit of all the other wells, only a few data points, and more specifically at some of the lowest observed values, have real values not within the prediction interval. The coverage of these intervals with respect to the observed values was, therefore, the highest for well 11 (82%) and lower for well 3 (75%) and for well 18 (79%).

ANN training model uncertainty – Monte Carlo

To evaluate the uncertainty of the ANNs training, the process described in section 'Uncertainty in ANN results' was performed 300 times, with each case having different training and testing datasets, as well as different initial random neural weights. For all wells used, the resulting box-plots of this process can be summarized in Figure 6, for the training and testing error, respectively.

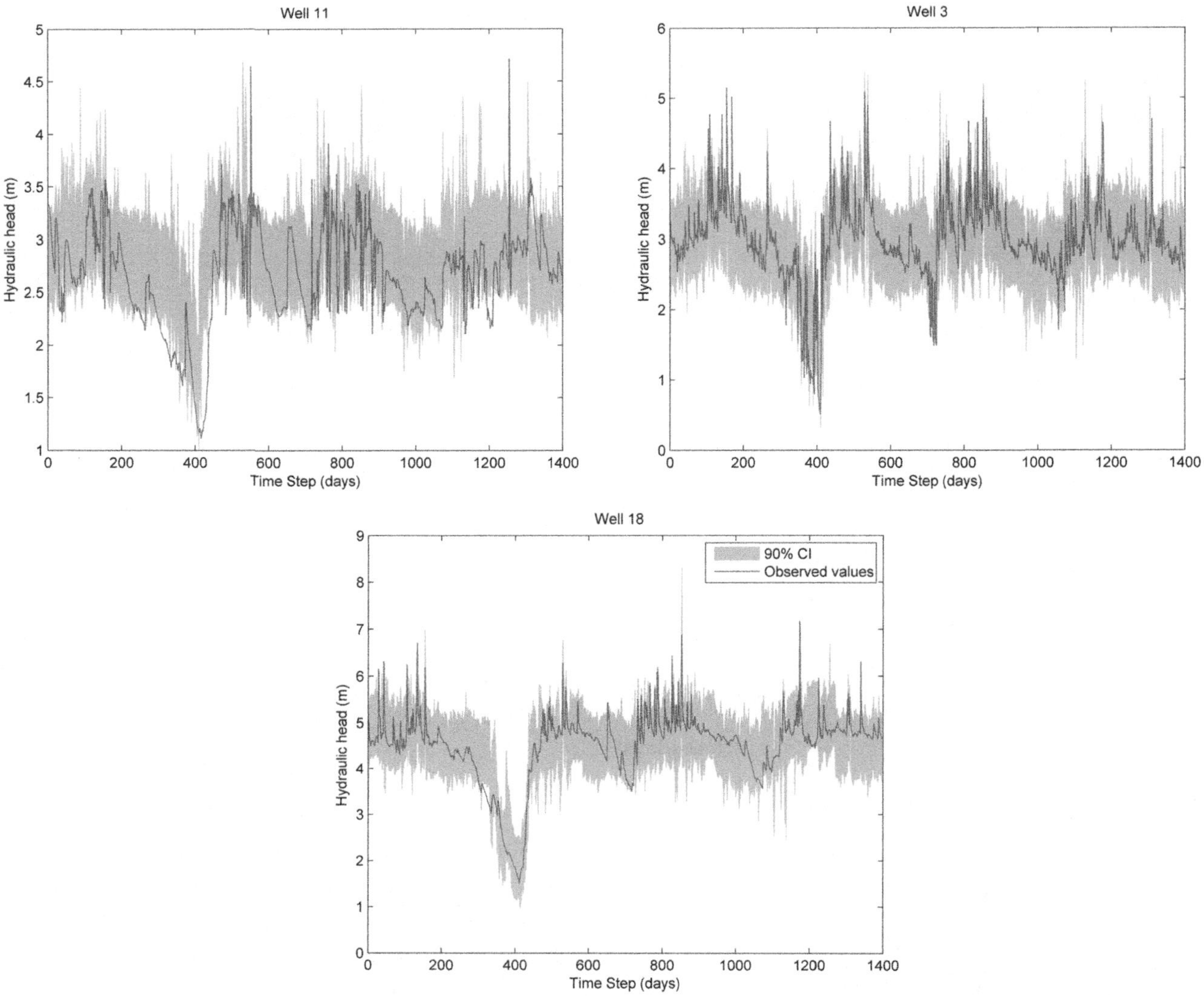

Figure 5 | 90% confidence intervals for ANN simulation.

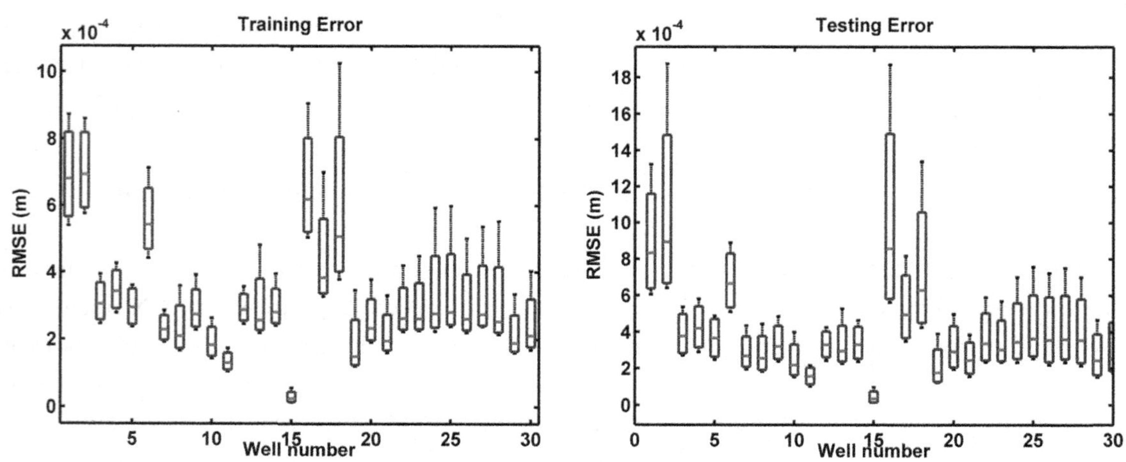

Figure 6 | Training and testing error range for training uncertainty for 30 wells.

The training error range, as expected, is smaller than the testing in most of the cases (26/30 wells). The results in terms of the representative wells are calculated and presented, together with those observed in the field values, in Figure 7.

For well 11, 94.5% of the observed values are within the 90% prediction interval. As the ANN performance deteriorates, this percentage shrinks to 90.9% for well 3 and 88.8% for the worst performance ANN category representative, well 18. The average width of the prediction intervals for each one of the representative wells are 0.153, 0.243 and 0.452 for wells 11, 3 and 18 correspondingly.

By comparing these prediction interval widths for the two different methodologies, it can be noted that the uncertainty derived for the training of the ANNs is only a small fraction of the model uncertainty calculated through the percentile methodology. This may also be attributed to the coarse nature of calculations in the percentile methodology. The second method creates narrower prediction intervals with higher coverage percentages, even for low performing ANNs, thus justifying the extra computational cost needed compared to the simple prediction confidence intervals of the previous section.

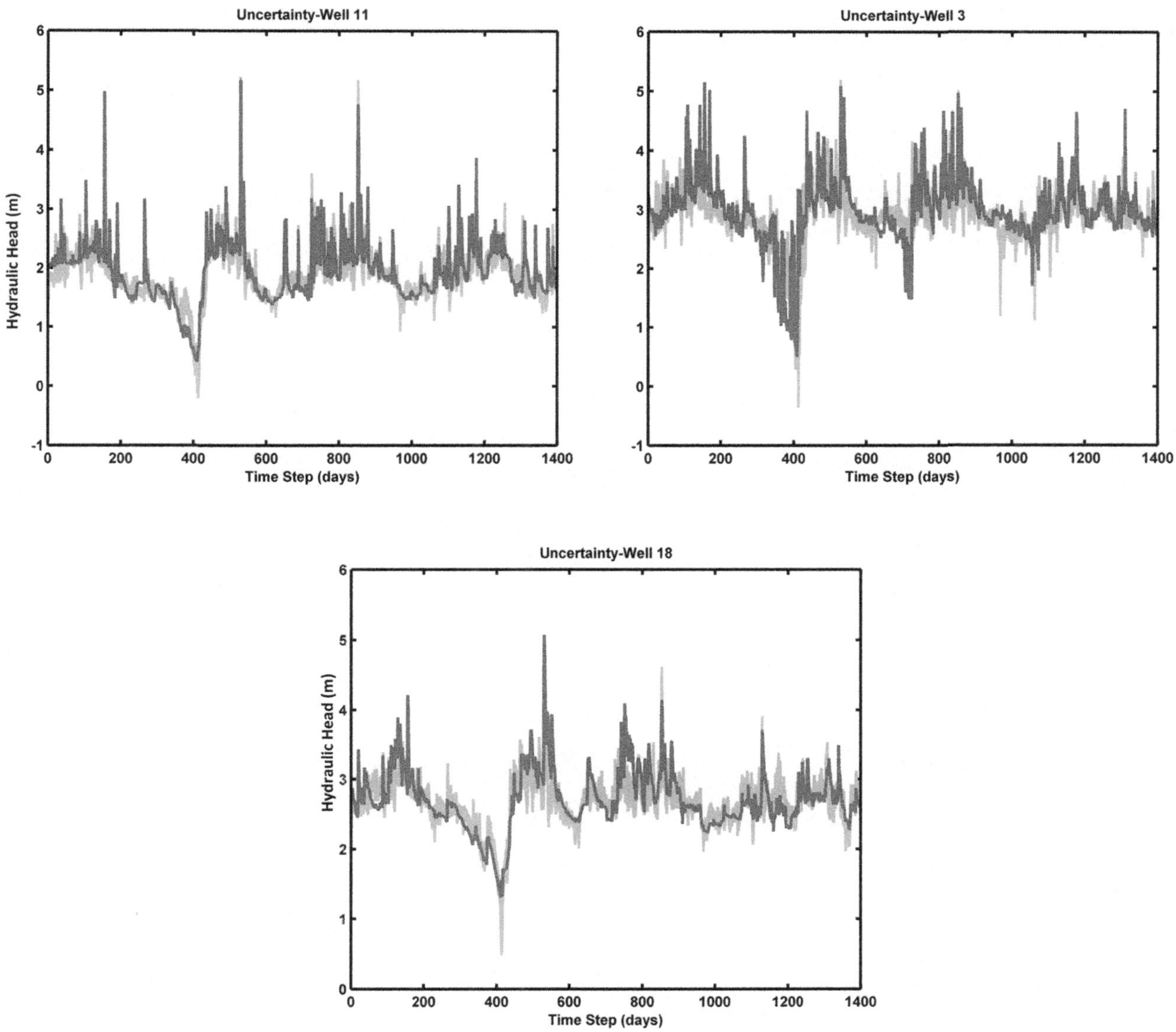

Figure 7 | Training uncertainty for three representative wells: in red, the observed in the field value and with blue, the prediction range. Please refer to the online version of this paper to see this figure in colour: http://dx.doi.org/10.2166/hydro.2019.137.

Uncertainty of the simulation algorithm due to ANN training uncertainty

For all 300 implemented trainings, the kriging part of the algorithm was also executed, resulting in a prediction interval for every studied prediction point. In Figure 8, the 90% prediction intervals for three random prediction points (namely 200, 400 and 600) are presented.

Depending on the prediction point and its proximity to observation points with good training error, the prediction intervals are either wider or narrower. More specifically, for the prediction points presented in Figure 9, the average width of the 90% prediction intervals are 0.25, 0.86 and 1.424 for prediction points 200, 400 and 600 correspondingly.

Looking at the location of these prediction points in the map, as the three prediction points fall closer, or further away from observation points with good training error, their prediction interval narrows or widens, respectively. Of the three prediction points, the one closest to observation points with good training error is the prediction point 200, while 400 lies a bit further away from the observation points. Point 600 is the one furthest away from observation data points.

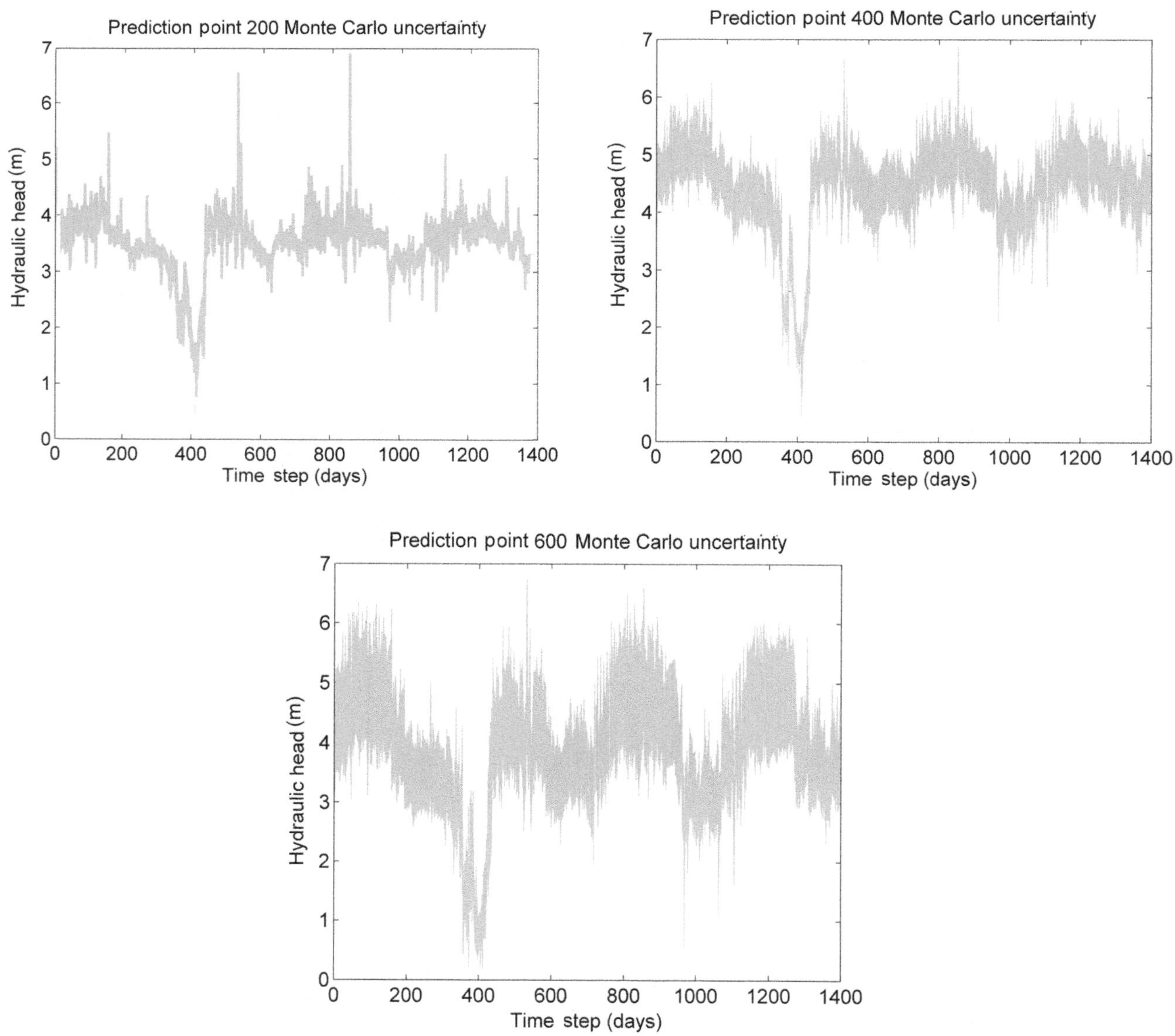

Figure 8 | Kriging 90% prediction intervals for four prediction points generated by different ANN trainings.

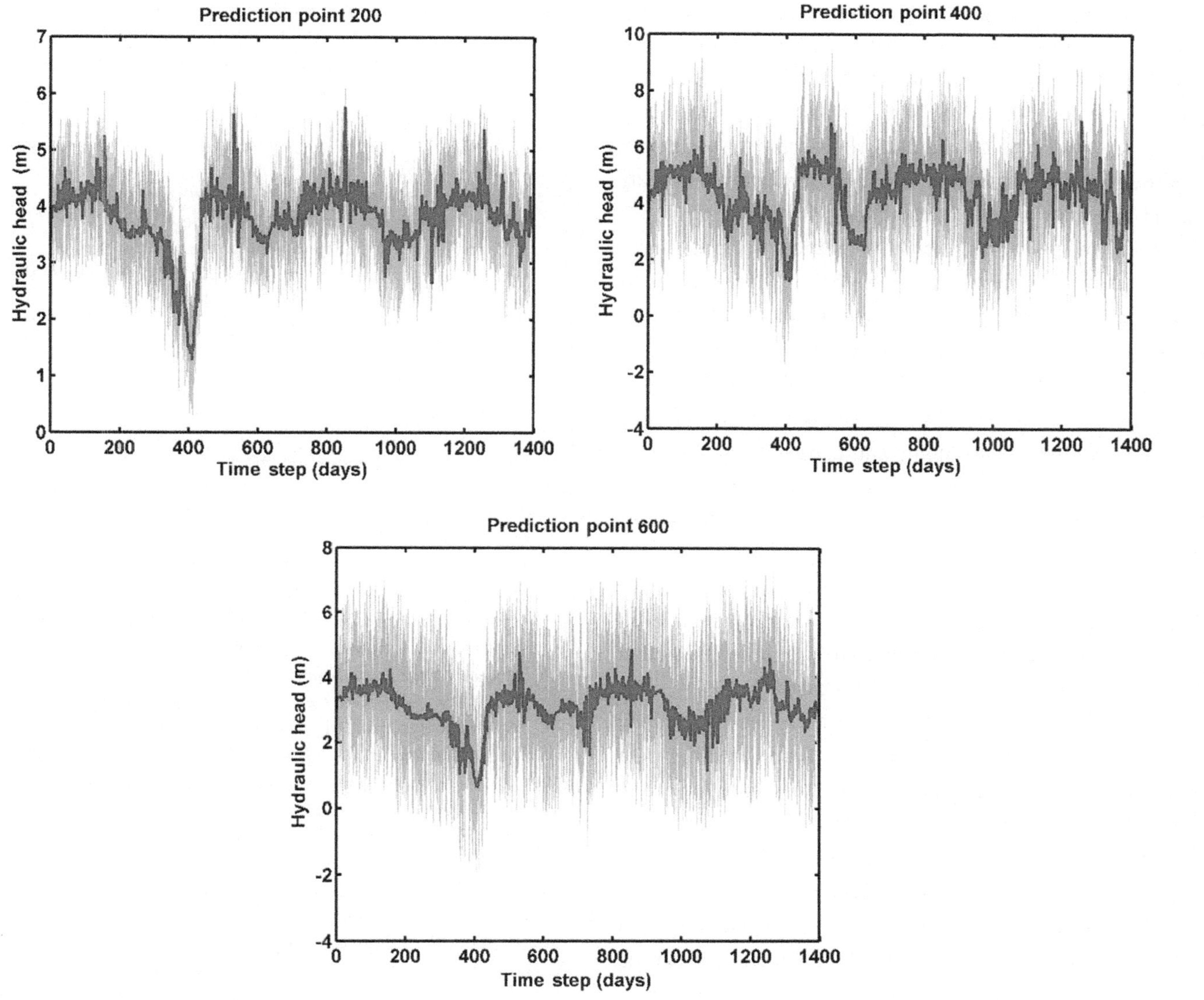

Figure 9 | Kriging parameter uncertainty for three prediction points. Please refer to the online version of this paper to see this figure in colour: http://dx.doi.org/10.2166/hydro.2019.137.

Uncertainty in hydraulic head change due to kriging parameters

Following the procedure described in the Methodologies section, the kriging parameter uncertainty is also calculated. For every prediction point and time step, 500 different data samples and hence 500 different variograms were constructed. The uncertainty was calculated for the algorithm with the use of 20 neighbors defined by the fuzzy logic system and for the exponential variogram. The 90% prediction intervals (blue range) together with the prediction derived from the initially simulated values (red lines) for the same three prediction points as before are presented in Figure 9.

In all cases, the uncertainty of the results, derived from using the described methodology, corresponding to the uncertainty due to the kriging parameter estimation is presented. The average prediction interval width for prediction points 200, 400, and 600 are 1.35 m, 2.32 m and 3.5 m correspondingly.

CONCLUSIONS

In the present study, an uncertainty analysis is performed on a combined ANN–fuzzy logic–kriging methodology for the hydraulic head simulation of an aquifer. The initial simulation of the hydraulic head in a complex study area, in Miami, Dade County, Florida, USA had an average RMSE training error 6.63×10^{-4}m, an average RMSE testing error 7.92×10^{-4} m and cross-validation RMSE of 0.962 m. The uncertainty analysis proved that the simulation algorithm used is both consistent and accurate, especially considering the complexity of the case study and the methodology involved.

Using the percentile methodology, the ANN uncertainty can be calculated. In this case, the 90% prediction intervals are wider, compared to the ones produced by all other methodologies presented in this study. This can be attributed to the coarse nature of the calculations in this methodology.

The Monte Carlo method for assessing the uncertainty attributed to the ANN training and consecutively for sensitivity analysis of the kriging results to the ANN training is also performed. The range of training and testing error varied, without, however, having a large effect on the 90% prediction interval of the hydraulic head, in locations where data were available. The 90% prediction interval of the hydraulic head is also depicted in three prediction points, having narrow intervals in all cases.

Kriging parameter uncertainty reflects the uncertainty attributed to the observed data used. Using artificial data and the Bayesian kriging methodology, this aspect is examined. The results in terms of predicted hydraulic head intervals are close to the simulated results using the real data.

All the above-mentioned uncertainty calculations indicate that the methodology used can provide consistent and reliable results under various conditions; hence, it can be used for groundwater-level simulation, especially in complex study areas where geological information is obscure, making conventional modeling unsatisfactory.

REFERENCES

Dehghani, M., Saghafian, B., Nasiri Saleh, F., Farokhnia, A. & Noori, R. 2014 Uncertainty analysis of streamflow drought forecast using artificial neural networks and Monte-Carlo simulation. *International Journal of Climatology* **34** (4), 1169–1180.

Delbari, M., Amiri, M. & Motlagh, M. B. 2014 Assessing groundwater quality for irrigation using indicator kriging method. *Applied Water Science* **6** (4), 371–381.

Dybowski, R. & Roberts, S. 2001 Confidence intervals and prediction intervals for feed-forward neural networks. In: *Clinical Applications of Artificial Neural Networks*. Cambridge University Press, pp. 298–326.

Eslamian, S. 2014 *Handbook of Engineering Hydrology: Environmental Hydrology and Water Management*. CRC Press, Boca Raton, FL.

Fine, T. L. 1999 *Feedforward Neural Network Methodology*. Springer Science & Business Media, New York.

Fish, J. E. & Stewart, M. T. 1991 *Hydrogeology of the Surficial Aquifer System, Dade County, Florida*. US Department of the Interior, US Geological Survey, Tallahassee, FL.

Jiang, Y., Nan, Z. & Yang, S. 2013 Risk assessment of water quality using Monte Carlo simulation and artificial neural network method. *Journal of Environmental Management* **122**, 130–136.

Journel, A. G. & Huijbregts, C. 1978 *Mining Geostatistics*. Academic Press, London, pp. 40–41.

Kasiviswanathan, K. S. & Sudheer, K. P. 2013 Quantification of the predictive uncertainty of artificial neural network based river flow forecast models. *Stochastic Environmental Research Risk Assessment* **27**, 137–146.

Maiti, S. & Tiwari, R. 2014 A comparative study of artificial neural networks, Bayesian neural networks and adaptive neuro-fuzzy inference system in groundwater level prediction. *Environmental Earth Sciences* **71** (7), 3147–3160.

Pilz, J. & Spöck, G. 2008 Why do we need and how should we implement Bayesian kriging methods. *Stochastic Environmental Research Risk Assessment* **22** (5), 621–632.

Sahoo, M., Das, T., Kumari, K. & Dhar, A. 2017 Space–time forecasting of groundwater level using a hybrid soft computing model. *Hydrological Sciences Journal* **62** (4), 561–574.

Schaefli, B. & Gupta, H. V. 2007 Do Nash values have value? *Hydrological Processes* **21** (15), 2075–2080.

Shrestha, D. L. & Solomatine, D. P. 2006 Machine learning approaches for estimation of prediction interval for the model output. *Neural Networks* **19** (2), 225–235.

Tapoglou, E., Karatzas, G. P., Trichakis, I. C. & Varouchakis, E. A. 2014 A spatio-temporal hybrid neural network-Kriging model for groundwater level simulation. *Journal of Hydrology* **519** (Part D), 3193–3203.

Tapoglou, E., Varouchakis, E. A. & Karatzas, G. P. 2018 Uncertainty estimations in different components of a hybrid ANN–Fuzzy–Kriging model for water table level simulation. In: *HIC 2018, 13th International Conference on Hydroinformatics*, Palermo. Vol. 3, pp. 2042–2050.

Trichakis, I., Nikolos, I. & Karatzas, G. P. 2011 Comparison of bootstrap confidence intervals for an ANN model of a karstic aquifer response. *Hydrological Processes* **25** (18), 2827–2836.

USGS 2014 *Geological Units in Miami-Dade county, Florida*. mrdata.usgs.gov/geology/state/fips-unit.php?code=f12086 (accessed 7 March 2016).

First received 11 December 2018; accepted in revised form 9 April 2019. Available online 29 April 2019

Using very high resolution (VHR) imagery within a GEOBIA framework for gully mapping: an application to the Calhoun Critical Zone Observatory

A. Francipane, G. Cipolla, A. Maltese, G. La Loggia and L. V. Noto

ABSTRACT

Gully erosion is a form of accelerated erosion that may affect soil productivity, restrict land use, and lead to an increase of risk to infrastructure. An accurate mapping of these landforms can be difficult because of the presence of dense canopy and/or the wide spatial extent of some gullies. Even where possible, mapping of gullies through conventional field surveying can be an intensive and expensive activity. The recent widespread availability of very high resolution (VHR) imagery has led to a remarkable growth in the availability of terrain information, thus providing a basis for the development of new methodologies for analyzing Earth's surfaces. This work aims to develop a geographic object-based image analysis to detect and map gullies based on VHR imagery. A 1-meter resolution LIDAR DEM is used to identify gullies. The tool has been calibrated for two relatively large gullies surveyed in the Calhoun Critical Zone Observatory (CCZO) area in the southeastern United States. The developed procedure has been applied and tested on a greater area, corresponding to the Holcombe's Branch watershed within the CCZO. Results have been compared to previous works conducted over the same area, demonstrating the consistency of the developed procedure.

Key words | ecognition, GEOBIA, gully erosion, landforms analysis, soil erosion

A. Francipane (corresponding author)
G. Cipolla
A. Maltese
G. La Loggia
L. V. Noto
Dipartimento di Ingegneria (DI),
Università di Palermo,
Viale delle Scienze, 90142 Palermo,
Italy
E-mail: *antonio.francipane@unipa.it*

INTRODUCTION

Overland flow erosion can be considered as a threshold process occurring only after resistance forces are exceeded (Huggett 1990; Francipane *et al.* 2012). Entity and evolution of this phenomenon depends on several factors, such as the hydrological regime, the geomorphological characteristics, the climate, and the land use of the basin (Pimentel *et al.* 1995; Nearing *et al.* 1999; Wilkinson & McElroy 2007; Francipane *et al.* 2015). All of these factors are related to each other and determine, to a different degree, the extent of the erosive process, changes on soil productivity, landscape evolution, and its variations in space and time (Henderson-Sellers 1994; Kirkby & Cox 1995).

Erosion typically starts as a series of subparallel rills parallel to the slope gradient. Slight variations in surface topography can produce greater depth of flow, resulting in increased erosive forces and in an accelerated erosion, which promotes the evolution of rills into more severe forms of erosion: the gullies (Ritter *et al.* 2011). According to Poesen *et al.* (2003), the total soil losses caused by gully erosion could rise up to 90% in different parts of the world, restricting land use and increasing risk to infrastructure. For all of these reasons, gully erosion and their identification in the territory have always attracted the interest of the scientific community (Valentin *et al.* 2005).

The first attempts to study gullies go back to the 1930s, when Ireland *et al.* (1939) tried to study and describe the conditions governing the development of gullies in the Piedmont of South Carolina. In the 1960s, Tuckfield (1964) described the processes that led to the formation of gullies in the New Forest in Hampshire (England) and Seginer

doi: 10.2166/hydro.2019.083

(1966) presented different methods to quantify gully erosion. Patton & Schumm (1975) tried to describe the relationship between critical valley slope and drainage-basin area and use it as a predictive model to locate those areas of instability within alluvial valleys where gullies could form. In 1977, the Soil Conservation Service (SCS 1966) used a multivariate analysis to predict gully growth, while Stocking (1980) used multiple regression analysis to predict gully-head retreat on 66 gullies in Zimbabwe. Kirkby & Bracken (2009) made an interesting synthetic review of gully morphology and genesis examining the conditions for gully formation, focusing on gully initiation, head and side-wall processes, and making a brief dissertation about implications for gully prevention and remediation.

The advent of digital elevation models (DEMs) led to the birth and diffusion of quantitative land-surface analysis (i.e., geomorphometry) (Grosse *et al.* 2012). For many years, the main problem related to the use of DEMs for land-surface analysis has been their relatively low spatial resolution. The issue seems to be overcome with the advent of very high resolution (VHR) imagery, which started with the launch of IKONOS in 1999 and which finds nowadays its maximum expression in LIDAR data, that allow the derivation of very high resolution digital surface models (DSMs) and bare ground DEMs. These products, which constitute a basic support for activities such as hydrological–hydraulic modeling and the identification of areas exposed to flood risk, have provided the opportunity to develop new methodologies for analyzing Earth surfaces (Tarolli *et al.* 2009) that have found widespread use also in identifying the scars of gullies in the landscape in the last decade (Jackson *et al.* 1988; Ritchie *et al.* 1994; James *et al.* 2007; Noto *et al.* 2017; Francipane *et al.* 2018).

So far, today, there are mainly two different procedures to deal with high-resolution DEMs: a per-pixel approach (pixel-based) and a per-object approach (object-based). The last one is often referred to as object based image analysis (OBIA or GEOBIA if it takes into account the geographical aspect as well). While the pixel-based approach uses the classical rules of map algebra applied to the pixel, the object-based approach uses the concept of the image object, which is defined as a group of neighbor pixels with similar characteristics. The main advantage of such an approach is that object characteristics (e.g., mean value, standard deviation, ratio, etc.) and features (e.g., shape and texture) can be calculated and used to differentiate land cover classes with similar spectral information. These extra types of information give OBIA the potential to produce classifications characterized by accuracies higher than those produced by traditional pixel-based method.

One of the earlier efforts in detecting gullies by means of OBIA techniques was made about ten years ago by Eustace *et al.* (2009). The authors developed a semi-automated method to predict the presence and volume of gullies for unsampled locations within the Fitzroy catchment in the Great Barrier Reef lagoon, on the east coast of Australia. They used data from 20 LIDAR transects acquired in 2007. Shruthi *et al.* (2015) conducted a study with the aim of quantifying temporal changes in gully system areas applying the OBIA to a 1-meter resolution DEM obtained from the IKONOS mission. The authors were successful in achieving the gullies by developing a set of improved rules based on the knowledge of gully landform and process related to their formation. D'Oleire-Oltmanns *et al.* (2014) put together the GEOBIA analysis and the expert knowledge to set up an ensemble of rules to delineate gullies. They used as input a multispectral QuickBird 2 satellite image and calculated the accuracy by comparing the results of their classification with the reference data obtained by field surveys. Yang *et al.* (2017) developed and applied a multidirectional hill-shading method to extract gullies with an OBIA approach. The approach allowed the authors to get boundaries of gullies with high location accuracy and solve the discontinuity and low location accuracy caused by the pixel-based methods. Rahmati *et al.* (2017) submitted evidence that the GEOBIA technique is less expensive than any other approach to detect gullies, such as soil erosion measurements, especially over the last few years because it leads to the highest values of accuracy. In their paper, gully erosion mapping was performed using a DEM, a SPOT-5 panchromatic satellite image, and a multi-resolution segmentation.

Starting from knowledge about morphology and geomorphological processes leading to gullies, this paper aims to develop a set of rules about morphometric aspects and processes related to the formation of these landforms for the identification of gullies within a GEOBIA framework. The framework is applied to the Calhoun Critical Zone

Observatory where several gullies have been observed and mapped by other studies (James *et al.* 2007; Noto *et al.* 2017) and where a very high resolution (i.e., 1 m) LIDAR DEM is available. The procedure, first calibrated for two gullies detected during a field survey carried out in June 2015, will be applied to the Holcombe's Branch watershed and results compared to those obtained by Noto *et al.* (2017). The framework has been developed within the software *eCognition developer 9.2* (Trimble Geospatial).

STUDY AREA AND DATASET: THE CALHOUN CRITICAL ZONE OBSERVATORY

The Calhoun Critical Zone Observatory (CCZO) is located in South Carolina (USA) and is part of the Critical Zone Observatory (CZO) project, which involves different institutions with the main goal to study and understand the processes that shape the surface of Earth and support terrestrial life.

The study site is on gentle slopes (<3%) that are covered by Appling and Cataula series soils (clayey, kaolinitic, thermic Typic Kanhapludults). These soils have a depth higher than 8 m to bedrock and are derived from granitoid gneiss, mica gneiss, and granite (Overstreet & Bell 1965). The soil has well-developed horizons with sandy loam in A and E horizons, clays in Bt horizons and deep saprolite C horizons (Fimmen *et al.* 2008). The upper 35-cm layer is coarse textured (68% sand, 15% clay), which ensures macroporosity to the soil (Richter *et al.* 1999). The groundwater depth is generally higher than 5 m, even though during winter and occasional summer rain events water can perch in the upper 1.5 m of the soil.

These soils supported crops of cotton, corn, and wheat during the 1800s through the first half of the 20th century, during which time, agricultural practices on Piedmont soils such as these included increased use of fertilizers and soil conservation practices (Sheridan 1979).

The CCZO is a particularly interesting site for studying gully erosion since after more than a century of intensive crop production, landscapes were abandoned and experienced severe erosion that has led to the loss of almost 20 cm of top soil with the formation of numerous gullies. Today, the gullies are mainly blanketed by reforestation, but the scars of these particular landforms are still present. More details about this area can be found in James *et al.* (2007).

Following the same work scheme of Noto *et al.* (2017), we identified a first area (SUBAREA-A) of about 1 km^2, containing two gullies surveyed during a field survey carried out in June 2015, and a second area (SUBAREA-B) with an extension of about 4.3 km^2, corresponding to the Holcombe's Branch watershed within the CCZO (Figure 1).

METHODS

The methodology builds on the use of four different morphometric indices able to detect the presence of specific landforms, which can be interpreted as part of a gully (bottom and edges): the Topographic Position Index (*TPI*), the terrain slope, the roughness slope, and the length to width ratio.

With reference to the data, LIDAR DEM was obtained by the study of Noto *et al.* (2017). The SUBAREA-A was used to develop and calibrate the GEOBIA procedure, while the SUBAREA-B was used to test the developed procedure. Figure 1 shows the LIDAR DEM for the SUBAREA-A (left part of the figure) and the SUBAREA-B (right part of the figure). Detailed information about the data and the two sub-areas can be found in Noto *et al.* (2017).

The following section is organized as follows. The section 'Morphometric indices' provides a brief description of the aforementioned indices. The section 'Gully detection within a GEOBIA framework' summarizes the OBIA procedure developed to identify gullies within the SUBAREA-A with particular reference to the two main phases of OBIA: segmentation and its calibration ('The *multi-resolution segmentation* algorithm') and classification ('The object-based classification').

Morphometric indices

TPI for gully bottom

The *TPI* (Guisan *et al.* 1999; Jenness 2006) is defined as the difference between the elevation of a central pixel, z, and the mean elevation, $\bar{z}_{\alpha}$, of its surrounding cells within the kernel of size α. Negative values mean the cell is lower than its

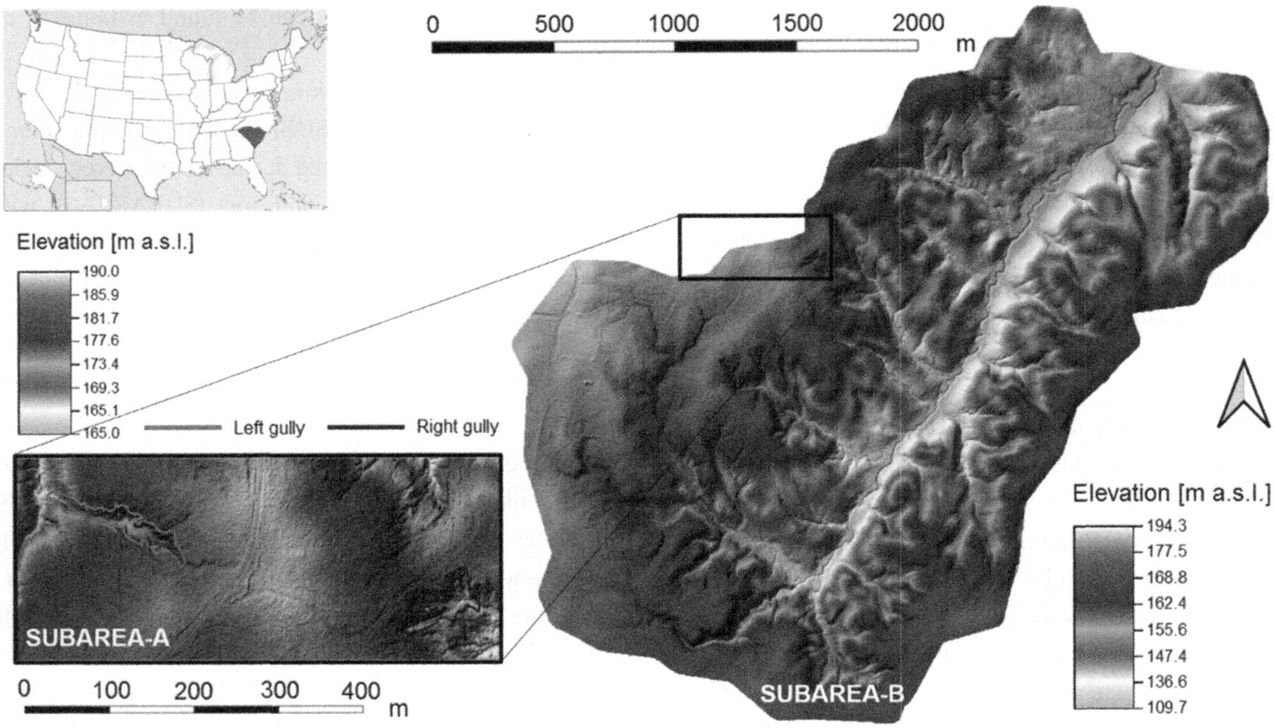

Figure 1 | LIDAR DEM of the SUBAREA-B (Holcombe's Branch watershed) of the CCZO (on the right) and focus on the SUBAREA-A (on the left) containing the two gullies surveyed and used for the calibration of the GEOBIA procedure.

surroundings. If z is significantly lower than $\bar{z}_\alpha$, which implies significant low values of *TPI*, then the cell is likely at or near the bottom of a depression (e.g., the bottom of a valley or a gully). In particular, we used the normalized *TPI* (*nTPI*) by dividing it by the mean value of elevation within the used kernel:

$$nTPI_\alpha = \frac{(z - \overline{z_\alpha})}{\overline{z_\alpha}} \quad (1)$$

Since the *TPI* is kernel-dependent, in order to decide the size of the kernel to use, we used different kernel size α (see 'Gully detection within a GEOBIA framework'). The $nTPI_\alpha$ was calculated in QGIS 3.4.0 (QGIS Development Team, http://qgis.org) starting from the existing GRASS *r.neighbors* function.

Terrain slope for gully edge

Terrain slope, which is the first spatial derivative of the terrain elevations, is one of the basic terrain parameters widely used in terrain analysis and landform classification. Slope has been here adopted as index of gully edge presence since the gully edges can be assimilated to abrupt changes of slope in the terrain (Shruthi *et al.* 2011; Liu *et al.* 2017). It was obtained in QGIS 3.4.0 from the LIDAR DEM through using the existing GRASS *r.slope.aspect* tool.

Terrain roughness for gully edge

Terrain roughness is a morphometric measure expressing irregularities of the soil surface, caused by factors such as soil texture, aggregate size, rock fragments, vegetation cover and land management. Since a DEM-derived roughness characterizes the local variance of surface gradients, distinguishing between smooth and rugged landforms might be a useful indicator of gully edge presence (Liu *et al.* 2017). According to Liu *et al.* (2017), here the terrain roughness is defined as the inverse of the cosine of slope angle and obtained using Equation (2) within the *raster calculator* of QGIS 3.4.0:

$$r = \frac{1}{\cos(slope)} \quad (2)$$

Since the roughness is directly derived from slope, in order to verify that the information provided by the terrain roughness is not redundant in detecting the edges of gullies, some tests for the SUBAREA-A have been carried out. The results of the tests, which are not shown here for the sake of brevity, demonstrated that considering this variable reduces the number of false positives, i.e., non-gully segments assigned to the class *gully edge*.

Length to width ratio for gully edge

Since gullies are usually narrow and long landforms, elements representing the edges are likely to have one of the two dimensions higher than the other one (Eustace *et al.* 2009; D'Oleire-Oltmanns *et al.* 2014). This information is contained within the length to width ratio, which gives an indication of the compactness of an element. In particular, values of length to width ratio greater than one means that the object is longer in one of its two dimensions. The index has been obtained in *eCognition developer 9.2*.

Gully detection within a GEOBIA framework

The gully detection procedure here developed can be summarized in the flow charts of Figure 2. The first step is the derivation of the first three aforementioned indices (*nTPI*, terrain slope, and terrain roughness). In particular, for the *nTPI* index, Evans & Lindsay (2010) suggest choosing the kernel size as a function of the typical width of the gullies to detect. In order to determine the best kernel size to detect gullies, *nTPI* was derived at three different kernel sizes, equal to 10, 20, and 30 m, respectively.

The second step consists of using the *nTPI* layer within *eCognition developer 9.2* as starting layer to apply the

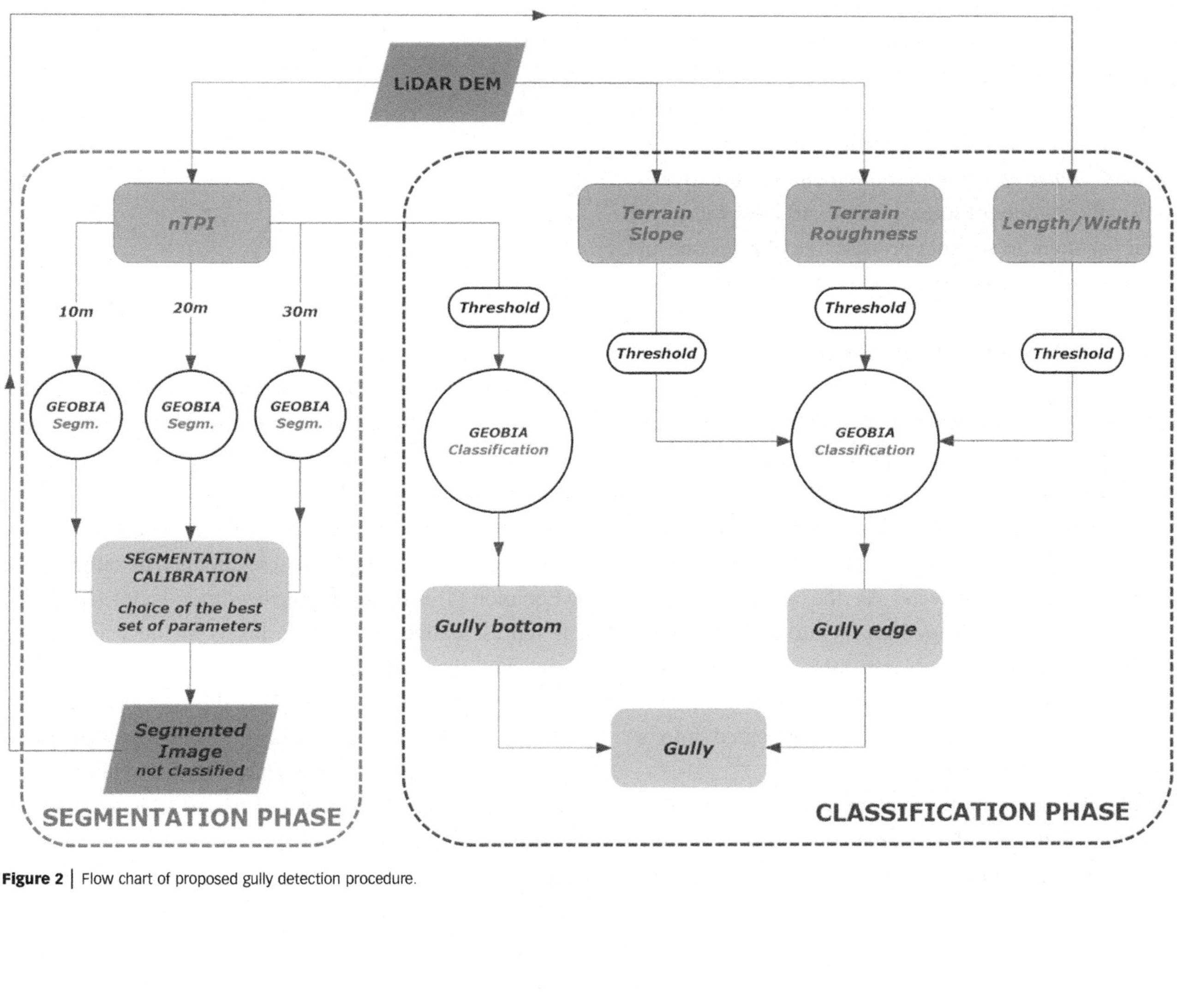

Figure 2 | Flow chart of proposed gully detection procedure.

segmentation process, which is the main process of *OBIA* (Hay & Castilla 2008), in order to derive homogeneous pixel clusters (also called segments or objects) that provide the base layer for the following analysis steps. Among the variety of segmentation algorithms, here it was decided to use the *multi-resolution segmentation* algorithm (Baatz & Schape 2000). It is a bottom-up approach that aims to merge neighbor pixels having a heterogeneity parameter less than a specific threshold, fixed by the user. Compared to the classical pixel-based approach, this process leads to a number of elements to process lower than that of pixels that make the original image and, for this reason, easier and faster to analyze. The optimal values of parameters, in order to obtain the segments from the original image, have been defined through a calibration procedure of segmentation detailed in the section 'The *multi-resolution segmentation* algorithm'.

A final phase of classification, based on different aspects of objects, such as spectral or textural attributes, allows the user to classify images and recognize a landform in the landscape (e.g., gullies in the present study). A *developed rulesets* section allows the user to set up an ensemble of rules aiming to automate the process and replicate it on different areas. A section called *Image Object Information* enables the user to query and get the values of some specific variables within an object and use them to set up a process tree of rules and/or algorithms aimed to detect a given object (e.g., gullies).

The *multi-resolution segmentation* algorithm

Segmentation creates segments or objects representing features that may be spectrally variable at the level of the single pixel. The most important parameter in order to identify homogeneous objects is the *scale parameter, e*, consisting of the maximum heterogeneity, or degree of fitting, between two neighbor pixels or objects. As the *scale parameter* increases, the number of merged pixels into a single object or the number of merged objects into a bigger object increases as well. On the contrary, as the *scale parameter* decreases, the original image will be fragmented into a higher number of small objects.

Actually, the similarity between two adjacent objects is evaluated as a function of a spatial and a spectral component (Happ *et al.* 2010). The first one depends on the spatial heterogeneity, h_{shape}, weighted by w_{shape}, and the compactness, h_{cmpct}, weighted by w_{cmpct}. For more details, see the first section of the Supplementary material, 'The *multi-resolution segmentation* algorithm'.

One way to optimize the values of the above parameters is to calibrate the segmentation procedure (Clinton *et al.* 2010; Liu *et al.* 2012, 2017). According to Liu *et al.* (2012), the overlay between the reference polygon (e.g., the actual gullies' boundaries) and the corresponding segments could generate under-segmented and over-segmented areas, apart from the overlapped areas (Figure 3). As is possible to see in Figure 3, the relationships between the reference polygons (continuous line in Figure 3) and the corresponding segments (dashed line in Figure 3) can be: one-to-many, one-to-one, and many-to-one.

According to the schemes depicted in Figure 3, it is possible to affirm that the goodness of the *multi-resolution segmentation* result, as far as concerning the geometric relationships, can be measured through the following three metrics:

$$OS = \frac{\sum |r_i - s_k|}{\sum |r_i|} \tag{3}$$

$$US = \frac{\sum |s_k - r_i|}{\sum |s_k|} \tag{4}$$

$$ED_1 = \sqrt{\frac{OS^2 + US^2}{2}} \tag{5}$$

In Equations (3) and (4), the term r_i stands for the reference polygon and the term s_k represents the segment that has to be compared with the reference polygon. In general, Equation (3) provides the *over-segmentation* factor (*OS*), because it sums up the areas included in the reference polygon but not in its corresponding segments. The output given by Equation (4) is the *under-segmentation* factor (*US*), since it is representative of all those pixels included in the segments, but that are outside the reference polygon. The output provided by Equation (5) is the *Euclidean distance* 1 (ED_1) and is the distance from the perfect geometric segmentation result. By analyzing Equations (3)–(5), it is easy to affirm that in the ideal case of perfect match between the reference polygon(s) and the segment(s) all the three metrics should be equal to zero. The over-segmentation

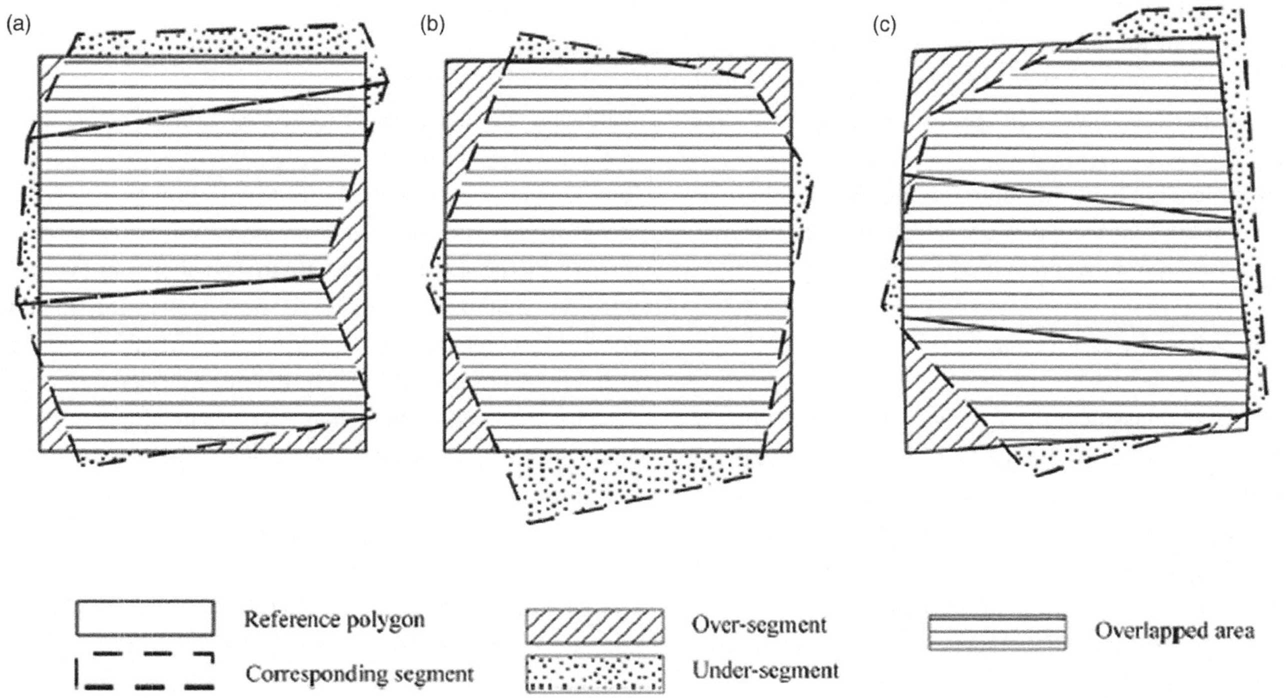

Figure 3 | Comparison between the reference polygon and the corresponding segments: (a) displays the one-to-many relationship, (b) the one-to-one, and (c) the many-to-one. Figure from Liu *et al.* (2012).

and the under-segmentation are two of the main limitations of object-based approach of images (Figure 3), which can affect the subsequent classification process in two ways: (i) under-segmentation results in image objects that cover more than one class and thus introduce classification errors because all pixels in each mixed image object have to be assigned to the same class; and (ii) features extracted from images affected by over-segmentation or under-segmentation do not represent the properties of real objects on the Earth such as shape and/or area (Liu & Xia 2010).

In order to take into account the arithmetic relationships as well, Liu *et al.* (2012) proposed the following new indices:

$$PSE = \frac{\sum |s_i - r_k|}{\sum |r_k|} \quad (6)$$

$$NSR = \frac{|m - v|}{m} \quad (7)$$

$$ED_2 = \sqrt{PSE^2 + NSR^2} \quad (8)$$

where PSE and NSR stand for *potential segmentation error* and *number of segment ratio*, respectively, while m and v represent the number of reference polygons and the number of corresponding segments, respectively. ED_2 is the *Euclidean distance* 2 and measures the distance from the perfect arithmetic segmentation.

In order to come up with a single index value that takes into account both the geometric and arithmetic segmentation results, the ED_1 and ED_2 have been combined into the key performance indicator, KPI (Goepel 2018), of Equation (9):

$$KPI = \sum_{i=1}^{k} w_i \cdot IC_i \quad (9)$$

where $k = 2$ (i.e., ED_1 and ED_2), w_i is the weight of IC_i, here assumed equal to 0.5 for both ED_1 and ED_2, and IC_i is calculated by means of Equation (10):

$$IC_i = 100 \cdot \frac{PI_i - base_i}{target_i - base_i} \quad (10)$$

where PI_i is the value of the ith index (i.e., ED_1 and ED_2), $base_i$ is the value indicating the worst performance of the ith index and $target_i$ is the value indicating the best performance of the ith index (i.e., zero for both ED_1 and ED_2). The *KPI* ranges between 0 and 100, with 100 indicating a perfect match in terms of geometric and arithmetic results.

More details about the multi-resolution algorithm can be found in the first section of the Supplementary material, 'The *multi-resolution segmentation* algorithm'.

The object-based classification

Object-based classification involves categorization of pixels on the basis of the spatial relationship with the neighboring pixels. Differently from the pixel-based classification, the object-based classification attempts to mimic what the human eyes do during visual interpretation based on the spectral properties (i.e., color), size, shape, and texture of objects obtained with the segmentation.

The process relies on the construction of one or more rules that can be used in cascade across a variety of layers to classify the segments resulting from the *multi-resolution segmentation* algorithm; such a procedure allows the user to produce a repeatable methodology. Depending on the purpose of the classification, it is possible to set different kinds of rules. As an example, some rules aim to classify objects on the basis of a threshold. In this case, the threshold value can be calibrated on the basis of literature indications or, in the absence of those, querying the values of some specific variables within objects falling within an interest area. Some other rules can be written up on the basis of adjacency or distance between segments.

The software *eCognition developer 9.2* allows the user to create different rules working with one or more layers, since they can be linked through logical *or* and *and* operators. Each rule can be independent from the others or linked to them in a cascade of rules, where each rule operates on the classified segments from the previous ones.

As it is possible to understand from the above considerations, a good classification requires a good a priori knowledge of the area and the types of land cover under investigation, which may not necessarily be available. Further, there is no definitive algorithm or parameters for the creation of image objects.

Method setup and calibration: application to the SUBAREA-A

Following the approach proposed by Clinton *et al.* (2010) and later applied by Liu *et al.* (2017) to identify gullies in the Yaojiawan catchment (Shaanxi Province, China), the proposed method has been previously calibrated on the two surveyed gullies of SUBAREA-A shown in Figure 1. Such an optimization process is very convenient, since it enables to compute segments fitting the actual boundaries of the two considered gullies in the most proper way.

The proposed methodology involves making choices of many different parameters, which may influence the resulting outcome in terms of accuracy of the identification of gullies. Main choices regard the *nTPI* kernel, α, and the segmentation parameters (i.e., e, w_{shape}, and w_{cmpct}). With reference to the *nTPI*, three different kernel sizes (i.e., 10, 20, and 30 m) were fixed a priori on the basis of a simple qualitative visual analysis and considering that the two gullies of the SUBAREA-A both have a width of about 20 m. With reference to the segmentation parameters, instead, a certain range of variability for the *scale parameter* (i.e., 3, 5, 10, and 20) and the weights related to the spatial heterogeneity and the compactness factor has been explored. In more detail, w_{cmpct} has been set equal to 0.2, 0.45, and 0.9, while w_{shape} is equal to 0.2, 0.6, and 0.9.

The *KPI* (see the section 'The *multi-resolution segmentation* algorithm') has been used to obtain the set of parameters that provides the best segmentation and the confusion matrix to evaluate the goodness of the final classification.

Multi-resolution segmentation

Based on morphology and the most common shape of gullies, the expected result of segmentation should have thin and long segments corresponding to the gully-affected areas and more compact segments outside them. Considering what was said in the section 'The *multi-resolution segmentation* algorithm', once the *scale parameter*, e, is fixed, these results should be obtained for low weights of the spatial heterogeneity and compactness factor.

In order to perform the *multi-resolution segmentation* algorithm, the only layer of the *nTPI* has been used.

In order to calibrate the process, a number of 108 possible combinations resulting from the considered ranges of variability of *nTPI* (3), *e* (4), w_{shape} (3), and w_{cmpct} (3) have been considered by setting up an ensemble of segmentation attempts and evaluating, for each of them, the goodness of the results in terms of *KPI*. The results of calibration, not shown here for the sake of brevity, demonstrated that the more suitable size for the kernel of the *nTPI* was 30 m (hereafter *nTPI*30). With reference to the two gullies of SUBAREA-A, a visual inspection of *nTPI*30 layer seems to ensure that pixels belonging to the gully-affected areas, either bottom or edges, and to the gully not-affected areas are likely to be included in the same kernel. In terms of *multi-resolution segmentation*, this means that all the pixels covering the bottom of a gully are likely merged into objects with strongly negative values of *nTPI*30; similarly, the edges of gullies are merged into objects still having negative values of *nTPI*30 but closer to zero than the object representing the bottom. More details about the 36 possible combinations and their results in terms of KPI can be found in the second section of the Supplementary material, 'Multi-resolution segmentation in SUBAREA-A'.

Figure 4(a) and 4(b) show the results of segmentation corresponding to the two surveyed gullies of SUBAREA-A relatively to $e = 10$, $w_{shape} = 0.9$, and $w_{cmpct} = 0.2$, while Figure 4(c) and 4(d) show the results of segmentation for

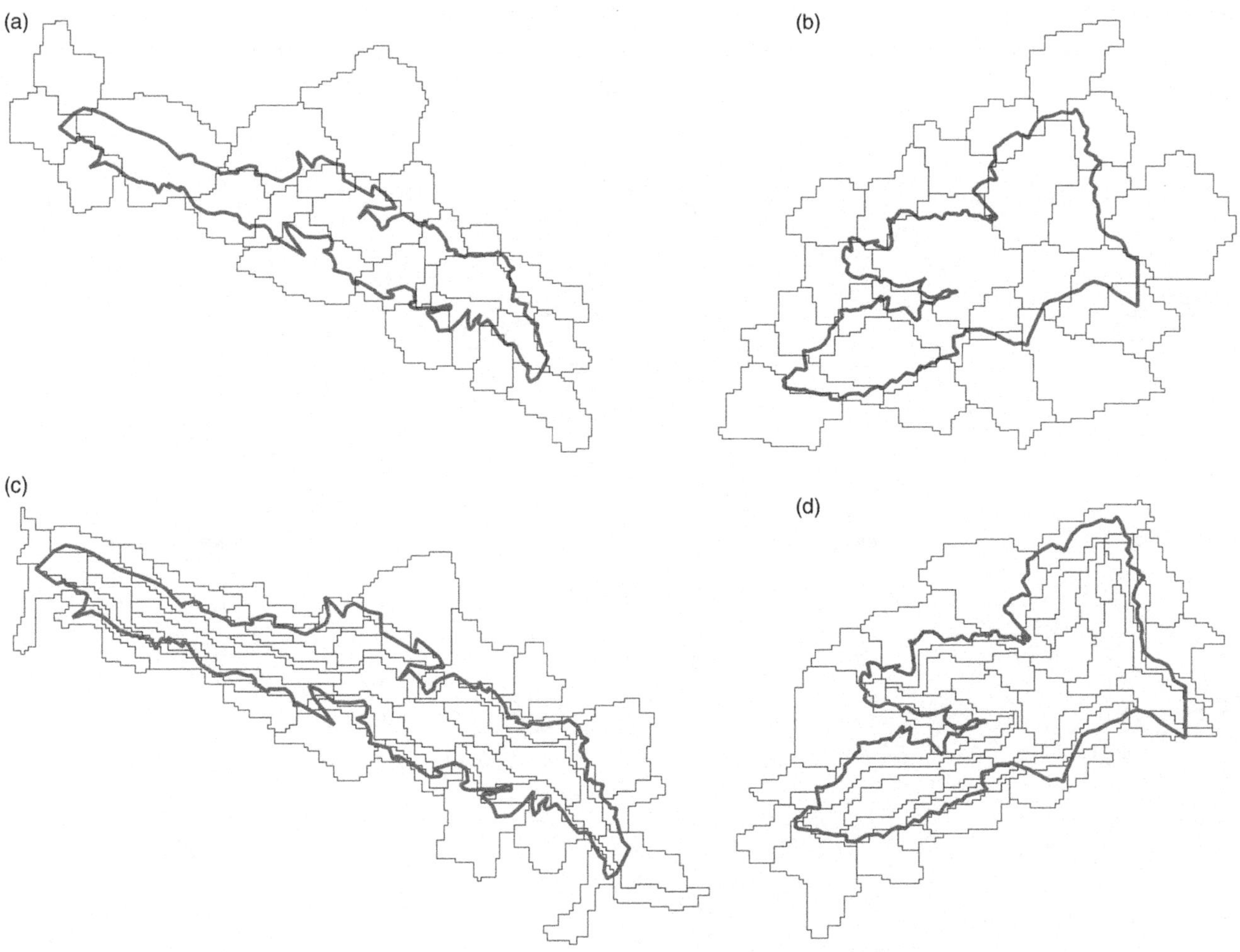

Figure 4 | Segmentation results for $e = 10$, $w_{shape} = 0.9$, and $w_{cmpct} = 0.2$ for the (a) left basin and (b) right basin and $e = 5$, $w_{shape} = w_{cmpct} = 0.2$ for the (c) left basin and (d) right basin. Black lines indicate segmentation.

$e = 5$, $w_{shape} = 0.2$, and $w_{cmpct} = 0.2$. The image segments displayed in Figure 4 are only those intersecting the actual gully-affected areas.

It is possible to notice in Figure 4 that, as the *scale parameter* increases, the size of objects increases as well. Looking at Figure 4(a) and 4(b) ($e = 10$), segmentations resulted in a low number of large objects leading to high values of *OS* and/or *US* and, consequently, higher values of ED_1. More importantly, there is a coarse segmentation inside the actual gullies that could result in a raw classification. The opposite considerations can be made by observing Figure 4(c) and 4(d), where $e = 5$. In this case, a higher number of long and narrow objects allows for a better description of geometry but a worse segmentation in terms of arithmetic results.

Considering the previous arguments, according to Baatz & Schape (2000), it has been decided to select the following values for the segmentation of SUBAREA-A: $e = 5$, $w_{shape} = w_{cmpct} = 0.2$.

Starting from this segmentation it has been possible to retrieve different information at the segment scale, among which is the length to width ratio, useful for the next step of classification.

Classification

Figure 5 shows the two surveyed gullies of SUBAREA-A overlaid to the spatial distributions of the four indices above mentioned: *nTPI30* (Figure 5(a)), terrain slope (Figure 5(b)), terrain roughness (Figure 5(c)), and length to width ratio (Figure 5(d)). Looking at Figure 5(a), as already discussed in the section 'Multi-resolution segmentation', it is possible to notice that the objects belonging to the bottom of gullies are characterized by negative values, while edges are identified by still negative values but closer to zero. By observing the slope and the terrain roughness, shown in Figure 5(b) and 5(c), respectively, it is clear that the higher values of both these variables correspond to gullies' edges.

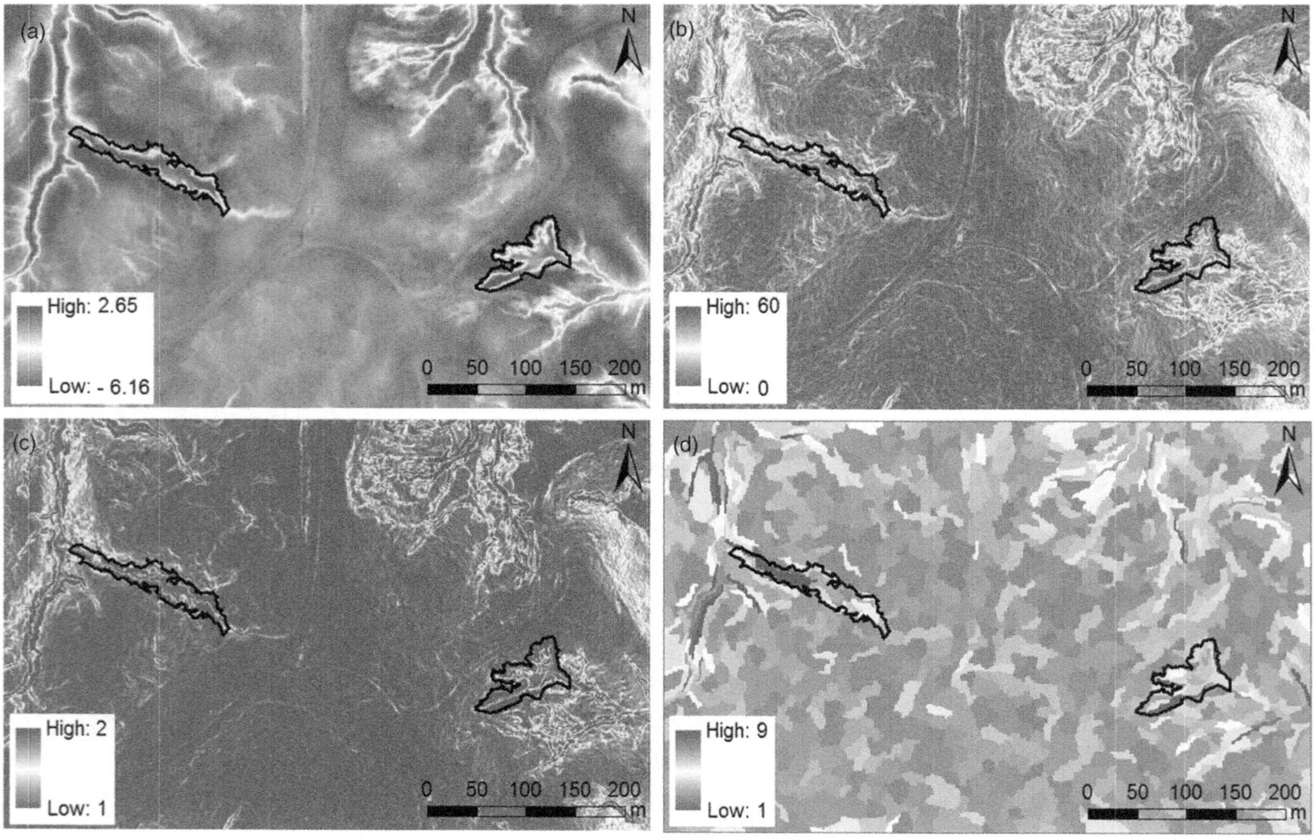

Figure 5 | Spatial distribution of the four indices used to detect gully bottom and edges: (a) TPI30, (b) slope, (c) terrain roughness, and (d) length to width ratio.

Finally, from Figure 5(d), it is possible to see that if the element is narrow and long, as gullies usually are, there are high values of length to width ratio.

Starting from information contained within the segmented objects that, differently from the classical pixel-based approach, can concern also the values of min, max, mean, standard deviation, etc., of a given variable, we tried to detect landforms connected to gullies. Investigation of previous layers allowed us to set up an ensemble of rules and algorithms to detect the bottoms and the edges of the two surveyed gullies of SUBAREA-A.

First of all, we identified the bottom of gullies by using the *nTPI* layer. As already said in the section '*TPI* for gully bottom', low values of *TPI* indicate a cell at or near the bottom of a valley that is likened to the bottom of a gully. That being said, by querying the mean value of *nTPI*30 for both those segments within and outside the surveyed gullies, it was possible to identify and calibrate the value of a threshold, T_1, distinguishing between the bottom and other landforms.

Once the bottom part of gullies within the SUBAREA-A was identified, for the identification of edges, a similar process led to the identification of three thresholds (i.e., T_2, T_3, and T_4) for the terrain slope, terrain roughness, and length to width ratio layers, respectively.

This process resulted in the writing of Equation (11), which is the first rule of the cascade process, for the identification of the bottom, and Equation (12) for the identification of edges:

$$Mean\ (nTPI30) < T_1 \tag{11}$$

$$\begin{cases} Mean\ slope > T_2 \\ Mean\ roughness > T_3 \\ Length/width > T_4 \end{cases} \tag{12}$$

Equation (11) affirms that, for each segment, if the average *nTPI*30 within the object is less than T_1 it is classified as a gully bottom. Equation (12) puts three different conditions together through an *and* spatial operator and, therefore, it represents an intersection. For this reason, only those objects where all of the three conditions are contemporarily satisfied are classified as gully edges. In particular, from previous analysis, the values of T_1, T_2, T_3, and T_4 resulted as equal to −2, 20°, 1.15, and 1.5, respectively.

All the segments classified as bottom or edges have been merged into a single class labeled *gully*, while all the unclassified objects have been assigned to a class labeled *non-gully*. The boundaries of the two surveyed gullies, highlighted in a continuous black line, overlaid to the classification results, are shown in Figure 6.

In order to verify the goodness of the achieved classification, the *Accuracy Assessment* tool provided by *eCognition developer* 9.2 has been used for the same two rectangular frames surrounding the two gullies of SUBAREA-A retrieved by Noto *et al.* (2017) (see Box-SX and Box-DX, Figure 2a of Noto *et al.* (2017)). The tool returns the *overall*, the *user*, and the *producer accuracies* for the classes *gully* and *non-gully* and the *K-index* for both the two classes and takes into account all of the pixels contained in an object. In particular, the overall accuracy is the percentage of pixels correctly classified over the total number of them in the confusion matrix.

Since the overall accuracy is very satisfactory (almost 95% for the left gully and 91% for the right gully), in order to test the procedure developed for the SUBAREA-A, the same rules here discussed have been applied to detect gullies in the SUBAREA-B. More details about the *Accuracy Assessment* tool and its application to SUBAREA-A can be found in the third section of the Supplementary material, 'Classification in SUBAREA-A' and Table S1.

Application of the developed procedure to the SUBAREA-B

As was said in the section 'Classification', since the accuracy assessment related to the SUBAREA-A is satisfied, the same classification rules have been applied to detect the gullies within the SUBAREA-B. Because of the greater extension and the more complex morphology of the SUBAREA-B, as compared to the SUBAREA-A, in addition to the developed rules for the SUBAREA-A, other rules have to be used in order to exclude some morphological features (e.g., some parts of hydrological network and the bank erosion areas) which, at the watershed scale, can be erroneously classified as gullies.

In particular, for the correction of those objects belonging to the hydrological network, all of those objects with a mean value of flow accumulation lower than a threshold,

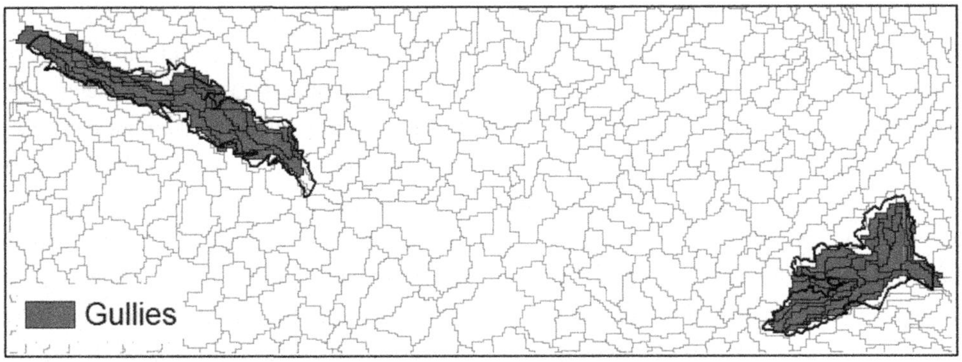

Figure 6 | Results of the classification procedure for the SUBAREA-A after applying Equations (11) and (12).

T_5, have been corrected as *non-gully*. The layer of flow accumulation for the Holcombe's Branch watershed was previously derived from the DEM in QGIS. For the correction of those objects belonging to the bank erosion areas, instead, we used a distance threshold, T_6, from those elements corrected as hydrological network at the previous step. In this case, all of those segments with a distance from the hydrological network lower than T_6 have been corrected as *non-gully*. This process resulted in the implementation of two new rules in *eCognition developer 9.2*:

$$Mean\ (Flow\ Accumulation) > T_5 \quad (13)$$

$$Distance\ to\ hydrological\ network < T_6 \quad (14)$$

In particular, investigation of objects belonging to the river network and to the bank erosion areas led to set $T_5 equal\ to\ 4,000\ pxl$ and $T_6 equal\ to\ 30\ pxl$.

In this case, since a spatial distribution of the digitalized actual gullies is not available, in order to assess the appropriateness of the developed procedure, the spatial map of results for the object-based procedure and the pixel-based procedure, developed for the same area by Noto *et al.* (2017), has been compared (Figure 7).

From an overview of Figure 7, it is possible to notice that the number of gullies detected with the object-based approach is higher than that obtained with the pixel-based approach. Moreover, while in some cases there is almost a perfect match between the gullies detected with the two approaches (see box B in Figure 7), in some other cases the match is only partial (see box A in Figure 7) and in some others there is not a match at all (see box C in Figure 7). Rarely, the pixel-based approach detects gullies that the object-based approach is not able to identify. However, observing the morphology of the Holcombe's Branch watershed in Figure 7, it seems plausible that the landforms identified by the object-based approach are likely gullies and there is a failure in identifying correctly these landforms from the pixel-based approach.

Starting from the distribution map of the gullies in Figure 7 we derived some indices of the gullies such as the area, the boundary length (i.e., perimeter), and the compactness index, and their empirical distribution function (*ecdf*, Figure 8). According to Noto *et al.* (2017), the compactness index was defined as the ratio between the perimeter of the gully and the root square of its area (multiplied by 4π). It ranges between 1.28 (most compact gully) and 5.31 (very elongated gully) and characterizes the shape of the extracted gullies; the more it is close to 1, the more the shape of the gully is close to a circle.

A total number of 143 gullies were identified with an area ranging between $20\ m^2$ and about $10{,}700\ m^2$ (Figure 8(a)), whereas the perimeter ranges from 24 m to about 1,950 m (Figure 8(b)). The compactness index (Figure 8(c)) ranges from a minimum of 0.04 (very elongated gullies) to a maximum of about 0.6 (more compact gullies). In order to compare the results of this study with the results of the pixel-based approach provided by Noto *et al.* (2017), Figure 8 shows also the *ecdf* for the area, perimeter, and compactness index returned by that study. As it is possible to notice from Figure 8, apart for the compactness index, which is perfectly overlapped to that of Noto *et al.* (2017), the object-based approach led to

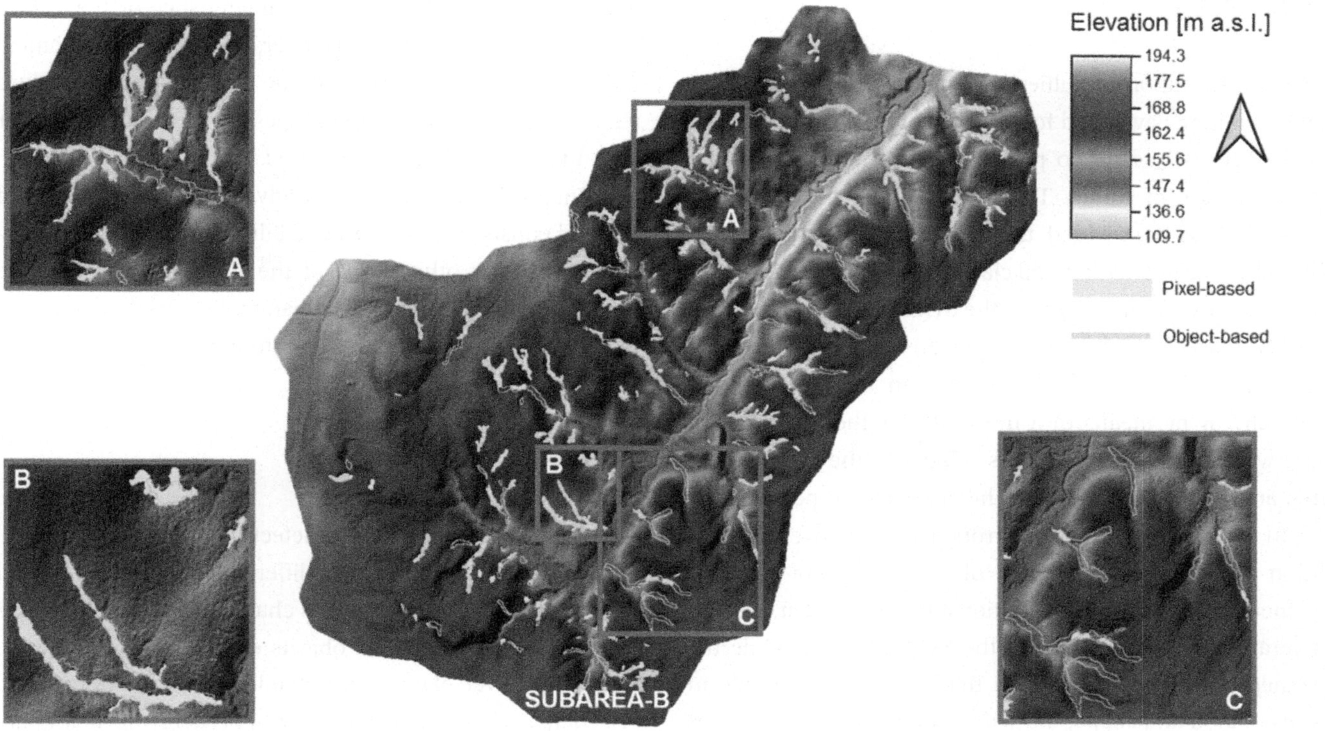

Figure 7 | Comparison between the gullies achieved through a GEOBIA technique and a pixel-based image analysis. Boxes A, B, and C show some focuses on the Holcombe's Branch watershed.

identifying more extended gullies (e.g., larger area and perimeter) than the pixel-based approach. While the difference in terms of upper limits is due to the two different approaches and, mainly, to the fact that the object-based approach works with elements, which are aggregation of pixels, where the minimum size of the element is not that of the pixel anymore, the different lower limit is mainly due to the fact that, differently from Noto *et al.* (2017), we did not eliminate from classification those gullies with an area smaller than a given threshold.

Some more statistical characteristics of extracted gullies can be found in the fourth section of the Supplementary material, 'Application of the developed procedure to the SUBAREA-B' and Table S2.

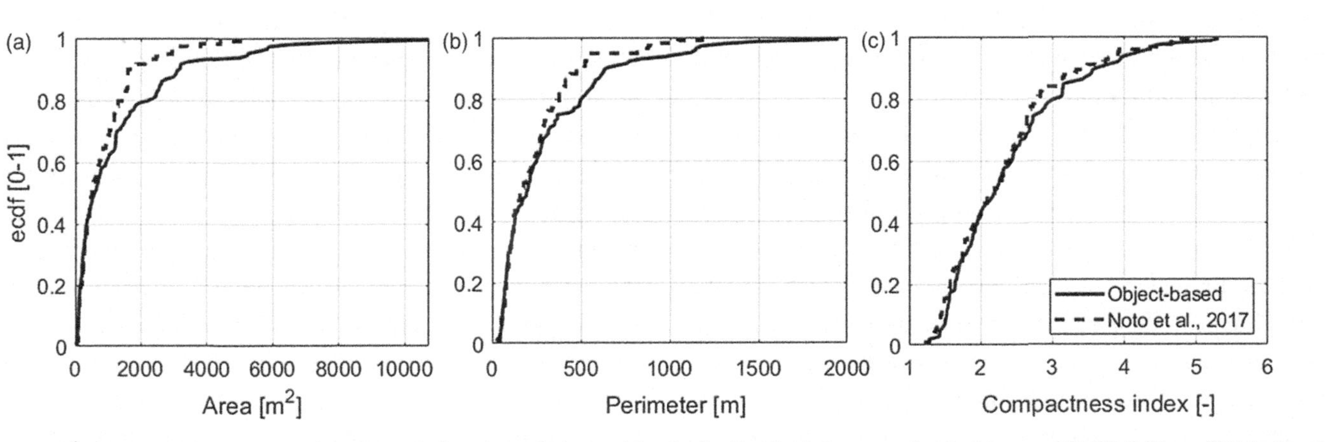

Figure 8 | Empirical cumulative distribution function (*ecdf*) of morphometric characteristics of gullies identified by the proposed methodology over SUBAREA-B: (a) area, (b) perimeter, and (c) compactness index.

DISCUSSION

Spatial distribution of gullies identified by the object-based approach here developed for the Holcombe's Branch watershed (Figure 7) seems to provide a good definition of the spatial pattern of gullies. The application of the procedure to SUBAREA-B provided the identification of gullies on the basis of segmentation and classification previously developed, calibrated, and tested for the SUBAREA-A. Differently than Noto *et al.* (2017), which removed the main river channel and the bank erosion landforms in the SUBAREA-B manually or by means of a threshold for the gullies' area, here we used a couple of rules related to the contributing area and to the distance from the channels, respectively.

In order to minimize the errors due to the over-segmentation and under-segmentation of images (Figure 3), some geometric and arithmetic relationships have been used to calibrate the segmentation on the SUBAREA-A. While the geometric relationship ensures the best match possible in terms of area and shape, the arithmetic relationship ensures the best segmentation in terms of number of objects (i.e., as few objects as possible). The geometric and arithmetic relationships have been combined in a unique index representative of the goodness of segmentation.

For the classification of objects, a set of rules based on parameters' thresholds and/or adjacency or distance from other objects has been used. Since a gully can be usually characterized by a bottom and some edges, which are defined by different characteristics, it has been preferred to set up a rule for the identification of the bottom and three rules in cascade for the identification of the edges. For the first case, we used the *nTPI30* parameter, while for the edges we used the terrain slope, the terrain roughness, and the length to width ratio.

For the SUBAREA-B, the physical consistency of the developed procedure has been evaluated by comparing the gullies' pattern provided by the proposed method with that of Noto *et al.* (2017). From the observation of the Holcombe's Branch watershed morphology (Figure 7), the GEOBIA procedure seems able to detect also gullies, or parts of those, that have not been detected by the pixel-based approach. For this reason, it is plausible to say that the developed procedure is a reliable tool to identify gullies once it has been calibrated and tested. Moreover, the comparison of *ecdf* for some characteristics of the gullies such as the area, the perimeter, and the compactness index (Figure 8) showed that the object-based procedure leads to gullies with bigger ranges of area and perimeter than those identified by Noto *et al.* (2017).

One of the most important advantages of the proposed method consists of the reproducibility of the developed procedure also for other areas of the Earth. In contrast, the main limitation is the fact that parameters need to be recalibrated in order to adapt to different gully patterns.

CONCLUSIONS

An object-based approach to detect landforms is presented in this study. The approach, differently from the classical pixel-based approach, uses the characteristics and features of groups of pixels, called objects or segments, to differentiate land cover classes with similar spectral information. As compared to the pixel-based approach, the possibility to consider this extra information gives GEOBIA the potential to produce land cover thematic maps with a higher accuracy.

GEOBIA has been applied to the Calhoun Critical Zone Observatory (CCZO), an area of South Carolina particularly interesting since it is characterized by the presence of numerous gullies because of very intense erosion activity due to the abandonment of terrains used for intensive crop production in the past century.

The software used to carry out the study is the *eCognition developer 9.2*, developed by Trimble Navigation Ltd, which provides a collection of tools for OBIA and GEOBIA, facilitating the classification of images in various application fields. The software allows the user to develop a procedure that starts from the segmentation of an image to create groups of homogeneous pixels (also called objects or segments) to which is assigned the same class during the following classification process. The main advantage of the developed procedure, as compared to a classical pixel-based approach, is that the number of elements to process is significantly lower than that of pixels that make the image and, for this reason, easier and faster to analyze.

Two areas, the SUBAREA-A and the SUBAREA-B, inside the CCZO have been used to calibrate and test, respectively, the developed procedure. The first area is characterized by the presence of two gullies detected during a field survey in June 2015. The surveyed gullies have been used as reference to calibrate the GEOBIA procedure and thus obtain the optimal segmentation and classification. The procedure, then, has been applied to the SUBAREA-B, corresponding to the Holcombe's Branch watershed, to test the goodness of the developed procedure.

Results showed that the GEOBIA procedure developed in *eCognition developer 9.2* is able to detect gullies with good accuracy even though, because of the absence of surveyed gullies for the SUBAREA-B, the results have been evaluated using as reference the results from a pixel-based approach developed for the same area by Noto *et al.* (2017).

Although the approach surely requires further testing in other gullied systems where good quality topographic data are available, the demonstrated capability of *eCognition developer 9.2* in developing OBIA or GEOBIA procedures to identify gullies in a given area of Earth, coupled with the increasing availability of LIDAR data, allows us to say that this automated technique has the potential for individuation of different landforms (e.g., landslides) and applications across a range of environments.

SUPPLEMENTARY MATERIAL

The Supplementary Material for this paper is available online at https://dx.doi.org/10.2166/hydro.2019.083.

REFERENCES

Baatz, M. & Schape, A. 2000 Multiresolution segmentation: an optimization approach for high quality multi-scale image segmentation. In: *Angewandte Geographische Informations-Verarbeitung, XII* (J. Strobl, T. Blaschke & G. Griesbner, eds). Wichmann Verlag, Karlsruhe, Germany, pp. 12–23.

Clinton, N., Holt, A., Scarborough, J., Yan, L. & Gong, P. 2010 Accuracy assessment measures for object-based image segmentation goodness. *Photogrammetric Engineering and Remote Sensing* **76**, 289–299.

D'Oleire-Oltmanns, S., Marzolff, I., Tiede, D. & Blaschke, T. 2014 Detection of gully-affected areas by applying object-based image analysis (OBIA) in the region of Taroudannt, Morocco. *Remote Sensing* **6**, 8287–8309.

Eustace, A., Pringle, M. & Witte, C. 2009 Give me the dirt: detection of gully extent and volume using high-resolution Lidar. In: *Innovations in Remote Sensing and Photogrammetry* (S. Jones & K. Reinke, eds). Springer, Berlin, Heidelberg, pp. 255–269.

Evans, M. & Lindsay, J. 2010 High resolution quantification of gully erosion in upland peatlands at the landscape scale. *Earth Surface Processes and Landforms* **35**, 876–886.

Fimmen, R. L., Richter, D. D., Vasudevan, D., Williams, M. A. & West, L. T. 2008 Rhizogenic Fe–C redox cycling: a hypothetical biogeochemical mechanism that drives crustal weathering in upland soils. *Biogeochemistry* **87**, 127–141.

Francipane, A., Ivanov, V. Y., Noto, L. V., Istanbulluoglu, E., Arnone, E. & Bras, R. L. 2012 TRIBS-Erosion: A parsimonious physically-based model for studying catchment hydro-geomorphic response. *Catena* **92**, 216–231.

Francipane, A., Fatichi, S., Ivanov, V. Y. & Noto, L. V. 2015 Stochastic assessment of climate impacts on hydrology and geomorphology of semiarid headwater basins using a physically based model. *Journal of Geophysical Research F: Earth Surface* **120**, 507–533.

Francipane, A., Mussomè, F., Cipolla, G. & Noto, L. V. 2018 Object-based image analysis technique for gully mapping using topographic data at very high resolution (VHR). In: *HIC 2018. 13th International Conference on Hydroinformatics* (G. La Loggia, G. Freni, V. Puleo & M. De Marchis, eds). EasyChair, Manchester, pp. 725–730.

Goepel, K. 2018 Implementation of an online software tool for the analytic hierarchy process (AHP-OS). *International Journal of the Analytic Hierarchy Process* **10**, 469–487.

Grosse, P., van Wyk de Vries, B., Euillades, P., Kervyn, M. & Petrinovic, I. 2012 Systematic morphometric characterization of volcanic edifices using digital elevation models. *Geomorphology* **136**, 114–131.

Guisan, A., Weiss, S. B. & Weiss, A. D. 1999 GLM versus CCA spatial modeling of plant species distribution. *Plant Ecology* **143**, 107–122.

Happ, P. N., Ferreira, R. d. S., Bentes, C., Costa, G. A. O. P. & Feitosa, R. Q. 2010 Multiresolution segmentation: a parallel approach for high resolution image segmentation in multicore architectures. In: *International Conference on Geographic Object-Based Image Analysis (GEOBIA 2010)*, Ghent, Belgium.

Hay, G. J. & Castilla, G. 2008 Geographic Object-Based Image Analysis (GEOBIA): A new name for a new discipline. In: *Object-Based Image Analysis: Spatial Concepts for Knowledge-Driven Remote Sensing Applications* (T. Blaschke, S. Lang & G. J. Hay eds). Springer, Berlin, Heidelberg, pp. 75–89.

Henderson-Sellers, A. 1994 Land-use change and climate. *Land Degradation & Development* **5**, 107–126.

Huggett, R. 1990 Book review. In: *Modelling Geomorphological Systems*, Vol. 15 (M. G. Anderson, ed.). Earth Surface Processes and Landforms, Chichester, pp. 292–293.

Ireland, H. A., Eargle, D. H. & Sharpe, C. F. S. 1939 *Principles of Gully Erosion in the Piedmont of South Carolina*. US Department of Agriculture, Washington, DC, USA.

Jackson, T. J., Ritchie, J. C., White, J. & Leschack, L. 1988 Airborne laser profile data for measuring ephemeral gully erosion. *Photogrammetric Engineering & Remote Sensing* **54**, 1181–1185.

James, L. A., Watson, D. G. & Hansen, W. F. 2007 Using LiDAR data to map gullies and headwater streams under forest canopy: South Carolina, USA. *Catena* **71**, 132–144.

Jenness, J. 2006 *Topographic Position Index (tpi_jen.avx) Extension for ArcView 3.x, v. 1.2, 1.2 ed.* Jenness Enterprises, Flagstaff, AZ, USA.

Kirkby, M. J. & Bracken, L. J. 2009 Gully processes and gully dynamics. *Earth Surface Processes and Landforms* **34**, 1841–1851.

Kirkby, M. J. & Cox, N. J. 1995 A climatic index for soil erosion potential (CSEP) including seasonal and vegetation factors. *Catena* **25**, 333–352.

Liu, D. & Xia, F. 2010 Assessing object-based classification: advantages and limitations. *Remote Sensing Letters* **1**, 187–194.

Liu, Y., Bian, L., Meng, Y., Wang, H., Zhang, S., Yang, Y., Shao, X. & Wang, B. 2012 Discrepancy measures for selecting optimal combination of parameter values in object-based image analysis. *ISPRS Journal of Photogrammetry and Remote Sensing* **68**, 144–156.

Liu, K., Ding, H., Tang, G., Zhu, A. X., Yang, X., Jiang, S. & Cao, J. 2017 An object-based approach for two-level gully feature mapping using high-resolution DEM and imagery: a case study on hilly loess plateau region. *China. Chinese Geographical Science* **27**, 415–430.

Nearing, M. A., Simanton, J. R., Norton, L. D., Bulygin, S. J. & Stone, J. 1999 Soil erosion by surface water flow on a stony, semiarid hillslope. *Earth Surface Processes and Landforms* **24**, 677–686.

Noto, L. V., Bastola, S., Dialynas, Y. G., Arnone, E. & Bras, R. L. 2017 Integration of fuzzy logic and image analysis for the detection of gullies in the Calhoun Critical Zone Observatory using airborne LiDAR data. *ISPRS Journal of Photogrammetry and Remote Sensing* **126**, 209–224.

Overstreet, W. C. & Bell, H. 1965 *The Crystalline Rocks of South Carolina: Relation of Rock Units in the Geologic Belts of the Piedmont and Blue Ridge Provinces of South Carolina*. Bulletin 1183, US Government Printing Office, Washington, DC, USA.

Patton, P. C. & Schumm, S. A. 1975 Gully erosion, northwestern Colorado: a threshold phenomenon. *Geology* **3**, 88–90.

Pimentel, D., Harvey, C., Resosudarmo, P., Sinclair, K., Kurz, D., McNair, M., Crist, S., Shpritz, L., Fitton, L., Saffouri, R. & Blair, R. 1995 Environmental and economic costs of soil erosion and conservation benefits. *Science* **267**, 1117–1123.

Poesen, J., Nachtergaele, J., Verstraeten, G. & Valentin, C. 2003 Gully erosion and environmental change: importance and research needs. *Catena* **50**, 91–133.

Rahmati, O., Tahmasebipour, N., Haghizadeh, A., Pourghasemi, H. R. & Feizizadeh, B. 2017 Evaluating the influence of geo-environmental factors on gully erosion in a semi-arid region of Iran: an integrated framework. *Science of the Total Environment* **579**, 913–927.

Richter, D., Markewitz, D., Trumbore, S. G. & Wells, C. 1999 Rapid accumulation and turnover of soil carbon in a re-establishing forest. *Nature* **400**, 56–58.

Ritchie, J. C., Grissinger, E. H., Murphey, J. B. & Garbrecht, J. D. 1994 Measuring channel and gully cross-sections with an airborne laser altimeter. *Hydrological Processes* **8**, 237–243.

Ritter, D. F., Kochel, R. C. & Miller, J. R. 2011 *Process Geomorphology*, 5th edn. Waveland Press, Long Grove, IL, USA.

SCS 1966 Procedure for determining rates of land damage, land depreciation and volume of sediment produced by gully erosion. US Dept. of Agriculture, Soil Conservation Service, Engineering Division, Washington, 1966.

Seginer, I. 1966 Gully development and sediment yield. *Journal of Hydrology* **4**, 236–253.

Sheridan, R. C. 1979 Chemical fertilizers in southern agriculture. *Agricultural History* **53**, 308–318.

Shruthi, R. B. V., Kerle, N. & Jetten, V. 2011 Object-based gully feature extraction using high spatial resolution imagery. *Geomorphology* **134**, 260–268.

Shruthi, R. B. V., Kerle, N., Jetten, V., Abdellah, L. & Machmach, I. 2015 Quantifying temporal changes in gully erosion areas with object oriented analysis. *Catena* **128**, 262–277.

Stocking, M. A. 1980 Examination of the factors controlling gully growth. In: *Assessment of Erosion* (M. de Boodt & D. Gabriels, eds). John Wiley and Sons Ltd, Chichester, UK, pp. 505–520.

Tarolli, P., Arrowsmith, J. R. & Vivoni, E. R. 2009 Understanding earth surface processes from remotely sensed digital terrain models. *Geomorphology* **113**, 1–3.

Tuckfield, C. G. 1964 Gully erosion in the New Forest, Hampshire. *American Journal of Science* **262**, 795–807.

Valentin, C., Poesen, J. & Li, Y. 2005 Gully erosion: impacts, factors and control. *Catena* **63**, 132–153.

Wilkinson, B. H. & McElroy, B. J. 2007 The impact of humans on continental erosion and sedimentation. *GSA Bulletin* **119**, 140–156.

Yang, X., Li, M., Na, J. & Liu, K. 2017 Gully boundary extraction based on multidirectional hill-shading from high-resolution DEMs. *Transactions in GIS* **21**, 1204–1216.

First received 19 April 2019; accepted in revised form 6 August 2019. Available online 3 October 2019

IWA Publishing's authorised EU representative for General Product Safety Regulations is Diane D'Arras, 15 rue Duret, 75116 Paris, France, e-mail: safety@iwap.co.uk.

Printed and bound by CPI Group (UK) Ltd, Croydon, CR0 4YY
17/09/2026
02218804-0001